Global environmental change is caused by, and affects, all humanity. *Leaving Eden* examines the causes and consequences of global change, with particular emphasis on the interaction of nature and human behavior. The book begins with a discussion of the natural Earth, focusing on the physical and chemical processes that have hitherto controlled our environment and climates. The author then turns to the effects of human disruption of these natural systems, arguing that human beings are now taking over the management of the Earth from nature. Issues such as the greenhouse effect, acid rain, deforestation, ozone depletion, chlorofluorocarbons, air pollution, nuclear waste, agricultural subsidies, and population management are addressed. Fossil fuels, nuclear energy, hydroelectricity, and solar power are compared and assessed. In addition to providing a comprehensive account of the development of the current environmental crisis, the book offers an extensive discussion of the ways in which the environment can be protected and restored.

Leaving Eden

E. G. NISBET
University of Saskatchewan, Canada

Leaving Eden
To protect and manage the Earth

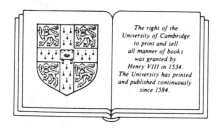

*The right of the
University of Cambridge
to print and sell
all manner of books
was granted by
Henry VIII in 1534.
The University has printed
and published continuously
since 1584.*

CAMBRIDGE UNIVERSITY PRESS

Cambridge

New York   Port Chester   Melbourne   Sydney

Published by the Press Syndicate of the University of Cambridge
The Pitt Building, Trumpington Street, Cambridge CB2 1RP
40 West 20th Street, New York, NY 10011, USA
10 Stamford Road, Oakleigh, Melbourne 3166, Australia

© Cambridge University Press 1991

First published 1991

Printed in the United States of America

Library of Congress Cataloging-in-Publication Data
Nisbet, E. G.
Leaving Eden / E. G. Nisbet.
p.   cm.
Includes bibliographical references and index.
ISBN 0-521-39311-6   ISBN 0-521-42579-4 (pbk)
1. Man – Influence on nature. 2. Pollution. 3. Global warming.
4. Environmental policy. I. Title.
GF75.N566   1991
363.7 – dc20   90-21823
                                                          CIP

British Library Cataloguing in Publication Data
Nisbet, E. G. (Euan George)
Leaving Eden.
1. Environment
I. Title
333.7

ISBN 0-521-39311-6 hardback
ISBN 0-521-42579-4 paperback

To my family,
for whom this book is written

This goodly frame, the Earth, seems to me a sterile promontory; this most excellent canopy, the air, look you, this brave o'erhanging firmament, this majestical roof fretted with golden fire, why it appears no other thing to me than a foul and pestilent congregation of vapours. What a piece of work is a man! how noble in reason! how infinite in faculty! in form and moving, how express and admirable! in action how like an angel! in apprehension how like a god! the beauty of the world! the paragon of animals! And yet, to me, what is this quintessence of dust? man delights not me; no, nor woman neither.

Shakespeare
Hamlet: Act II, Scene II

Contents

Preface

Our home is small and limited. The argument of this book is based on self-interest, that it is to our advantage to protect and to manage the Earth. Western society today exists without governing faith or purpose, and can only be swayed by the perception of advantage. My personal belief is that self-interest should not be our only guide: our first duty is spiritual; our task on Earth is to enjoy our home and to obey the commandment to love our neighbor. The biblical writings are filled with joy in creation and with love for the poor and the oppressed. We are told not to muzzle the ox, not to overuse nature. Our society is heedless of this. Its actions are based largely on short-term self-interest and the maximization of profit.

For all scientists, truth is the only arbiter of research, though truth may be elusive. Society, in contrast, often values interest more highly than truth. Wrong actions may be taken deliberately, simply because there is immediate profit in them. Truth frequently fails against self-interest. Truth alone seems unable in our society to change the world, not even if the Four Horsemen of the Apocalypse were to gallop by. Truth, however, is surely unassailable when allied with self-interest. Here lies my purpose, which is to convince not on the basis of the truth that we are devastating our planet and its beauty, but by an appeal to the self-interest of nations. The book first details the scientific concerns about the state of the planet and then outlines some possible consequences of the changes brought about by the policies of humanity. If my argument is correct, it is in the strong self-interest of nations to turn away from their present behavior. If they do not, they risk disruption, insurrection, war, famine, and pestilence. Moreover, there is profit in good environmental behavior, if the profits are counted over a decade, not over a year.

Oliver Cromwell once pointed out that "it is interest that keeps peace." Environmentalists may well win a battle or two in defense of the Earth, but a lasting solution can be achieved only if there is strong interest in sustaining it. We need to devise a global economy that recruits self-interest on the side of environmental peace. It is not an impossible task. On a smaller scale, this is how the democracies operate, using coalitions of self-interest groups to sustain constitutional peace.

We stand at a unique moment in human history. Though unaware, we now manage the Earth. We have the power to make or unmake the planet. We can see the future. Before the battle of Sedan in 1870, the French general, Ducrot, surveyed the end of what was called the Liberal Empire, a great, prosperous state. His despairing comment as he rolled up the map on his nation could fit us all today: "Nous sommes dans un pot de chambre, et nous y serons emmerdés." Our environmental laws and regulations today, for the most part, are simply exercises in putting up umbrellas as the first dollops fall into the chamber pot. Yet it is by no means too late to climb out of the pot. Perhaps it might even profit us to do so.

Saint Paul, in his letter to the Romans, commented that in a rightly based society all things work together for good. This is a doctrine of optimism, more difficult to believe and to act upon than easy despair or cynicism. The same letter carries the hope that the whole creation may be freed from the bonds of corruption. In understanding the Earth we will take a few temporal steps toward that freedom. We may also be able to bequeath that freedom, and peace, to our children.

Acknowledgments and apologies

Science in the late twentieth century is set in a reductionist structure. It is both difficult and dangerous to attempt a work of synthesis. Rutherford wrote up his results in books that were accessible to any educated person, but today science is communicated only within interest groups, from peer to peer, by scientific papers, abstracts, and talks at meetings of specialists. Only rarely, if the results are important or wished to be seen to be important, is the research more widely communicated, often in a partial or inadequate context, by means of a press release. Within science, the barriers between disciplines are almost as high as those that bar science as a whole from the broader intellectual community. Those barriers are very difficult to climb. A work of synthesis such as this is perilous, because it is nearly impossible to reach the research literature in so wide a range of fields as global change covers. Inevitably, the synthesis must be guided by colleagues. Errors will be common, either by being out of date and not aware of the latest paper in the *Journal of Tittlebats,* or from a lack of that intuition possessed by all good scientists when judging work within their field, or simply by getting things wrong and not understanding the arcane language of another discipline. Atmospheric chemists make mistakes when talking to foresters, foresters do not understand the economics of the commons, economists are baffled when faced with atmospheric chemistry. Geologists comprehend none of the above.

Perhaps this book should have been written by a committee. To that committee, and to all specialists and reductionists, I apologize. I ask them to forgive with gentle hearts my errors and omissions, which I hope are not too great; the Earth is a whole, and we need to see that whole, even if the vision is imperfect. I ask them also to write a better book that more accurately guides our decision makers. We need that book: this is not it, but my hope is that it will help to bring about a climate favorable to such a synthesis.

A work of synthesis is not only difficult, it is also not part of the pattern of a career in modern science, and I thank my colleagues who have willingly guided me in so perverse a project. In particular, Mike Keen of the Bedford Institute of Oceanography provided strong criticism and support. I also thank the community of soil scientists, hydrologists, biologists, and atmospheric modelers at the University of Saskatchewan, who have so bravely attempted to create a group outlook on global change that is unique in its integration. Various atmospheric chemists at NCAR and NOAA in Boulder and at DSIR, New Zealand, have been immensely helpful, welcoming an outsider with kindness and teaching an amateur with patience, as have Cathy Law and John Pyle in Cambridge and Jim Lovelock at Coombe Mill. Similarly, the gentle help given me by the limnologists and aquatic biologists in the Department of Fisheries and Oceans in Winnipeg has been invaluable. Canada's leading generalist, Bill Fyfe, has encouraged my global-change interests and given much enthusiasm to those around him. S. H. Alexander's attack on my poor writing style has saved the reader from torture as has Peter-John Leone's (Cambridge University Press) rigorous and uncompromising editing. Kathleen Zylan, Sophia Prybylski (both of CUP), and W. M. Havighurst turned a jumble of paper into a book, very professionally. For criticism, advice, or general encouragement, I am most grateful to Rodney Davenport, Sir Crispin Tickell, A. Barrie Pittock, Charles Caccia, Henry S. Caplan, W. David Hopper, John Hartwick, John Stewart, Hugh Hendry, Peter Fowler, Paul Lowman, Tim O'Riordan, G. H. Miller, E. Ripley, Simon Rippon, and especially Roger Jones. Gordon Wells is thanked for his help with space imagery, and Lindsay Embree for preparing diagrams. Hazel, Hermann, and Aline tolerated a crowded shared office filled with unruly piles of books and papers, and, for all their help, I thank the Earth Science Department in Cambridge, especially Jane Shears, Sam Lal, and Andy Buckley. I thank Angie Heppner, who has created a manuscript out of a pile of sheets scribbled upon by a fountain pen with a bent nib. Author's royalties from this edition all go to the Zimbabwe Scientific Association; however, the Association is in no way responsible for any opinions expressed or errors made, has exerted no editorial control, and is not aware of the views discussed.

Finally, my thanks to C.M.R.F., descendant of Prometheus, who has given up so much for this and borne so much of the load. Perhaps the work will be worth it, for the next generation.

Some decades ago, the British writer C. P. Snow drew attention to the rift between the sciences and the liberal arts. In his view, the intellectual community had divided into two cultures, each battling for the leadership of society. Today it is apparent that neither culture has prevailed: both have been swept aside by a dominant new culture of social management, as represented by the economists, psychologists, lawyers, and business administrators. Science, the humanities, and also religion, the older source of leadership, have all become the domestic animals of modern society, pets fed their scraps with varying degrees of indulgence but no longer playing any important role in the direction of humanity. Religion, which has nourished and guided our world view over the centuries despite its internal conflicts, is now unheard and almost mute; the voices of science and the humanities are also little regarded and becoming faint and shrill. In the corridors of power the social managers reign unchallenged.

To these social managers, my apologies. Among those natural scientists who deal with environmental issues the commonest attitude toward economists, lawyers, and business administrators is a deep aversion, even often the prejudicial contempt of an old aristocracy for a new. Scientists like myself, especially those in field disciplines who have traveled widely in poor nations, find most economic texts or management plans almost impossible to read, because the basic axioms so often seem to be wrong, possibly describing another planet but with no validity on ours. It is perhaps not accidental that one of the most influential of the economic appraisals of the actions of humanity, Garrett Hardin's thoughts on the tragedy of the commons, was published in a scientific journal. However, as a consequence of the rift between environmental scientists and social managers, our prejudice in the scientific community may have blinded us to the real successes of economic thought. It is for this reason that I apologize: any scientist attempting to put forward a unified picture of global change inevitably risks falling victim to prejudice against social managers and to ignorance of their valid contributions.

To the politicians, too, I must apologize. Many of us can tell stories of politicians who have not listened. Environmental issues are too commonly misunderstood, or exploited for political gain, even if the result is harmful to the community or nation. There is rarely a long-term view, and few politicians are willing to study the issues in the sort of depth that they are prepared to accord to economic and fiscal policy. In consequence of this type of behavior, there is now in the scientific community a widespread mistrust of political leaders. But this too is prejudice on the part of the scientists: my own experience of giving evidence to a parliamentary committee is that the minister could spare us virtually no time, but attending that same committee were many thoughtful backbenchers who asked searching and intelligent questions whether or not they believed us. To them, then, my apologies if I have suggested politically impossible steps in the latter chapters of this work: I look to them to suggest better.

Science has itself changed as it has lost influence. In the Western nations we no longer attract the best minds in the student population: they have left for more rewarding occupations. In exchange, we recruit our graduate students from the poorer countries – wonderfully gifted people, some of them, and an asset to cherish, but their world view is different, and their loss an enormous burden to their homelands: China, India, Africa. Some of them, though, do share the deeper vision that characterized the early leaders of modern natural science: that insight into the reality below the surface which Rutherford and his school possessed. Among the modern leaders of science, the professors and heads of research teams, that insight is now of little value. In response to the political and managerial onslaught, the demands for productivity and commercial results, we have retreated into what has been termed the excellence of mediocrity: practical, safe science, seeking small solutions to little well-defined problems of immediate piecemeal benefit, and to this path we guide our students. As we shorten the range of our vision, we become ignorant of the ground around us, and grow unaware of the doings of fellow scientists in other disciplines. Ignorance is the breeding ground of isolation and then of prejudice: each of us in science has prejudice against other disciplines, as less pure, or less factual, or less mathematical, or less relevant, or whatever. To my scientific colleagues in other disciplines, then, my apologies where I have omitted or misrepresented the significance of their work through prejudice or unjustified ignorance.

The diversity of these apologies illustrates the width of the global problem. Lord Rayleigh, who was one of the founders of the modern dairy and milk industry, and also a most distinguished physicist when he had the time, commented of his predecessor as Cavendish Professor, J. C. Maxwell, that he had always endeavored "to keep the facts in the foreground." Grand ideas or deep emotions will not of themselves rescue our planet: we need to base our world view on facts. But which facts? Each community that makes up society, in the rich countries and in the poor, needs to search out those facts and to examine its basic axioms of behavior and understanding. My hope is that this book can stimulate that work, however slightly.

1
Introduction

Nature is at its end: we need to put something in its place. During the past decade the validity of this statement has been debated within a small part of the scientific community, without unanimous agreement. More recently, the discussion has moved beyond that small group, into the political forum. As the debate has broadened it has become less rational, more intuitive; opposing viewpoints are stated forcefully, sometimes with beauty, often in anger. Emotional vision can be deeper than rational sight, but both emotion and reason need clarity and understanding in order to comprehend the fabric of nature, the forces bringing it to an end, and the consequence for humanity and the living world. Over the past centuries, humanity has assumed that nature is infinite, a free common good. If that assumption is wrong, reason tells us that the consequences will be severe and, for many of us, life-taking; emotion tells us that we may also lose beauty, and with it some of the substance of life itself. The debate is important and unresolved. To enter into it, we need to define nature, the challenges that may have ended it, and the future constructs that may replace it.

The statement that nature has ended begins with the idea of "nature," but the scientific understanding of nature is still very limited. All modern science, physical and biological, is an investigation of nature, and the best known of all scientific journals is called, simply, *nature*. The vigor of our investigation hides much ignorance, as it is usually much easier to uncover detail than generality. By funding a professor to study a point mutation in a gene, or by setting a student to analyze the composition of a lava flow, society can reliably obtain results, but if we ask why the atmosphere is about one-fifth oxygen, not one-sixth or one-third, we will find no answer in the textbooks and very few scientific papers on the subject. Only brave young innocents or the bypassed elders of science dare to tackle a problem that is so poorly constrained. In consequence, we have no clear knowledge of the system that regulates the natural environment, the world about us, which to most people is "nature." We understand weather fairly well, we have analyzed the air, we have counted the trees and classified the birds, but we do not yet fully know what deeper process regulates the composition of the air, the temperature of the planet, or the distribution of species.

Because our knowledge of nature is so poor, it is difficult to justify the proclamation of the end of nature. This point is often raised in the debate about the Earth: we cannot be certain about the death of what we did not know, and even if it has died, we should not mourn its passing. The human world is divorced from nature and the boardrooms of Tokyo or New York should not concern themselves with the fate of a distant ghost that may or may not have been destroyed by their actions. To this the rational answer is that the distant ghost may yet control the air of Manhattan; the emotional and more oblique answer is that the band on the Titanic also continued playing.

If nature has ended, that end has come swiftly. Today, as I write this during a visit to Cambridge, the college fellows play their lunchtime game of bowls on a soft green lawn, as they have done for centuries with the same ancient wooden balls. All, seemingly, is continuity, without termination or even change. The day is very warm for early February, and the spring flowers are opening. Above the buildings, flags stand out in the strong wind. There have been great storms recently, but perhaps this warmth and energy is part of the stochastic variability of weather, not evidence of permanent change. The flags fly to celebrate the anniversary of the accession of Queen Elizabeth II, a symbol of changeless continuity. Her accession took place in 1952, when nature seemed unchallenged: less than four decades later we debate nature's end. Just as the British Empire has vanished in those few decades, so also the wild empire of nature may have melted away. The elements of nature remain – we can breathe the air, rain falls from the sky – but power over the environment may have passed out of nature's hand. The spring flowers may not be opening at nature's call.

The old saying is that nature abhors a vacuum. If the power of nature has ended, then it is humanity that has become the regulator of the Earth's surface. In recent years, a continuing scientific debate, fostered especially by the American Geophysical Union, has considered the problem of nature's design, whether there are controlling laws that optimize the natural management of the environment. If there are no such laws, then all is accident and the death of old nature is simply another nasty accident, a loss of diversity but nothing seriously to perturb the body of humanity, though it may sadden our soul. If, however, there are laws that govern the stability of the environment, we need to discover them soon, so that we can manage the system after the death of our victim. It is also possible that nature is not yet dead, but has unknown laws and corrective mechanisms that will turn upon any destabilizing agent and remove it.

Humanity is disparate, in factions, and in rapidly increasing number. It is optimistic to assume that so divided an agent is able to replace the hand of nature, yet it is pessimistic to imagine that there will be no attempt.

In what follows, these themes – nature, its end, its replacement – are explored at length. Chapter 2 deals with the operation of the natural world, especially the atmosphere, and outlines some of what we know or suspect about the management of the planet. In Chapter 3 the forces that are causing change are discussed, with an assessment of the immediate consequences of change (Chapter 4), to discover whether or not they justify the reports of the death of nature. Chapters 5–7 examine the second part of the opening statement: if nature has ended, how shall it be replaced, and how is humanity to manage itself? Finally, the less rational but deeper aspects of the problem are briefly addressed: how should we live in a changed and still-changing world?

Reading list

Friday, L., and R. Laskey, eds. (1989). *The fragile environment: the Darwin College lectures.* Cambridge: Cambridge University Press.
McKibbin, B. (1990). *The end of nature.* New York: Viking Penguin.

2
The natural Earth

But ask the animals and they will teach you, or the birds of the air, and they will tell you; or speak to the Earth, and it will teach you.

Job 12: 7–8.

Stability is not natural, and there has never been such a thing as a perfectly stable natural environment. Throughout geological time the environment has been changing, sometimes slowly, sometimes dramatically. Life has adapted constantly to those changes; many of the changes were directly caused by life.

Humanity – modern *Homo sapiens* – evolved about 100,000 years ago, and immediately began altering the natural world. The effect seems to have been a wave of extinction of animals and plants: mammoths and mastodons, the lions of Judah and Greece, and the chestnuts of America. By the year A.D. 1950 the fauna and flora of the planet and hence the global environment were permanently changed, though not yet to the extent that climate was measurably altered.

2.1 The history of the natural Earth

The atmosphere regulates the temperature of the Earth's surface and has done so throughout geological time. Most models of the evolution of the Sun and solar system imply that the Sun was significantly fainter when it was young, billions of years ago, and has brightened over time. If these models are correct, the reason the early Earth did not freeze was that atmospheric gases acted as a warm

blanket, increasing the surface temperature by as much as 50°C.
with a bright Sun, the effective temperature of the Earth, which is
ture of a hypothetical airless body that would radiate heat at the sam
Earth does, is well below freezing. A planet at this temperature wou
port life. Fortunately the natural water vapor, together with carbo
methane, and other trace gases in the air, act to raise the temperature to a global
average around 15°C, which is rather more comfortable. The gases blanket the
Earth and keep the surface warm.

Nearly four billion years ago, the young Sun was probably only roughly three-
fourths as bright as the modern Sun. Despite this, geological evidence shows
that the oceans were liquid, because of the blanketing effect of the atmosphere
which probably contained much more carbon dioxide (CO_2) than today. Scientists have little sure knowledge about the state of the early atmosphere; one suggestion is that it was dominantly carbon dioxide and nitrogen together with water
vapor. At no time over the last four billion years is it likely that either methane
(CH_4) or ammonia (NH_3) was a major component of the atmosphere, although
trace amounts are sustained in the air by the emissions of living organisms.
Sometime, probably in the period 4.2–3.8 billion years ago, life began on Earth.
The oldest firm geological record of life dates from about 3.5 billion years ago,
by which time there is evidence that a complex and varied bacterial community
existed. Life had already started to use most of the important biochemical reactions, such as those involved in photosynthesis, and the geological evidence
shows that life had begun to process the atmosphere, managing the oxygen and
carbon dioxide, and nitrogen too.

For most of the Earth's history, life was single-celled and bacterial. Bacteria
multiply to exploit all available resources, and so life was abundant, but left little
fossil record. Around a billion years ago multicelled organisms evolved, and
about 570 million years ago, at the beginning of the Cambrian period, the first
animals with hard parts such as shells appeared. Over the next 100 million years
the oceans were filled with a complex and competing chain of animal life. Colonization of the land began, first with simple plants, insects, snails, and so on.
By the Permian and Triassic periods, roughly 200 million years ago, a complex
ecological web had been set up on land, with a wide variety of animals and
plants, and a diverse and complex community occupied the seas. Many of the
larger Permo-Triassic land animals were mammal-like reptiles, including our
own ancestors. During much of the Permo-Triassic time these mammal-like reptiles dominated the land. In the next two geological periods, however, it was not
the mammals but the dinosaurs that became the largest and most numerous of
the land vertebrates. Dinosaurs flourished and occupied much of the Earth. Then
something happened: the dinosaurs suddenly disappeared at the end of the Cretaceous, about 60 million years ago.

Life is both fragile and versatile. It is fragile: at the end of the Permian in the
seas, and again at the end of the Cretaceous on land and at sea, a diverse and
seemingly stable ecosystem collapsed. The complex fabric of animal and plant
life that previously had dominated the landscape and the seas disappeared. What-

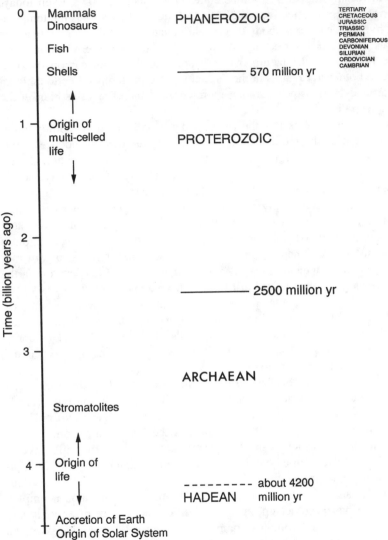

Figure 2.1. The geological time scale, showing the major events in the history of life. (Modified from Harland et al. 1989.)

ever the cause of the catastrophe at the end of the Cretaceous, when the dinosaurs died so did many of the plants. There is an extremely close relationship between life on Earth and the climate and atmosphere of the planet. The animals and vegetation changed, and so did the climate. Life is also versatile: the collapse at the end of the Cretaceous was followed by a period of ecological poverty, but then by further evolution to the richness and diversity inherited by humanity.

The interrelationship between climate and life has been recognized by a number of scientists, including the Russian V. I. Vernadski, who developed the notion of a *biosphere,* a term first made popular by the geologist E. Suess, to describe the realm in which life interacts pervasively with the environment. The biosphere is the thin skin on the Earth's surface in which life exists, varying from a few millimeters to a few hundred meters thick, though its influence pervades the planet. There is now some degree of consensus amongst atmospheric scientists that the biological controls exerted by the biosphere on climate are immense, pervasive, and critical. The composition of the air and the climate of the planet depend on life's history and present state.

The history of life on Earth, as deduced from the geological record, demonstrates that the biosphere is an extraordinarily complex interactive system which, like any human society, bears within itself the history and traditions of former times, and which can, on occasions, collapse disastrously. If enough links are broken, the whole fabric can unravel. This has happened several times in the past. Following each collapse, over the next tens of millions of years the fabric has been rewoven, but the community present at the time of the collapse never returns.

In the present century a large number of species has gone from the biosphere. This loss of species is equaled in the geological record only by the most massive extinction events, such as that at the end of the Cretaceous.

2.2 The physical controls on the environment

2.2.1 Energy

Virtually all the energy received by the atmosphere comes from the Sun. It is simple to show this: when one lies on a beach it is the sunny side of the body that is warmed. Some energy does come from the Earth's interior, but this contribution is small and is not normally important to the climate, except when volcanic eruptions change the reception of solar energy by placing dust and gas in the air.

The Sun is hot, and as a result the energy that arrives from the Sun comes as light characteristic of hot bodies, rich in radiation of short wavelengths, especially in that part of the spectrum which we can see with our eyes as the colors of the rainbow. The Earth returns to space an amount of energy equal to that which it receives from the Sun, to balance the incoming energy. If it did not, the temperature would climb steadily. Because the Earth is much cooler than the Sun, the energy emitted by the Earth is of much longer wavelength, in the infrared part of the spectrum. We cannot see this energy, but we can feel it as heat.

The temperature of the surface of the Earth depends on exactly how the balance of incoming and outgoing energy is achieved. On the natural Earth, some of the incoming solar radiation is sent straight back to space – either backscattered by air, reflected by clouds, or reflected directly from the surface. A larger fraction of the incoming radiation is absorbed by water vapor, dust, and gases such as ozone in the air, by clouds, and by the surface. The fate of this absorbed energy, especially the energy absorbed by the surface, controls the surface tem-

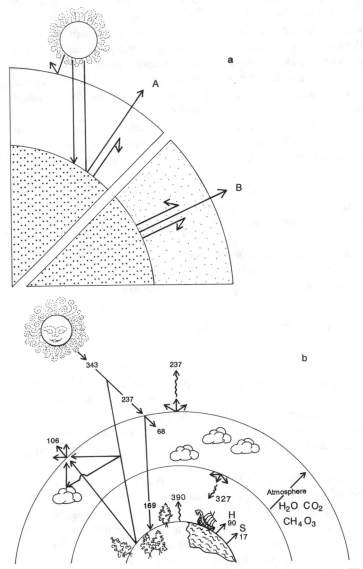

Figure 2.2. Energy inputs and outputs of the Earth and atmosphere. The diagrams show details of the complexity of the greenhouse effect. (a) An outline sketch of the greenhouse process. The greenhouse gases absorb the heat that leaves the Earth and reradiate it back to the surface, acting as a blanket. (The process is much more complex than this, as the lower diagram shows.) A: Sunlight falls on the planet and is absorbed or reflected at various stages as it penetrates the atmosphere. It is reradiated back into space at longer wavelengths. Some of this reradiated energy is absorbed by greenhouse gases in the air; if the concentration of greenhouse gases rises, as in B, more reradiated energy is absorbed in the lower atmosphere, causing temperatures there to rise before the energy is finally reradiated to space. (b) The global energy balance. Averaged over a year, the

peratures of the Earth. Some of the heat from the surface is directly radiated back to space, as infrared or heat radiation. Other heat is radiated from the surface to be absorbed by water vapor (in clouds or simply as vapor) and carbon dioxide in the air. Of the rest of the heat leaving the surface, part is carried up as heat that can be felt (e.g., in hot air), but much is carried in water vapor. An analogy is boiling a kettle: when we add energy, the water heats and then vaporizes into steam. It takes a large amount of energy to turn liquid water into vapor. This energy is then held in the vapor, and is called *latent heat*. On the Earth's surface, liquid water is vaporized and the vapor is carried into the air. As this happens, the heat is transferred from the surface into the latent heat in the water vapor in the air. We can feel the Sun's heat or the heat of a hot stone but we do not directly feel the heat carried up by water vapor (which is why it is called latent heat). The latent heat transfer is very important in controlling the surface temperature of much of the Earth's surface. The transfer of heat into water vapor happens all over the Earth, but especially over the tropical oceans. The rainforests also emit vast quantities of water vapor. To put it into physiological terms, it is as if the tropical parts of the Earth are sweating. Just as a human being sweats to keep cool on a hot day, so the forest emits water. J. E. Lovelock has used the term *geophysiology* to describe this process. There is a close analogy between the way in which we regulate our body temperatures, by sweating or by clothing and blankets, and the way in which the Earth operates.

2.2.2 The greenhouse effect

Some of the gases in the Earth's atmosphere play an especially important role in the energy balance of the planet because of the particular ways in which they absorb radiation. Some of these properties are shown in Figure 2.3. Ozone (O_3) is a strong absorber of light with a relatively short wavelength, in the ultraviolet. Carbon dioxide (CO_2) absorbs light at rather longer wavelengths, especially in the infrared region. Water has a complex absorption spectrum, absorbing light of various wavelengths. It also, of course, forms clouds. As light enters the atmosphere the energy is selectively absorbed by the atmosphere. Similarly, the

Caption to Figure 2.2 *(cont.)*
Sun's radiation brings 343 watts per square meter (Wm^{-2}) to the top of the atmosphere. For comparison, imagine all the output of three and a half typical 100-watt light bulbs shining on a square meter. Solar radiation amounting to 106 Wm^{-2} (or one light bulb per square meter) is reflected, which is why the Earth is a bright object to astronauts standing on the Moon. The other 237 Wm^{-2} are radiated back to space as long-wave or heat radiation, so that the planet is in radiative balance. The energy is processed in various ways as it descends to the surface and as it returns to space. The atmosphere absorbs some (68 Wm^{-2}). The surface receives 169 Wm^{-2} together with 327 Wm^{-2} of energy radiated down from the atmosphere to the surface. The surface radiates upward 390 Wm^{-2} of heat. The surface also gives off 90 Wm^{-2} of latent heat (H) and 17 Wm^{-2} of sensible heat (S). (Modified from Ramanathan 1988.)

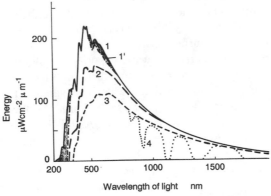

Figure 2.3. The passage of radiation from the Sun through the Earth's atmosphere. The horizontal axis is the wavelength, in nanometers, of the Sun's light, from ultraviolet (left-hand end), through the visible spectrum of the colors of the rainbow, to infrared (right-hand end). The vertical axis, in microwatts per square centimeter per micrometer of wavelength, is the flux of energy at each wavelength. Curve 1 shows the radiation arriving into the outermost atmosphere. Curve 1′ shows how much radiation emerges below the stratospheric shield of ozone. The shaded regions, especially around 200 to 300 nm wavelength in the ultraviolet part of the spectrum, indicate how much radiation is absorbed by the shield. The remaining radiation passes through the rest of the atmosphere, with further absorption and scattering by air molecules (Curve 2) and dust particles (Curve 3). Curve 4 shows the light that eventually reaches the Earth's surface, mostly in the region between 400 and 1000 nm. This region includes the visible spectrum, on which life depends and to which it is adapted. (From Deutscher Bundestag 1988, after earlier sources including Lacis and Hansen 1974.)

energy that is reradiated from the surface back into space is also selectively modified by atmospheric absorption as it passes through the air.

Sunlight that arrives at the top of the atmosphere is successively filtered by a variety of atmospheric gases and processes. The first major filter is absorption by oxygen and ozone in the high atmosphere (the *stratosphere*), which removes the energetic short-wavelength ultraviolet light. This energetic light is dangerous to most organisms, and so the ozone filter protects life on the surface below. As the light descends into the lower atmosphere, it is scattered by air molecules and chemicals in the air. The scattering process gives us a blue sky. In the lower atmosphere water vapor and other components of the air absorb light. This heats the air and reduces the intensity of the light that eventually reaches the surface.

In order to maintain a balance the Earth must return radiation to space. To expand the argument above, as much radiation must leave as enters. The temperature of the surface and of the various layers of the air depends on how this balance is achieved. A hot radiating body, such as the Sun, emits much short-wavelength light, such as ultraviolet light. This short-wavelength light is energetic. In contrast, a cool radiating body, such as the Earth, produces light of longer wavelength, in the infrared part of the spectrum, that is less energetic.

This infrared light (which we feel as heat) leaves the surface and the atmosphere and radiates back into space. If for some reason the radiation returning to space is blocked, the energy is trapped. The system must warm until eventually the balance is restored.

The light that leaves the surface of the Earth and the atmosphere to return to space is selectively absorbed by several gases, especially carbon dioxide, methane, and water. This absorption warms up the lower atmosphere or *troposphere,* which eventually reradiates the energy upward. In general terms, the higher the absorption the higher the temperature of the lower atmosphere, although the process is complex and much modified by other factors such as clouds and dust.

The consequences of the radiative properties of these gases are commonly called the *greenhouse effect.* In an ideal garden greenhouse made of glass, sunlight (short-wavelength light) can enter, since glass is transparent to ordinary visible or short-wavelength light. The heat energy that has entered cannot leave again by radiation, because glass is opaque to the long-wavelength infrared or heat radiation given off by the ground surface. As a result, the interior of the greenhouse warms up. The same applies to the atmosphere. The air is generally colder than the ground (except at the very top of the atmosphere). This means that any heat energy that is emitted by the ground or the sea will be trapped to some extent as it passes back upward. The trapping of heat, or infrared energy, is mainly done by water vapor and clouds, and also by carbon dioxide (CO_2) and methane (CH_4) as well as other trace gases. For example, carbon dioxide is transparent to most sunlight, which passes through to fall on the ground or on low-level clouds, but the returning infrared radiation is blocked by the carbon dioxide, which absorbs the radiation, becomes warmer, and warms the atmosphere below and around it (Figs. 2.4 and 2.5).

If the atmosphere did not exist, or had no greenhouse effect, the Earth's temperature would be around $-18°C$. Its actual average temperature is about $15°C$. The difference is produced by the atmospheric greenhouse, which warms the planet by about $33°C$ naturally. If it were not for this, the seas would freeze and life would be extinct. If the greenhouse effect were too small, the planet would die; if it were too large, the surface would heat up uncomfortably.

It is clear from Figure 2.4 that an increase in the concentration of gases such as CH_4 and CO_2 will change the behavior of the atmosphere as a whole and warm the lower atmosphere, or troposphere, where these gases are most abundant. Natural ozone, on the other hand, acts oppositely, since much of it is found at high altitude in the stratosphere. The net effect of natural ozone in the atmosphere is to warm the stratosphere and, by filtering sunlight, to cool the troposphere. At present, all these gases are present in trace amounts, but any changes will alter the radiative behavior of the atmosphere, especially as it affects the important wavelengths of light around the visible spectrum. An increase of the greenhouse absorption of the troposphere will mean that in order to maintain an energy balance the temperature of the Earth's surface and lower atmosphere must rise. The other major gases in the atmosphere, not shown in Figure 2.5, are radiatively insignificant in the temperature ranges found at low altitudes, al-

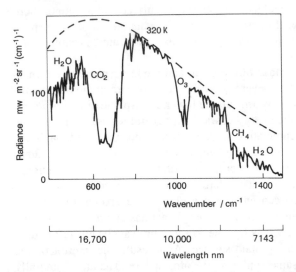

Figure 2.4. Thermal energy leaving the Earth, compared to the energy that would be expected to leave from a blackbody at 320 K (47°C). In this satellite measurement from above the Sahara, wavenumber (number of light waves per cm) is plotted on the horizontal axis, with equivalent wavelengths in the long wavelength or infrared region shown below. Note that the plot is reversed compared to Figures 2.3 and 2.5. The vertical axis is an index of energy actually emitted to space and detected by the satellite. The dashed line (320 K) shows roughly the energy radiated by the hot Saharan surface. The difference between the two lines is the greenhouse absorption of the atmosphere. Note the importance of H_2O, CO_2, O_3, and CH_4 in blocking outward radiation, and the atmospheric window between wavenumbers of 800 and 1000 cm^{-1}. (Modified from Nimbus 4 results in Hanel et al. 1971.)

though they can be important at the top of the atmosphere. When changes to the climate are being considered, it is CH_4, N_2O, O_3, CO_2, and H_2O, as well as the man-made chlorofluorocarbon (CFC) gases, that are of critical importance.

Domestic greenhouses do not really work in this idealized way, however; much of their effect comes from their ability to trap warm air near the ground. So too is the atmospheric greenhouse not ideal. Water vapor, for instance, is the strongest greenhouse gas, but it is not completely transparent to sunlight. Moreover, it tends to form clouds, which are enormously important in controlling the temperature of the atmosphere, outweighing the direct effect of CO_2 by a hundredfold. Clouds regulate the radiation falling on the surface, and they also control the radiation as it returns to space. A simple demonstration of this is in the contrast between a cloudless day in fall, sunny and warm in the daytime but frosty at night, and a damp cloudy day in fall, which may be much cooler in the daylight but does not produce frost at night.

Clouds are still poorly understood, but the available evidence suggests that on the whole they cool the Earth. A cloud has two effects. At its bottom, a cloud absorbs radiation given off by the surface – ground or sea. The bottom of the cloud is usually cooler than the Earth's surface. At the top of the cloud, which is even colder, radiation is given off to space. The net balance is that clouds act as blankets, protecting the surface by trapping heat. Deep clouds, which are especially common over the monsoon regions of the Indian Ocean and Indonesia, exert a large greenhouselike effect. However, clouds also have a second effect. They are white and so they reflect short-wavelength sunlight from their tops back into space. This reflection back to space has a cooling effect. The biggest net cooling effect from clouds comes over the middle- and high-latitude oceans of the Northern Hemisphere. Widespread clouds reduce the radiative heating in these areas by 100 watts per square meter, as people in Ireland or Newfoundland well know.

The greenhouse effect of clouds is many times the effect that would be given by doubling the CO_2 level of the air. The cooling effect of clouds is even greater, one and a half times as large as their greenhouse warming. The net overall effect is that clouds cool the planet by $10°-15°C$ in contrast to a hypothetical Earth without clouds. Clouds thus make the Earth cooler than it would otherwise be. This overall cooling effect depends strongly on the planet's weather systems, and the balance could shift dramatically if weather patterns changed. Because the effects are so much greater than the greenhouse effects of other gases, climate can presumably change substantially if the distribution of cloudiness of the planet is changed, for instance by changing vegetation patterns, especially in forest areas, or by changing the weather of the North Atlantic region.

The individual roles of various greenhouse gases are discussed in more detail later. Apart from water, CO_2 is the most important, and methane (CH_4) is also significant. If extra molecules of these gases are added to the air, the greenhouse effect is increased. The addition of a molecule of CH_4 has a much greater effect than the addition of a molecule of CO_2. This is, in part, because there is far less CH_4 than CO_2 in the air. That part of the atmospheric window where CO_2 is active is already partly closed by CO_2, while the CH_4 window is still fairly open. Recently, the man-made chlorofluorocarbons have been added to the air. They are very powerful greenhouse gases, even in comparison to CH_4. Each molecule of CFC has a greenhouse effect that is thousands of times greater than a molecule of CO_2.

Over the past four billion years the greenhouse effect of the atmosphere has probably slowly decreased. In the early part of the Earth's history the Sun may have been faint but the greenhouse effect strong, so that liquid water existed on the surface despite the low solar input. Venus, much closer to the Sun, probably also had liquid oceans. Over time, the Sun brightened and the greenhouse effect of the Earth's atmosphere must have decreased; if this decrease in the greenhouse effect had not happened, the surface temperature of the planet would have risen markedly, perhaps boiling the oceans. On Venus, in contrast, there was no reduction in the greenhouse effect. There, the oceans were eventually lost to space, and a runaway greenhouse effect warmed the surface to its present-day temperature of 470°C, an inferno.

During the last one billion years of the geological record, the temperature of

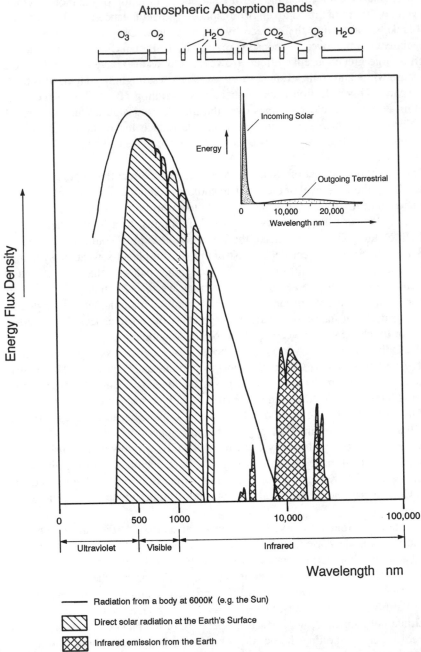

Figure 2.5. Another summary of the Earth's input and output of energy, this time by wavelength of the radiation. The inset diagram shows the distribution with wavelength of the incoming solar radiation, which is mostly in the short-wavelength or visible part of the spectrum, and the outgoing terrestrial radiation, which is mostly infrared, or heat.

the Earth's surface has fluctuated, often dramatically, but never beyond the bounds of habitability: the oceans have sustained complex living communities. In the Permian period there was a major glaciation; in contrast, the Cretaceous period was warm even in the Arctic. More recently, during the Tertiary period, temperatures have fluctuated greatly. About 40 million years ago, which is geologically fairly recent, there were forests in the high Arctic and alligators in Arctic streams; in contrast, 15 to 10 thousand years ago large parts of the northern continents, including virtually all of Canada, were covered with ice, and today Antarctica and Greenland remain icebound.

Through all this change life has acted to influence climate. In periods when conditions are warm, abundant growth and the formation of limestone, coal, and oil removes CO_2 from the air, reducing the greenhouse effect; in colder times there is less plant life and less carbon is extracted from the air. Volcanoes erupt irregularly but, over a geological time scale, are steadily adding to the CO_2 content of the biosphere; simultaneously, living organisms extract carbon from the biosphere, fixing it as coal and oil, or as limestones, some of which are eventually returned to the interior.

2.2.3 *The biological influence on the reflectivity and color of the Earth*

The reflectivity of a planet is very important in controlling its surface temperature. The *albedo* of a body or surface is the ratio of reflected radiation to incident radiation. This ratio depends on many factors. In general, the Earth's surface only reflects a small part of the incident radiation. Clouds, in contrast, typically reflect more than half of the radiation falling on them. As discussed above, the reflectivity of oceanic regions is very dependent on the extent of their cloud cover. On land, plants with glossy leaves are highly reflective, or can vary their reflectivity. Snow reflects; water bodies absorb much of the incident light. As the vegetation of the planet changes, so does the albedo.

An example of the importance of the reflectivity and color of a planet is the daisy world model of A. J. Watson and J. E. Lovelock. They considered a planet such as the Earth, but inhabited only by black and white daisies. In the model, over the planet's history the Sun warms up, as shown in Figure 2.6. If the planet were simply a lifeless earthen ball, it would warm up in sympathy with the warming of the Sun. On a planet inhabited by the daisies, however, the black

Caption to Figure 2.5 (*cont.*)
The main diagram shows this in detail, but this time with both axes plotted logarithmically. The solar radiation, as it comes in, is partly absorbed (see Fig. 2.3) by the various absorbing gases. On the right is shown the outgoing radiation, which escapes through the cross-hatched regions in the infrared part of the spectrum. At the top of the diagram are shown the main atmospheric absorption bands according to wavelength. Ozone, oxygen, and water affect the incoming radiation, while ozone, carbon dioxide, water, and various trace gases affect the outgoing radiation leaving the Earth. (Modified from Barry and Chorley 1987.)

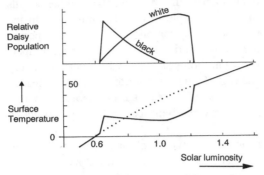

Figure 2.6. The daisy world model of A. J. Watson and J. E. Lovelock (1983), which
supposes a hypothetical Earthlike planet covered entirely with daisies. Top diagram
shows the change in relative abundance of black and white daisies with solar luminosity,
which increases with time. Bottom diagram shows the variation in surface temperature
with increasing luminosity, over time. Solid line: daisy planet. Dotted line: dead planet.

daisies, which absorb much but reflect little of the incident radiation, would be
favored in the early days of a faint Sun. The daisies would blacken the planet,
absorbing heat on the surface and warming it. As the Sun brightened over time,
the black daisies would become uncomfortably hot and would begin to suffer
from heat stress. In contrast, cooler white daisies would become competitive.
The planet would slowly turn gray as white daisies became more abundant, and
then white, cooling the surface. Over billions of years of the planet's history the
surface temperature of the planet would remain steady, despite the warming of
the Sun. Eventually, the planet would become wholly white. If the Sun continued to brighten it would finally become so bright that the temperature on the
planet would at last rise (there being no daisy that is whiter than white) and
conditions would become impossible for the daisies. They would die and the
planet would return to its inorganic warming curve.

Even in a system as simple as a two-daisy planet, life can and will modify the
climate to suit itself. This is Darwinian selection. The competition between the
black daisies and the white daisies has the paradoxical effect of allowing the two
species of daisies cooperatively to manage the environment of the planet. There
is evidence that plankton regulate sea-surface temperature by varying the color
of the ocean. It is possible that over the aeons the Earth has evolved in this way.
Has life been the factor that saved the Earth from the fate of Venus?

2.2.4 The circulation of the air

The energy falling on the Earth from the Sun, together with the physical features
of the surface − such as mountains − and the Earth's rotation, all drive and
control the air circulation. The troposphere is heated from below by the ground,
and is cooled from within as it emits radiation. As a result, it is turbulent and is
very effectively stirred up. To us this process appears as weather: wind, storms,

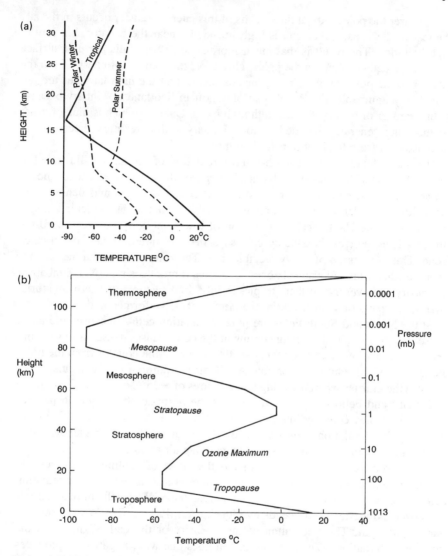

Figure 2.7. The variation with height of the distribution of temperature in the atmosphere. (a) For various latitudes and seasons. (b) Generalized distribution over the planet. (From Barry and Chorley 1987.)

rain, and drought. The stratosphere is not like this. It is a much more peaceful place. The boundary between the two regions is called the *tropopause,* and the height of the tropopause depends on the surface temperature. In the tropics it is as high as 17 km (10 miles), while in the polar regions it can be as low as 8 km (5 miles).

The circulation of the air and of the waters of the oceans serves to even out the temperature of the planet, cooling the equatorial regions and warming the areas nearer the poles. About three-fifths of this energy transfer occurs in the air, the rest in the oceans. The sun is high in the sky near the equator, but low in high latitudes. The result is that much more solar energy falls per unit surface area on the equator than on the poles. However, the transfer of energy from the equator to the poles by the air and the sea means that the high latitudes are not completely uninhabitable. People can play golf in Scotland in January because of the transfer of heat by the sea, although North Dakota, which is much further south, is not generally regarded as good January golfing territory, because it is more isolated from the heat transfer in winter.

The major features of the north-south circulation of the air are illustrated in Figure 2.8. Near the equator, solar heating causes the air to rise. The air moves toward the poles, carrying energy with it. As it cools, it becomes denser; near 30° latitude, the dense air sinks and returns to the equator. This circulation system is called the Hadley cell. The region of low-level convergence toward the upward-rising part of the Hadley cells is called the Intertropical Convergence Zone. This is a region of heavy precipitation. The zone moves with the seasons over the equatorial and tropical zones. In Africa it produces rain over Zimbabwe in January and over the northern tropics in July. In Amazonia, western equatorial Africa, and parts of southern Asia, the rainforests have formed under it.

At the North and South Poles are polar circulation cells. Because cold air is denser than warm air, the strong cooling at the poles causes dense air to sink and drives a circulation system in which air flows toward the equator from the pole, along the surface. Outbreaks of cold Arctic air are familiar to Canadians, especially in the prairies where morning temperatures of −35°C or lower are common in January and February. The third part of the north-south circulation lies between the Hadley cells and the polar cells. These are the Ferrel cells. They are driven by the circulation in the other two cells, with dry descending air at around 30° of latitude (above the deserts), and ascending air at 60°.

The movement of air in these cells has the effect of cooling the tropics and warming the regions closer to the poles. There is also an east-west circulation cell, not shown in Figure 2.8, that is known as the Walker cell. In this cell, air rises over Indonesia and descends over the eastern Pacific. The Walker cell can vary in strength. The fluctuation of the intensity of the cell is known as the Southern Oscillation. When the oscillation goes one way, negative, it produces weather conditions, known as El Niño events, that can produce major droughts in the Southern Hemisphere, wet weather in deserts, and warm winters on the prairies. When the oscillation goes strongly the other way, positive, the weather conditions that result are known as La Niña.

A model of the variation in the Walker circulation during an El Niño–Southern Oscillation (ENSO) event is shown in Figure 2.10. The top panel shows the December–February flow along the equator in normal times. Rising air and heavy rain occurs over the Amazon basin, in central Africa, and over Indonesia and the western Pacific. The bottom panel shows the circulation during a strong ENSO event, which produces a catastrophic drought over Africa and changes sea temperatures in the Pacific.

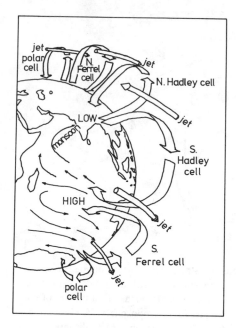

Figure 2.8. The basic circulation pattern of the atmosphere over Asia and the Indian Ocean during the Northern Hemisphere summer. (Modified from Meehl 1987.)

The air circulation cells control the major vertical motion in the atmosphere. On the surface, the winds we observe and that bring rain or drought are strongly directed by the Earth's rotation. This produces an easterly component to the low-latitude trade winds (the bottoms of the Hadley cells). The Earth's rotation also imparts a westerly drift to the path of the surface air of the Ferrel cells, giving a counterclockwise circulation of depressions in the Northern Hemisphere and a clockwise circulation of depressions in the Southern Hemisphere.

In general, the circulation system of the planet is robust. It is difficult to change the broad pattern of the cells. In detail, though, it may be more fragile: the location of the boundaries between cells may change. The vigor of the Intertropical Convergence Zone in Africa is very variable, for instance. Some years the rainfall is intense while in other years the convergence is weak and little rain falls. The Walker cell also varies greatly. It may be possible to modify the latent heat transfer below the uprising limbs of these cells. If we were to warm the poles substantially, especially the North Pole, the polar circulation might change drastically. Such warming happened in the Cretaceous and during part of the Tertiary, when one or both of the poles were ice-free and forests grew in high latitudes.

2.2.5 *The circulation of the seas*

The movement of water in the oceans carries heat from equatorial regions toward the poles. In the tropics and subtropics, solar heating exceeds outgoing infrared

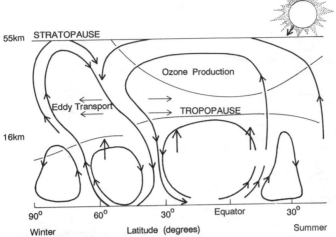

Figure 2.9. The circulation in the troposphere and stratosphere. The tropopause is the boundary between the lower atmosphere, the troposphere, which is turbulent, and the overlying higher air, the stratosphere, which is not. The stratopause is at the top of the stratosphere; above it lies the highest part of the air. In the troposphere the circulation cells are those shown in Figure 2.8. The height of the troposphere shifts abruptly around the latitude of the jet stream (see Fig. 2.8). In the stratosphere, ozone is made by sunlight acting on oxygen. This ozone blocks ultraviolet light, protecting us from skin cancer. In blocking the ultraviolet, it acts as a greenhouse gas, and plays an important role in controlling the temperature structure of the atmosphere. (Modified from Ramanathan, Barkstrom, and Harrison 1989.)

radiation, giving a net warming. Near the poles the opposite is true, with outgoing radiation exceeding incoming, to give a net cooling. Together the atmosphere and the ocean carry excess heat from the equator to the poles, to balance the planetary heat budget. A large part of this heat transport is via the oceans, which may carry up to 40% of the total transferred heat.

The ocean currents are driven by the equator-to-pole heat transfer, either directly by the winds or by temperature-salinity variations at deeper levels. The most familiar of all currents is probably the Gulf Stream, first studied scientifically by Benjamin Franklin in an attempt to improve the transatlantic postal service. This current carries significant quantities of heat northward into the North Atlantic, moving the heat across the surface of the ocean. A similar system operates in the North Pacific.

Currents in the intermediate and deep levels in the oceans are also very important in transferring heat. These deeper currents are driven partly by cooling (which makes water denser and thus causes it to sink) and partly by salt enrichment (which has the same effect). In the modern ocean, waters dense enough to fall to the bottom form in only two regions, the Southern Ocean around Antarctica and the northern North Atlantic. The Antarctic bottom water moves north across the base of the Atlantic, Indian, and Pacific Oceans, to equatorial and

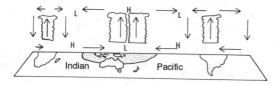

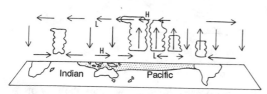

Figure 2.10. The Walker circulation. Top diagram shows the circulation during a strong positive Southern Oscillation, with high pressure over Tahiti and low over Australia. Warm water occurs in the East Indies and cooler water in the Pacific off the coast of South America. Rain falls over the East Indies, Africa, and central and eastern South America. Lower diagram shows an El Niño event, with negative Southern Oscillation. Warm water crosses the Pacific. Rain falls on the Pacific coast of the Americas, but rainfall is less than normal over central and eastern South America, Africa, and Australia. In both diagrams, clouds show region of convection; sea-surface temperature warmer than 27°C is stippled. (From Frederiksen and Webster 1988.)

northern tropical latitudes, and is balanced by a counterflow southward of warmer water at intermediate depth.

The North Atlantic deep water has an extremely important role in the global circulation, and in consequence a major impact on climate, especially in northern Europe. Relatively saline warm water moving northward at a depth of about 800 meters rises to the surface around Iceland, to replace surface water swept aside by winter winds. The water is at about 10°C and is exposed to cold air. The water gives up heat to the atmosphere, and in doing so cools and becomes denser, sinking again, to create a deep-water mass that flows southward. The annual total release of heat by this process is about 5×10^{21} calories or about one-quarter to one-third of the direct impact of solar energy to the surface of the North Atlantic. This heat is largely responsible for the warmth of the winters of northwestern Europe: people in the Scottish Hebrides can admire palm trees in January while at a similar latitude the inhabitants of Yellowknife in northern Canada experience temperatures of −40°C.

The current created by the sinking water in the northern North Atlantic has a volume that is about 20 times the combined flow of the world's rivers. This current of deep salty cold water flows south through the Atlantic, then south of Africa, across the southern Indian Ocean, south of Australia, and finally north through the Pacific toward Alaska. It has been named the Atlantic Conveyor. In effect, it is the cold-water return of the global climate control, analogous to the cold return in a domestic heating system.

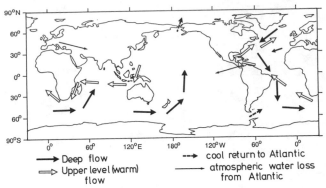

Figure 2.11. The salt-heat conveyor belt in the oceans. Solid arrows mark the flow of deep, salty, cold water, open arrows the upper-layer return flow. The deep flow begins in the North Atlantic, especially around Iceland, and moves south around Africa and Australia and into the Pacific. The upper-layer warm return flow may begin in the warm seas around Indonesia and includes the strong flow out of the Gulf of Mexico. (From various sources, summarized in Street-Perrott and Perrott 1990.)

One of the forces driving the Atlantic Conveyor is the salt content of the North Atlantic Ocean. This ocean loses water vapor from its surface, which is blown across Eurasia and Central America. Much of the water falls as rain that runs back to the Atlantic, but some crosses Panama or into southern Asia and is lost to the Atlantic. There is a net transport of fresh water in this way, via the atmosphere, from the Atlantic to the Indian and Pacific Oceans. The result is that the Atlantic is saltier than it would otherwise be, and this salty water is dense, driving the Atlantic Conveyor current. Because water vapor is fresh water, while the current is only slightly saltier than its surroundings, the current moves about 50–70 times as much water as the water-vapor deficit. Consequently, the current is very sensitive to changes in weather and water-vapor transport: more transport of water vapor across Central America (e.g. in hurricanes) or Eurasia would intensify the current, while less transport would switch off the current. In the ice ages the current seems to have been switched off, and consequently the transfer of heat from the equator to the pole must have been much less efficient. The sensitivity of this current to small changes in the transport of water vapor may be an important factor in global climate change.

2.2.6 Rainfall

The Sun delivers energy to the Earth's surface. When this energy is transferred to water, the water evaporates and enters the air as vapor. This water vapor stays in the air until it loses energy again, when it falls back to the surface as rain or snow. In Figure 2.2b the surface is shown as giving off 90 watts per square meter (Wm^{-2}) of latent heat. This heat is moved as water vapor. It is a very important part of the energy transfer that drives the world's weather.

The ways that water moves around the Earth are enormously complex. Intense evaporation takes place in the warmer parts of the oceans and over the rain-forests. The most important regions of the oceans for evaporation are near the equator, especially in the East Indies in a belt running east and west of Indonesia. Important northern summer evaporation also occurs in the western Pacific be-tween and to the west of the Philippines and Japan, and in the Caribbean and neighboring parts of the North Atlantic.

Over land, massive evaporation takes place in the rainforest regions. These regions are supplied with water that comes originally from the oceans, but is often recycled several times (especially in Amazonia). Large clouds in these regions rise to high altitudes. Through these clouds, the rainforests play a role in transferring latent heat to the atmosphere that is much greater than would be expected from the forests' relatively small area. Water loss from the surfaces of plants, especially from leaves, occurs during transpiration. When the vapor pres-sure of water in the leaf is greater than the water-vapor pressure in the surround-ing air, the water leaves the leaf and as it does so it cools the leaf; simultaneously more water can rise from lower in the plant, bringing nutrients. In a rainforest the rain (with any nutrients in it) is often caught and recycled several times before it finally escapes to a river and the sea. Each time, energy from sunlight falling on the leaves is transferred into the atmosphere via the latent heat in the emitted water vapor.

The curve in Figure 2.12 shows the amount of water that the air can hold at sea level, plotted against temperature. Clearly, warm air can hold much more water than cold air. The relative humidity is simply the amount of water that a sample of air actually holds, when compared to what potentially it can hold.

When warm water vapor in warm air rises, entering a region of lower pressure and temperature, the air expands and cools. As it does so, the relative humidity increases and the air becomes saturated and can no longer hold the vapor: clouds form. As the water vapor condenses, its latent heat is released. However, it is difficult to begin the formation of a condensed droplet of water in completely clean air, unless the air is highly supersaturated. If a preexisting nucleus is pres-ent, the water can condense around it. Completely clean air is extremely rare, and a wide variety of cloud condensation nuclei exists in the air, including dust, smoke, sea salts, and an assortment of chemical compounds. In general, the air is full of potential nuclei for water droplets, mostly dust or sea salt; the com-pounds given off by living organisms may also be important in controlling the formation of clouds, especially at certain times of day in forest areas or over marine algae. The role of the rainforest in controlling climate is discussed further in 2.4.2.

2.3 The chemical controls on the environment

2.3.1 The air

One of the basic facts drilled into children at elementary school is the pressure of the atmosphere at sea level. In my school, we were taught 14.7 pounds per square inch, but it is much easier to remember the figure in metric units: around

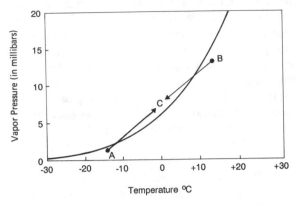

Figure 2.12. Variation with temperature in the ability of air to hold water (dew-point curve). Any air mass to the left of the curve is supersaturated with water (i.e., it contains more water than it can stably hold) and is likely to precipitate. If a cool air mass at A comes into contact with a warm mass at B, the mixing will produce a supersaturated air mass C which is likely to precipitate. (From Barry and Chorley 1987.)

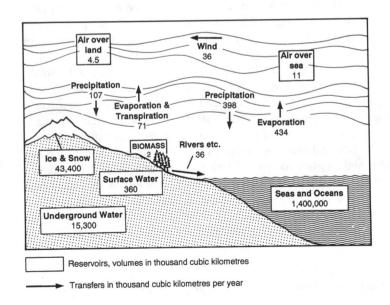

Figure 2.13. The global water cycle, showing estimates of the various reservoirs in which water is stored on Earth and the transfers of water each year between them. (From *Global change in the geosphere-biosphere,* National Research Council of Canada, 1986.)

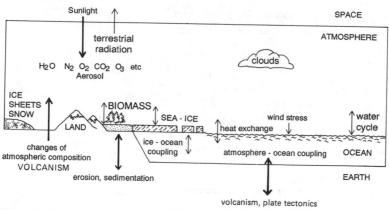

Figure 2.14. The climate system. (From various sources.)

one kilogram per square centimeter. This number, when multiplied by the surface area of the Earth, which is $4\pi r^2$, where the Earth's radius, r, is 6.38×10^8 centimeters, gives the mass of the atmosphere, about 5.1×10^{18} kg, or five million billion tons.* Thought of in total the answer is a large number with many zeros, but one kilogram per square centimeter sounds much more modest. That is the weight of the air above us, which manages our environment, keeps off radiation, and gives us life. My school, which was at high altitude, had about one-fifth of the air below it.

Imagine a small leaf, one square centimeter in area, transpiring water and chemicals through its surface: over a year the total weight transpired will be significant compared to the one kilogram of air above it. Imagine a gasoline engine burning, say, two kilograms of fuel an hour, or a power station burning a billion kilograms (one million tons) of fuel a year. All these, and humans and all breathing animals, over time affect the whole atmosphere.

After being taught the pressure of the air, we were next taught its composition: 78% nitrogen, about 21% oxygen, about 1% argon, and 0.03% carbon dioxide, with variable amounts of water vapor and many trace components. That childhood memory is now wrong: the CO_2 level is now best rounded off as 0.04%. Table 2.1 lists the composition of the air in more detail.

One of the most interesting things about the Earth's atmosphere is that it instantly proclaims the presence of life. A little green man in space could immediately discover that there are living creatures here simply by looking at the spectrum of light reflected from the Earth, which would allow him to analyze the atmosphere. Our atmosphere contains much oxygen. That in itself is not especially unusual – Mars too has oxygen, though less than Earth. What is odd is that our air also contains gases such as methane and ammonia that should not coexist with oxygen. Everyone knows that methane, or natural gas, burns in oxygen, and the little green man would instantly conclude that something, prob-

* The ton referred to throughout this book is the metric ton, equal to 1000 kilograms.

Table 2.1. *Components of dry air (by percentage)*

Nitrogen (N_2)	78.084
Oxygen (O_2)	20.946
Argon (Ar)	0.934
Carbon Dioxide (CO_2)	0.035 rising
Neon (Ne)	0.00182
Helium (He)	0.000524
Methane (CH_4)	0.00017 rising
Krypton (Kr)	0.000114
Hydrogen (H_2)	0.00005
Plus many trace components.	

Note: For more detail, see A.2 in the Appendix.
Source: Compilation in Lutgens and Tarbuck (1986), modified.

ably life, must be maintaining the supply. Similarly, we can tell that there is no life on the other planets of the solar system (or at least, no life as we know it) because on them the mixture of atmospheric gases is not odd, and does not contain improbable components.

The comparison between the atmospheres of Earth and Venus is now worth exploring in more detail. Venus is almost Earth's twin. It is nearly as large, and probably of roughly similar composition. However, the atmospheres are very different. The Earth is covered by air, with a surface pressure of one atmosphere, and water with a pressure roughly equivalent to 350 atmospheres (the oceans are about 3.5 kilometers deep). The surface of Venus is covered only by air – there is no free water at all – and that air is nearly all carbon dioxide, with a surface pressure equal to around 90 Earth atmospheres (less on the mountain tops) and a surface temperature around 470°C. The planet is dry. This seems quite different from the Earth, but when one looks more deeply there are close parallels between the two planets.

The first parallel is in the ancient presence of water on the two planets. There is evidence in the atmosphere of Venus, from the amount of the hydrogen isotope deuterium (heavy hydrogen), that Venus once had water. What seems to have happened (the interpretation remains controversial, and there are other explanations) is that early in the history of Venus the surface became very hot, and water vapor managed to rise to the top of the atmosphere where hydrogen was lost to space. The small amount of hydrogen now left is very rich in deuterium, the heavy isotope, which, being more massive, was less easily lost. It is likely that Venus once had oceans that were kilometers deep, like the Earth.

The second parallel is in the large amount of carbon dioxide present on the surface of each planet. On Venus, this CO_2 is in the air. There is an enormous amount of carbon dioxide on the Earth's surface too; so much, in fact, that the amount on Earth is comparable to the amount in the air of Venus. The difference is that the carbon dioxide on the surface of the Earth is locked up in limestones and other carbonate sediments and in oil and coal. Over the past three and a half billion years, as Earth's volcanoes have degassed carbon dioxide, there has been a steady regulation of the amount of the gas in the air. The process of sequestration has been fine-tuned by the biosphere as it constantly cycles carbon and oxygen. The excess has been put safely away, mostly in the rocks as limestones and dolomites, or returned to the deep interior. Were we today to liberate all this carbon dioxide in sedimentary rocks, the Earth's atmosphere would come to resemble that of Venus. Venus has been less fortunate than Earth: the carbon dioxide in its air has not been controlled by making limestone. The noted American geochemist, Wally Broecker, once wryly asked whether "God put Venus there as a symbol of what can go wrong if a planet is mismanaged."

The Earth's atmosphere is almost entirely controlled by a complex system of biological and geological processes. Only the rare gases, such as argon, are to some extent excluded from this management system as they are nearly inert chemically and can be regarded as the accumulation of aeons of geological release. Each of the other gases – oxygen, nitrogen, and so on – has a story to tell.

Oxygen makes up slightly over a fifth of our air. This proportion has not been constant over geological time, and there is some evidence from the types of sediment deposited in the past that oxygen levels were once much lower, though it should be pointed out that the interpretation of this evidence is contested. Most geologists consider that the atmosphere was once very poor in oxygen and that blue-oxygen levels slowly rose during Precambrian time, as organisms such as cyanobacteria carried out photosynthesis and liberated oxygen in consequence. A minority of geologists, including some field geologists who work on older rocks, has disputed this and has argued that even three and a half billion years ago the atmosphere was not necessarily oxygen-poor, and that in the earliest days of the Earth's history the air may have been rich in oxygen. Both schools of thought agree, though, that the oxygen in the modern atmosphere is biologically controlled by the metabolism of plants, animals, and especially bacteria.

Nitrogen is, at first thought, an almost inert gas. Yet this is not correct. Lightning can make the nitrogen in the air combine with oxygen. This process is known as *fixing* the nitrogen. If all life were to disappear from the Earth, lightning would slowly fix the nitrogen in the air and over millions of years it is possible that the atmospheric nitrogen content would decline as fixed nitrogen, nitrate, was collected in the sea, unless other inorganic processes acted to the contrary. On the real Earth, nitrogen is managed by a complex set of controls. Some of these, mainly bacterial but also inorganic processes such as lightning and the aurora, act to fix nitrogen. Other bacteria again release the nitrogen to the air. Fixed nitrogen, as nitrates, is essential to living organisms. Both in the sea and on land, the chain of life depends on the availability of nitrate.

Between the actions of the nitrogen-fixing bacteria and lightning, and the nitrogen-releasing bacteria, comes a balance that sets the nitrogen content of the atmosphere at 78%. This is an excellent level at which to hold the balance. The biosphere is adapted to this air. Nitrogen is a fire-retardant and protects plants. If there were less nitrogen in the air, and more oxygen, most of the world's trees would catch fire. On the other hand, if there were more nitrogen, fire would rarely occur, and perhaps animals such as ourselves would find breathing a more difficult process.

The major components of the atmosphere reflect the cumulative result of millions of years of biological processing. For many of the trace gases the situation is quite different – they often have atmospheric lifetimes of only a few years to a few millennia and are continually being created and destroyed, in an environment that is constantly recycling its chemical components.

2.3.2 Geological recycling and the environment

The external environment of our planet includes a variety of components: atmosphere, oceans, surface sediments. Within the store of chemicals that makes up the environment is the inventory of hydrogen, oxygen, carbon, nitrogen, sulfur, calcium, phosphorus, and all the other elements from which our surface biosphere is created. The inventory has accumulated over the aeons and has been complexly cycled and recycled over the past four and a half billion years by geological processes. Some of these processes have over time released more material from the Earth's interior into the exterior environment, while other processes have removed material from the surface and replaced it into the interior.

The Earth's surface moves in large slabs, or spherical caps, known as plates (Fig. 2.15). The process that determines the movement of these plates is known as *plate tectonics*. Plate tectonics is the cause of continental drift and sea-floor spreading (as well as many TV documentaries). It constrains the Earth's thermal budget, controls the movement and chemistry of the Earth's surface, and has structured the continents and built the oceans.

A rough analogy to plate tectonics is the activity at the top of a boiling pot of oatmeal porridge, although, unlike most porridge, the Earth's surface is rigid. Warm material rises from the interior of the pot, reaches the surface along well-defined lines of upwelling, and cools. As it cools it degasses water and becomes slightly more rigid. It moves away from the line of upwelling until it becomes colder and denser and falls back down into the interior. In plate tectonics, the lines of upwelling are the *mid-ocean ridges* and the lines of descent into the interior are called *subduction zones*. Each of the major oceans has a mid-ocean ridge, at which the ocean floor is splitting apart at velocities of between 1 and 10 cm per year. The subduction zones are mostly around the rim of the Pacific Ocean, although subduction also takes place in the Caribbean, the western South Atlantic, the eastern Mediterranean, and the northeastern Indian Ocean. At the subduction zones the cool old oceanic floor falls back into the interior of the Earth. Above the zones are lines of volcanoes: the volcanoes of Japan, Indonesia, and the western coast of the Americas all occur over subduction zones. Those volcanoes play an important role in the long-term cycling of the chemicals on the Earth's surface. They can also explode and dramatically alter the climate.

Table 2.2. *Summary of the factors influencing the surface temperatures of the terrestrial planets*

	Surface pressure of atmosphere (kg/cm^2)	Distance from Sun $(10^6 km)$	Effective temperature $(°C)$	Greenhouse warming $(°C)$	Actual surface temperature $(°C)$
Mercury	0^a	58	+160	0	+160
Venus	90^b	108	−40	+510	+470
Earth	1.03	150	−18	+33	+15
Mars	0.007^c	228	−61	+6	−55

Note: Effective temperature is temperature implied by infrared radiation from surface. Actual surface temperature includes greenhouse increment.
^{a}No atmosphere.
^{b}Mostly CO_2 (96%), minor N_2 (4%), trace H_2O, CO.
^{c}Mostly CO_2 (95%), minor N_2 (3%), Ar (1.6%), O_2 (0.13%).
Sources: Lewis and Prinn (1984), Henderson-Sellers (1983), and others.

Boiling oatmeal porridge is an instructive and healthy way to study continent formation. On the surface of the boiling porridge is a scummy (but edible) film that collects over the downgoing flow as the porridge boils and overturns. The continents of the Earth are analogous to this scum. Continents are made of granitic material, which is less dense than the material in the Earth's interior or the bulk of the old ocean floor. Because they are light, the continents are not usually subducted.

Geologists have identified a series of cycles by which the chemical components of the external environment are processed over time. The most rapid of these geological cycles is at the mid-ocean ridges. Seawater passes in huge volumes through the new volcanic rocks at the mid-ocean ridges, so that over about ten million years a volume of water equal to the total volume of the oceans passes through the ridges. In earlier geological time the rate of exchange was even faster. This flow of water through the ridges is known as a *hydrothermal system,* and it serves as one of the most important of the Earth's interior cooling systems – a gigantic radiator on the front of the Earth's engine, like the radiator in front of a car's engine. Not only do the hydrothermal systems cool the new plates, they also exchange chemicals between the seawater and the rock, and they serve as one of the most important controls on seawater chemistry. By influencing the chemistry and acidity of the sea, they have a profound effect on the air, because the air exchanges chemicals with the sea.

The second cycle begins here and involves the whole plate system. Chemical components such as water and carbon dioxide are introduced into the newly created ocean floor by the hydrothermal systems, which add water and CO_2 to the basalt lava that makes up the oceanic crust. More water is trapped within the

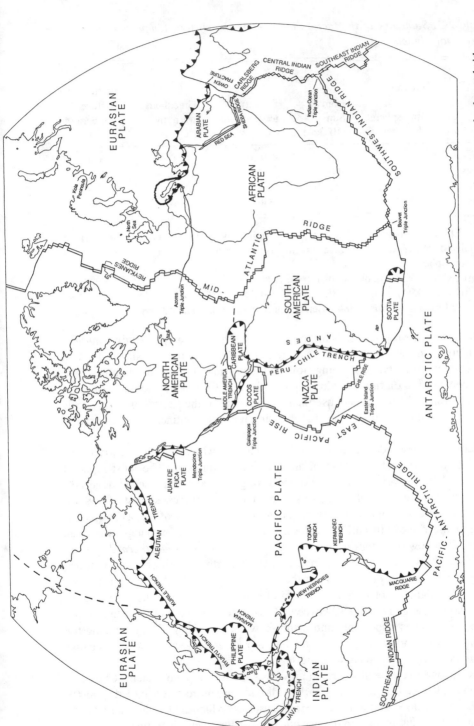

Figure 2.15. The Earth's tectonic plates. Most of the world's earthquakes occur along plate boundaries. Most volcanoes are also located either along mid-ocean ridges or above subduction zones. Plates spread apart at mid-ocean ridges on rises; they are consumed at subduction zones at trenches. In the figure, teeth indicate the over-riding plates at subduction zones. (From Fowler 1990.)

sediments that sink to the ocean floor, and carbon dioxide is deposited in carbonates that fall to the bottom as shells and skeletons of dead creatures. All this is carried, as the ocean floor moves, to the subduction zones, where some is scraped off and plastered against the continents, and the rest is carried down into the interior by the subduction zones. Here the pressure is greater and the temperature is higher, and the water and carbon dioxide are driven off and rise into the material under the edge of the overriding plate. The added water acts as a flux and causes the rock to melt partially. This melt, which includes the water and carbon dioxide, is less dense than the remaining rock and rises up to the top, where it produces volcanoes and igneous intrusions, regenerating the continents. The water and carbon dioxide are ultimately degassed again into the external environment from the volcanoes. These volcanic gases enter the atmosphere, and are eventually trapped as rain or as carbonate in the skeletons of living organisms, respectively, and repeat the cycle which takes on the order of 100 million years. The rocks of the volcanoes and igneous intrusions are also eventually eroded and washed to the sea, to reenter the cycle.

The time scale of the cycling that involves the deep interior of the Earth is even longer than this. The deep interior was the source of much of the water and air of the planet, and even today it is still slowly losing volatiles to the surface. One example of this loss is the anomalous amounts of an isotope of helium, ^{3}He or helium 3, at the mid-ocean ridges. This gas can have come only from the deep interior. Water and carbon dioxide are also lost from the deep interior at places such as the mid-ocean ridges and at volcanoes that tap deeply into the interior, such as Hawaii. These gases are continually but slowly leaking out from the interior. They are also being returned to the interior, because some of the water and carbonate that moves down the subduction zones does not escape back to the volcanoes, but is instead carried right down into the deep interior. We do not know whether the Earth's surface is at present losing or gaining atmospheric gases – that is, whether more is carried down into the deep interior at subduction zones than is brought up at mid-ocean ridges. One rough estimate is that there is a net loss of one part in two billion of the oceans to the interior each year, but this is not certain. To sum up, there is a very complex set of cycles, of progressively longer time scale, that govern the interaction of the external environment with the interior of Earth. In all of these cycles the effects of water are critically important.

We are accustomed to thinking of geological processes as being very slow, and therefore not of interest on a human time scale, but it should be remembered that many of these slow processes operate sporadically, sudden bursts interrupting long periods of peace. For instance, on plate boundaries such as the San Andreas fault all may be quiet for decades, until the fault suddenly moves during a few seconds of activity. As far as the atmosphere is concerned, volcanic eruptions that are powerful enough and directed enough to influence the weather by placing an enormous amount of aerosol high in the atmosphere occur roughly every decade or two. Some eruptions can trigger or accentuate massive climate change over the whole planet. An example is the extraordinary series of changes

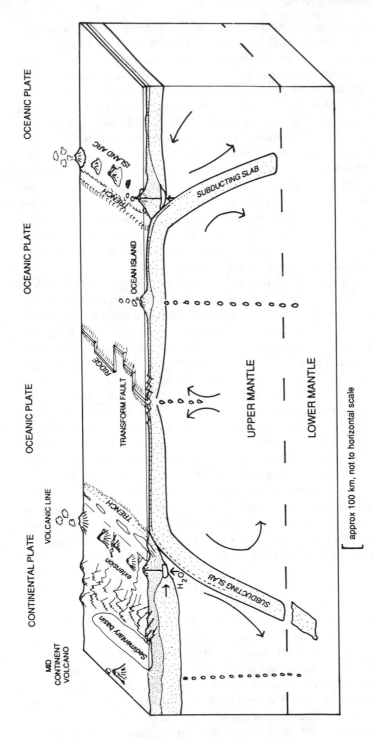

Figure 2.16. The geological system in cross section. (From Fowler 1990.)

to the global climate in 1982–3 that followed the eruption of sulfur gases into the stratosphere by the southern Mexican volcano El Chichón. The stratospheric sulfur aerosol that resulted may have enhanced the development of the 1982–3 El Niño event, making it one of the strongest ever recorded, and causing widespread drought in the Southern Hemisphere. This was a relatively small eruption. The bigger eruption of Tambora in 1816 – the year without a summer – may have drastically changed the chemistry of the stratosphere. Some eruptions have been much larger than this. A really large eruption could have a major impact on the environment, possibly even causing a mass extinction. Similarly, mid-ocean ridges can belch gas at irregular intervals. Geological processes can sometimes alter the surface environment of the whole planet, and rapidly.

More generally, the close link between climate and the geological processing of planetary surfaces is illustrated by the difference between Earth and Venus. On Earth the surface temperature of the equatorial regions has over billions of years remained much as it is today, probably in the range 25°–30°C. This is despite the apparent degassing of carbon dioxide from volcanoes, and despite the probable slow brightening of the Sun over the last four billion years. As the Sun has brightened and as carbon dioxide has degassed, the greenhouse effect has declined and the surface temperature appears to have remained steady or varied only within narrow limits. On Venus, in contrast, roughly the same amount of carbon dioxide has been degassed over time, but it has been stored in the atmosphere. Today the planet is a hot, lifeless desert.

2.3.3 The carbon cycle on the surface of the Earth

The carbon cycling discussed above is the long-term system, acting over millions or billions of years. Humanity, and most of the biosphere, is more immediately concerned with the very much more rapid surface cycling of carbon, especially carbon dioxide. Many processes can be distinguished, from very short-term processing, taking a few days to a few years, to medium-term systems taking some thousands of years.

In the very short term – days to years – carbon is rapidly processed around the biosphere. Plants trap carbon dioxide from the atmosphere and use it to build leaves, fruits, stems, and trunks. The Earth breathes each year, drawing in atmospheric CO_2 in the northern spring and exhaling it in the northern autumn. A striking feature of Figure 2.17 is the sawtooth nature of the CO_2 variation in the Northern Hemisphere. Each northern spring, during the spring growth spurt, the CO_2 content of the air declines by as much as 6–7 parts per million by volume (ppmv). The sawtooth pattern records this event when forests, fields, and even tundra bogs all burst into life. Growth is rapid, and the CO_2 content of the air is drawn down as carbon is captured by biological organisms. Later in the year, the growth slows and the planet exhales CO_2 as leaves fall and bacteria oxidize organic matter. By the northern winter the CO_2 content of the air has regained its former value. However, some carbon captured by plants is not recycled immediately. Some goes into the soil as organic matter, some is stored for a century

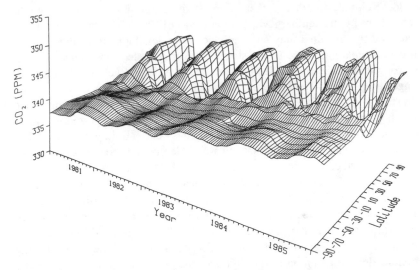

Figure 2.17. The seasonal cycling of carbon dioxide in the atmosphere. Note the strong seasonality in the Northern Hemisphere. The steady growth in CO_2 levels, shown well at 90°S latitude, is discussed in Chapter 3. (From U.S. National Oceanic and Atmospheric Administration/Climate Monitoring and Diagnostics Laboratory, with thanks.)

or more in tree trunks. A small amount goes into the rocks to form oil or coal eventually. Fire returns large quantities of carbon from the plants to the air and leaves some on the ground in burned charcoal.

Some of the slightly longer-term processes are also biologically mediated. Living organisms in the sea remove carbon from the environment to form cells. Other organisms form shells of calcium carbonate. When they die the carbonate is rained down, eventually, to the sea bottom, where it forms a carbonate mud and, finally, a limestone. Sedimentation is a great sink of carbon, but it operates slowly in relation to a human lifespan. When massive amounts of CO_2 are added to the atmosphere (e.g., from a huge volcanic eruption), it takes several thousand years or more before the precipitation of carbonate on the ocean floor is able to return the atmosphere to the earlier CO_2 level.

There is much debate about how long CO_2 is held in the air and how it exchanges with the plants and the ocean water. The mean lifetime of CO_2 in the air seems to be relatively short – perhaps a few years – before it is transferred into the uppermost layer of the sea or into plants. However, studies of radioactive isotopes and pollutants show that the sea does not mix very rapidly, and much of the CO_2 is recycled back into the air, rather than being carried down to deeper layers in the sea. If the sea warms, for whatever reason, odd things can happen. Warm water holds less CO_2 than cold water, all other things being equal, and warmer oceans may degas CO_2 into the air. The final removal of carbon dioxide from the surface system is slow. Eventually, CO_2 enters the deep sea and is precipitated as carbonate or organic carbon, over thousands of years.

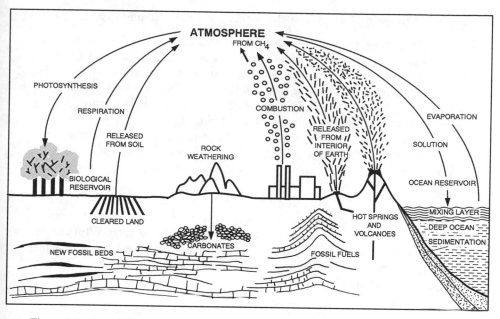

Figure 2.18. The CO₂ system in the atmosphere. (Modified from various sources.)

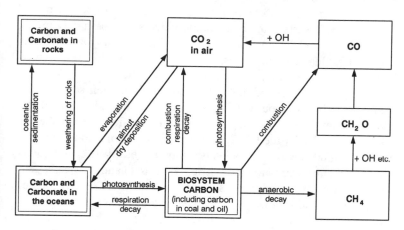

Figure 2.19. The carbon cycle. (Modified from various sources.)

The other carbon gas that is important in the natural greenhouse is methane. Unfortunately, the Earth's methane cycle has now deviated so far from prehuman norms that it can no longer be described as natural. Any attempt to describe the natural methane cycle is now an exercise in atmospheric paleontology: accordingly, atmospheric methane is mainly discussed in a later part of this book, where human influence is considered.

2.4 The processing factories of the Earth's air-conditioning system

So far, the Earth's air-conditioning system has been described as a whole, in terms of physical input and output of energy and processing and cycling of the major chemical components. But an air-conditioning system is made of many parts, each playing a distinctive role: fans, coolant, heat exchangers, air purifiers. The Earth's system is more complex but similar. It contains many distinct components that collectively control the environment.

The most obvious atmospheric processing system involves water: water evaporates from the sea surface; some falls back, some is blown over land; rain falls; streams flow down to sea, carrying with them eroded material that enriches the oceans.

Or so it seems. In fact, almost every aspect of the water cycle is biologically modulated in some way, and so too are the cycles of most of the other atmospheric trace gases. The biota of the planet exert a complex and powerful control on the global air-conditioning system: the Sun drives the process, but only in the way that electric current drives a building's air conditioner, while control and design are under biological constraint.

2.4.1 Oceans and the water cycle

The oceans are vast, covering nearly three-quarters of our planet. Water evaporates off the sea whether or not life exists, so long as there is an ocean, but in many parts of the ocean's surface the evaporation process may be highly modified by marine plankton. For instance some plankton emit a particular chemical known as dimethyl sulfide (DMS). It has been suggested that this emission of DMS may help to control the timing of the nucleation of water droplets, forming clouds in the air, above the algae. Clouds keep the algae and their surrounding water cool and control evaporation: if this idea is correct, it is almost as if the algae are shading themselves. This suggestion is controversial, but whether or not the clouds are controlled by the DMS, they certainly contain sulfide from the algae. Some of the clouds and the associated sulfide are blown over the continents, carrying water and sulfur onto land where it falls as a natural, weakly acid rain. Another product of marine algae is a compound known as methyl iodide, which releases iodine that also blows over the land and helps to return iodine to the continents. Bromine compounds may also be cycled in this way: they may be important in controlling ozone levels in the stratosphere. Nitrogenous gases are also produced in quantity.

This complex concoction of gases is present in the clouds of water vapor that float over parts of the oceans. There is much cycling to and fro. In the southern oceans and in places where the air is relatively free from cloud condensation nuclei (small particles that help to nucleate water droplets), the individual drops in clouds grow very large and can fall even from thin clouds; elsewhere, drops are smaller and do not fall as easily, and clouds grow thicker. In many parts of the ocean, there may be a close relationship between the structure of clouds and the activity of the marine algae below. The role of marine animals has not yet been mentioned, as it is poorly understood, but there is obviously a complex chain of life depending on the marine plankton, reaching right up to the great sharks and whales. Disruption of the population of large animals could alter the population balance amongst the smaller animals and plankton and, conceivably, even change the way clouds form.

The clouds of water above the oceans help to control the temperature of the planet, since they determine the whiteness of the surface they cover. The relationship between clouds and temperature is very complex (see 2.2.2) but, overall, clouds cool the planet. If for some reason the planet warmed, more water would evaporate, more clouds would form, and the surface again might cool somewhat, all other things being equal: a negative feedback effect that helps to give the Earth a steady temperature. However, this depends strongly on the type of cloud. If cloud patterns changed in any major way, so too would the surface temperature distribution. If the clouds of the northern oceans became more like tropical clouds, the cooling effect would be greatly reduced and the planet could warm rapidly: a runaway is conceivable.

Water vapor also controls the temperature of the planet in a more subtle way, by helping to control methane and other trace greenhouse gases in the air. The destruction of methane is brought about by hydroxyl (OH) radicals in the air. OH is the chief chemical scavenger of the atmosphere and is the main controller of trace gases in the air. It is produced by the action of light on the water evaporated off the sea. Water vapor and the production of OH in the air vary according to season and altitude; as a result there are major seasonal and latitudinal fluctuations in the cleaning power of the atmosphere.

All of this shows the intricate complexity of the processes that affect the air and the clouds above the seas. The temperature of the oceans controls the evaporation rate and hence the clouds, yet the temperature is itself controlled by the clouds, which in turn depend in part on the abundance of cloud condensation nuclei. In the air the water forms droplets nucleated around dust, salt, sulfur gas molecules, and organic molecules. Droplets form clouds, and the clouds carry with them the condensation nuclei, and in those nuclei are many chemicals that are returned to the land. If there are many nuclei of the right size in the clouds, the clouds will be thick and carry much water; if there are few nuclei the clouds will be thin and will easily rain out large droplets back to the ocean. About a quarter of the Earth's surface is covered in stratus, stratocumulus, altostratus, and altocumulus clouds, many of which may have been made in this way. Overall, clouds are the most important day-to-day control on the planet's temperature.

The water vapor blows across the land and rain falls where the air masses interact, controlling the distribution of forests and deserts.

2.4.2 The rainforests

Two of the most important and least understood components of the Earth's air-conditioning system are the rainforests in Southeast Asia and in Amazonia. They

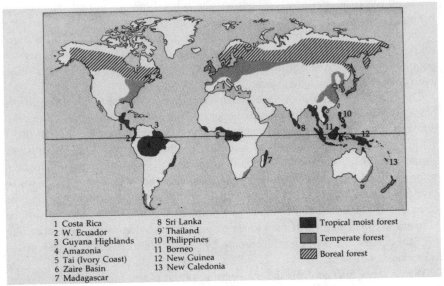

Figure 2.20. World distribution of various types of forest. (From Myers 1989.)

are both equatorial and they occupy antipodal points: if one were to dig a hole through the Earth's core starting from Borneo, the hole would come out in the central Amazon. Both are poorly understood, and it is very insecure to generalize about such diverse regions. Nevertheless, the role of forest, especially rainforest, in the physical and chemical processing of the air is very significant.

Flowering plants can lift water about 40 meters from the roots to the canopy of a tree. This governs roughly the height of most modern forests; eucalypts and some modern members of an older plant family, the gymnosperms, can lift water higher, to more than 100 meters. In a rainforest, the trunks of the trees serve to lift the leaves, so that a dense overlapping canopy can be built to catch light, water, and nutrients, while air can be processed across the interface as the leaves wave and turn in the sunlight, each leaf in effect controlling its surface temperature. The dense canopy is able to intercept most of the sunlight energy available. By varying transpiration the forest can maintain a steady temperature.

In the dense Amazon forest, of all the rain that falls, only about one-quarter runs off in rivers (Fig. 2.21). The rest is either caught on the leaves of the trees (one-quarter) and re-evaporated to the air, or enters the plant matter and is transpired via the leaves (one-half). These figures vary with local conditions, but typically between one-half and three-quarters of the rain that falls onto the canopy of the forest is returned to the air, to fall again. In other words, the forest recycles much of its water, making its own rain. If the forest were removed, rain would still fall, but there would be less of it. Moreover, the rain would have a different character.

Transpiration by the forest has some analogies to sweating in human physiology. As the leaves transpire water into the air, they place energy into the air as

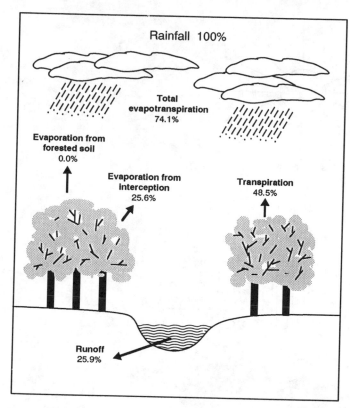

Figure 2.21. Water balance in a model basin in the Amazon rainforest. (Modified from Salati 1987.)

latent heat in water vapor, and in so doing they keep the forest cool. The amount of energy put into the air is large. The air rises, carrying up the vapor until it condenses as clouds and releases the carried-up energy into the air. This process lifts the energy high and thoroughly mixes the troposphere above the forest, through the condensation of vapor in tall cumulus clouds. Without the forest the energy would be transferred as sensible heat – heat that can be felt, as hot air – which would warm only a relatively thin atmospheric layer immediately above the ground.

The dense forest in the East Indies probably behaves in much the same way. The East Indies are islands in a tropical sea, and would probably be very wet even in the absence of forest, but nevertheless the dense Southeast Asian rainforests may be a critical factor in making this region the center of the most intense electrical activity in the planet's atmosphere. This is the place on Earth where thunderstorms are perhaps most common, and it is here that intense transfer of latent heat high into the troposphere takes place. In other words, the dense clouds that form over and around the Indonesian and Malaysian land surface constitute one of the main heat exchanges that drive the global air conditioner.

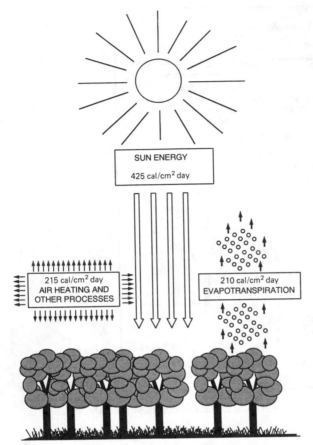

Figure 2.22. Energy transfer each day above forest in the Amazon Basin. (From Salati 1987.)

Much of this effect may be simply the result of geography, land in a warm sea, but the forest may play a major role in making the exchange so intense.

On a planetary scale, evaporation from tropical oceans and forests helps to transfer energy from the equator toward the poles. On land, water vapor, as it evaporates, keeps the forest cool and so helps to cool the tropics. Oceanic evaporation plays a similar role, although the activity of thunderstorms over the oceans is not nearly as concentrated and intense as over the rainforests. Rising vapor carries latent heat into the higher troposphere. Heat is released again when the warm and still moist tropical air interacts with the cooler air to the north or south. The effect of the whole process is to keep the equatorial region cooler than it would otherwise be, and to warm up regions closer to the poles, giving the Earth a more even temperature than it would have if oceans and forests did not exist. The size of the latent heat transfer is shown in Figure 2.2. Of course, much of this transfer is not above rainforest, but the concentration of heat transfer above

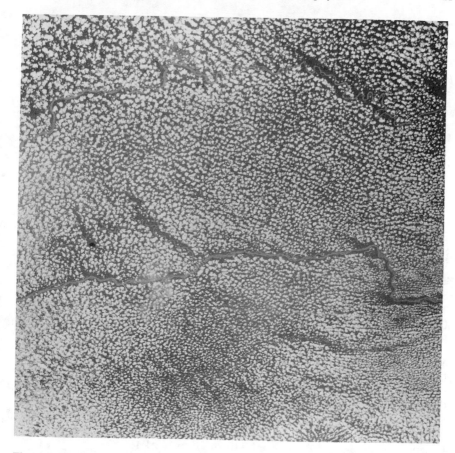

Figure 2.23. Cloud development in Amazonia. In this photograph of the Rio Meta region, Colombia, note the even distribution of clouds (white patches) that have formed at regular intervals in convective cells above the forest. Where forest is absent, as above large rivers, the clouds are absent too. The lines of the large rivers can thus be seen clearly in the photograph. (From Wells 1989, NASA Hasselblad photo.)

the rainforest – the power of the fan – makes it very important in influencing the way the system as a whole behaves.

Trees, like algae, may use trace emissions to control the rate and timing of cloud nucleation. Rainforests emit a variety of cloud condensation nuclei. It is possible that these nuclei help to shape clouds above the trees, often in small local convective systems of 5 km or less in diameter. These small-scale systems are linked into much larger convective motions of the air, as lines of squalls move across the forest. The aerosols produced by the forest are eventually carried out to oceanic areas, where they supply nutrients such as phosphorus to the marine life. Over Amazonia there is a complex exchange of gases and trace chemicals between the air and the forest, as moisture-laden air enters from the

Table 2.3. *Relationships between ecosystem types, showing area (in millions of sq. km) and percentage of the Earth's surface occupied by each, their primary productivity (in billions of tons of carbon per year), and their productivity/area ratios*

Ecosystem type	Area (A) and percentage of Earth's surface	Total net primary productivity (P)	P/A
Terrestrial ecosystems			
Tropical rainforest	17.0 (3.4%)	15.3	0.90
Tropical seasonal forest	7.5 (1.5%)	5.1	0.68
Temperate evergreen forest	5.0 (1.0%)	2.9	0.58
Temperate deciduous forest	7.0 (1.4%)	3.8	0.54
Boreal forest	12.0 (2.4%)	4.3	0.36
Savanna	15.0 (3.0%)	4.7	0.31
Woodland and shrub land	8.0 (1.6%)	2.2	0.28
Temperate grassland	9.0 (1.8%)	2.0	0.22
Swamp and marsh	2.0 (0.4%)	2.2	1.10
Tundra and alpine meadow	8.0 (1.6%)	0.5	0.06
Desert scrub	18.0 (3.6%)	0.6	0.03
Rock, ice, and sand	24.0 (4.84%)	0.04	0.002
Marine ecosystems			
Lake and stream	2.5 (0.50%)	0.6	0.24
Continental shelf	26.6 (5.4%)	4.3	0.16
Open ocean	332.0 (66.9%)	18.9	0.06
Upwelling zones	0.4 (0.08%)	0.1	0.25
Algal bed and reef	0.6 (0.12%)	0.5	0.83
Estuaries	1.4 (0.28%)	1.1	0.79

Source: From Holland (1984).

Atlantic, and the water cycles and recycles through the forest, eventually to blow or to run back to the sea.

Beneath the canopy, rainforest soils are notoriously poor in nutrients. In many cases it is almost as if the forest has lifted all available chemicals out of the ground up into its canopy, where it lives by using sunlight to exchange water and assorted other chemical components with the air. Many trace gases such as carbon monoxide are processed by the canopy; other trace gases such as methane are processed in the swamps and waterlogged soils that appear seasonally at the foot of the trees in many areas. The forest system is also a powerful sink that removes tropospheric ozone, and it is a source of sulfate and ammonia. Forests are chemical factories cleansing and processing the air for the rest of the planet.

The role of forests in sustaining the chemistry of the environment is not confined to the day-to-day control of trace gases. Forests are also major stores of carbon, in their tree trunks and other biomass. One of the chief limitations on organic growth is the supply of phosphorus. In the oceans, living organisms typically have a carbon:phosphorus ratio of 100:1. Plankton cannot, and do not need to, develop tree trunks. In contrast, land organic matter has a ratio of 1000:1 so more carbon can be stored for a given supply of phosphorus. This means that forests, although of comparatively small area, store far more carbon than the oceans. Enormous quantities of carbon are locked up in the trees. The carbon in tree trunks represents an important part of the short-term (1–1000-year) carbon store of the Earth's surface.

There may be a close link between the oxygen, carbon dioxide, and phosphorus levels of the exosphere. If, for instance, the oxygen level rose for some reason, there would be more fires. These release phosphorus from the land biomass into the streams. They also consume oxygen and raise the CO_2 level of the air. The phosphorus and carbon dioxide would stimulate marine life. More marine life would eventually mean an increase in the rates of burial of dead plankton on the sea bottom. This burial would reduce the total phosphorus of the exosphere and also reduce the carbon dioxide level of the air, thereby restoring the system toward its original state.

The importance of the rainforest is shown in Table 2.3. The rainforest occupies only 3.4% of the Earth's surface, yet its total biological productivity rivals that of the open oceans which make up two-thirds of the surface. This productivity is closely related to the immense importance of the rainforest. If the forest is replaced with tropical savanna grassland, the productivity of the region will decline very markedly: the consequences for the climate system of the planet may be severe. The concentrated energy transfers above the forest are the intense fans of the global air conditioner, passing air over each leaf, which is a complex chemical factory keeping our air clean for us. The forest moves the air and cleans it at the same time.

2.4.3 Boreal forest

The Earth possesses millions of square kilometers of northern, or *boreal* forest, especially in Canada and the Soviet Union. Much of this forestland is relatively flat, and waterlogged ground and swamps are common in it. Far less water falls on these forests than on the tropical rainforests (in many places, the precipitation over boreal forest is less than a quarter of the rainfall on tropical forest), but evaporation is minimal during the winter, so that the water is available for summer growth. The boreal forest is briefly able to burst into biological activity from June to September, during which time it rivals the tropical rainforest in its productivity.

Northern forest is of great significance in the annual transfer of the carbon resources of the biosphere. Each year there is a spring growth spurt in the Northern Hemisphere, much of which occurs in the boreal forest, which is so important that it affects the carbon cycling in the atmosphere of the whole planet. This short-term transfer of carbon contrasts with the longer-term storage by the rainforest. The major difference, however, between the two types of forest is that northern forest has fires, while rainforest normally does not. In the northern forest, fire is so common that the plants have evolved a complex fire ecology to take advantage of the opportunities of competitive reseeding after major fires. The fires can be immense; a single fire can burn thousands of square kilometers, and the smoke can travel across continents. Early this century, huge tracts of Siberia burned; in past centuries there have been enormous Canadian fires. A recent fire in Yellowstone National Park produced smoke that could be seen as a thin layer at 10,000 feet over very large areas of western Canada.

The burning of forest has a complex effect on the carbon dioxide and oxygen levels of the air. The British scientist Andrew Watson has shown that fires cannot be started, even in dry wood, when the oxygen level is below 15%. On the other hand, when the oxygen level is above 25%, even very wet rainforest would burn. In the northern forests of Canada and Eurasia (and in the tropical dryland) plants compete by fire. Some trees have foliage full of oil, and burn fiercely, destroying the neighboring species. These plants that burn easily tend to have seeds that in some way are fire-resistant, which then grow in the rich ash after the fire. Other plants resist fire. The net effect is to set oxygen at a median level, so that fire occurs, but not with total destruction.

Fire, paradoxically, can also act to increase oxygen levels in the air. When a tree burns it is oxidized: oxygen is combined with carbon from the tree to make CO_2. The fire lowers the oxygen content of the air, until the CO_2 is recycled by other plants. However, usually the tree is not completely oxidized, but partly turns to charcoal, which is very resistant to alteration and represents carbon removed from the biosphere. Much of this charcoal lies on the soil, where it is probably eventually oxidized by bacteria. Some is washed to the sea and buried as sediment that may be lost from the system until it is eventually returned by geological processes hundreds of millions of years later. This represents a net removal of carbon from the biosphere, effectively allowing the oxygen level to rise. If, however, the oxygen level of the atmosphere were to rise too high – say to 25% – the burning would be more intense, and less or no charcoal would be produced: there would be a net removal of oxygen from the system. In sum, the fire ecology of the northern forest seems to play a critical role in maintaining the oxygen and carbon dioxide levels of the air.

The natural northern forest contains abundant swamps and shallow lakes. These wetlands produce methane. Much of this gas is oxidized to CO_2, but some reaches the air. In general, the extensive wetlands and swamps of the boreal forest make it one of the most important natural sources of methane, which is a very important greenhouse gas. Up to 20% of the total world production of methane may come from northern forests, making them important in the regulation of the Earth's greenhouse temperature increment.

Northern forests, like the tropical forests, seem to manage their own rainfall to some extent. The forest transpires. The smell of a pine forest may be the smell of cloud condensation nuclei; some of the long organic molecules given off by the trees may help in creating clouds over the forests. Furthermore, the northern conifers are green, even under snow: the dark trees absorb sunlight and warm the environment.

2.4.4 Other land regions: tropical dry forest, savanna, tundra

The tropics and subtropics of Africa, Australia, and also India once had extensive dry forest, fringed by parklands of scattered trees and grass. Along the outer edges of the parkland are extensive semidesert regions of scrub and poor grassland. In these regions lives the extraordinarily diverse large animal population of Africa, which until recently extended to southwestern Asia. Dryland Africa and Australia, like the boreal realm, have complex fire ecologies that influence and process the air.

Much of the world's treeless grassland is relatively modern, sustained by periodic droughts severe enough to kill trees, by fire, and by human activity. Here, too, a fire ecology persists. The grassland soil is one of the major carbon stores of the biosphere, with carbon held as root material or dead organic matter.

In the far north of the planet is another seasonal processing factory of the atmosphere. The Arctic tundra is highly productive for a very brief period of the year. Peat bogs may for a few weeks exceed the rainforest in their activity. The permafrost is also the site of one of the world's more important stores of the greenhouse gas methane, which is kept frozen in gas hydrates in the ground but could be released in a thermal runaway.

2.5 The Gaia hypothesis

There is, as yet, no scientific agreement on what controls the very complex interactions that together sustain our atmosphere. Very broadly, there are two extreme viewpoints and a spectrum of opinion between. One extreme is that there is no control: the atmosphere is simply chance, and life has evolved to accept and utilize what is there. This viewpoint sees the atmosphere as wholly accidental, controlled by the constraints of physical and inorganic chemistry, with living organisms simply exploiting their environment. Very few scientists now hold this extreme view.

The opposite extreme, probably still a minority view, is that life controls the planet. If so, the surface environment of the Earth is a complex maze of feedbacks, checks, and balances, in which living organisms play the directing role. In this view the Earth's surface, including the atmosphere, the land surface, and the sea, is seen as a system that, within the broad range of possibilities afforded by the basic constraints of solar radiation and geological inventory, has evolved over the aeons under the dominance of life. The word *system* is stressed – the complex network of feedbacks is comparable to, but much more complex than, an engineer's cybernetic system. A simple analogy would be a house thermostat; a more subtle analogy would be the myriad of interactions that make up a modern industrial economy; a yet closer analogy is the set of controls that runs the colony

of cells we call a human being. We can regard the biosphere as having a physiological system, a *geophysiology,* that controls its operation.

This hypothesis, which was proposed by J. E. Lovelock working with L. Margulis and others, is known as the *Gaia hypothesis.* It is that the physical and
chemical condition of the surface of the Earth, of the atmosphere, and of the
oceans, has been and is actively made fit and comfortable by the presence of life
itself. The idea is best expressed in Lovelock's books, *Gaia – A New Look at
Life on Earth* and *The Ages of Gaia.* The name *Gaia,* that of the ancient Greek
Earth goddess, was suggested by the novelist William Golding. Gaia represents
a radical new view of life on Earth. The hypothesis draws attention to the cooperation between living organisms to manage the planet. This idea is not necessarily incompatible with Darwinian evolution; indeed, the supporters of the Gaia
hypothesis claim that it is a logical corollary of natural selection.

The Gaia hypothesis is not accepted by most scientists. The majority opinion
is somewhere between the two extremes, though recently there has been much
evidence in favor of the Gaian interpretation. The matter is not yet settled. My
own opinion and that of many others is fully in support of the Gaia notion as
stated scientifically. The scientific proviso is important: it excludes the zoo of
mystical accretions to the hypothesis. But science is not democratic: opinion is
not by itself adequate to ensure that an idea is accepted. Note must be taken of
contrary opinion. The debate is interesting and productive. Proof may come. If
it does, the Gaia hypothesis may assume a status similar to the idea of natural
selection, Gaian cooperation growing out of Darwinian struggle: the two ideas
are different faces of the same coin. Whether or not the hypothesis is ultimately
accepted, it has made the scientific community firmly realize that there is an
intimate connection between life and climate on Earth.

2.6 The interconnectedness of nature

If one wishes to study fluctuations in the population of small rodents along parts
of the Arctic coast of Siberia, it is necessary to conduct a regular census of the
rodents – a difficult and expensive business. If, however, a rough supporting
estimate of population trends is needed, it is easier to fly to Cape Town and to
count Siberian migratory birds in a coastal lagoon. The reason is that the same
predators eat both the rodents and the birds: when the rodent population crashes,
as it does from time to time, the predators turn to the birds, and at the next
migration few birds arrive in the Cape.

Southern African farmers, like farmers in most countries, use chemicals to
attack pests. The chemicals kill frogs and insects and find their way into white
storks. Fewer storks return to Europe. In Europe the same thing happens, so
fewer storks return to Africa. Not many years ago, in parts of the eastern Cape
Province of South Africa, the trees at dawn were white and sagging with their
load of thousands of roosting storks. Today those trees are empty, and there are
European nations where storks are almost extinct. Similarly, the population of
many familiar North American birds depends on the state of the Central Ameri-

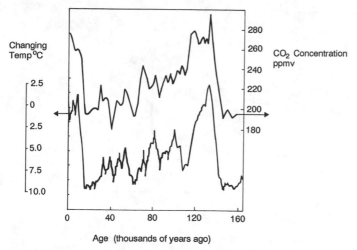

Figure 2.24. CO_2 and temperature variation over the past 160,000 years, from the Vostok ice core, Antarctica. Left hand scale, which refers to lower curve, shows change in temperature above and below present mean, which is at the 0 level. Right hand scale shows CO_2 variation, referring to top curve. The ice age ended approximately 13,500 years ago: note the sudden rise in CO_2 and temperature. The end of the previous ice age, about 140,000 years ago, is also marked by a sudden rise in CO_2 and temperature. (Redrawn from Jouzel et al. 1987. Reprinted by permission from *Nature,* vol. 329, pp. 403–7. Copyright © 1987 Macmillan Magazines Ltd.)

can rainforest: changes in Guatemala reduce the population of songbirds in the back gardens of the United States, which are becoming audibly less pleasant each year.

The long-ranging travel of migratory birds is well known. Distant biological systems are closely connected. Less well known are the complicated movements in the atmosphere. A volcanic eruption in Central America may set off a chain of events that induces catastrophic drought in Australia or floods in California. Borneo's forest may indirectly influence rainfall in China and India. The atmosphere is complex and connected to the biosphere. There is no independence in nature.

2.7 The possibility of change

When a child pushes a glass across a table, each small push moves the glass an extra inch or two. That is, until the edge is reached. In the past few years, it has been discovered that the atmosphere too has edges. Geologists have always been aware that the Earth is continually changing, and they have also been aware that major disasters can occur, such as the catastrophe at the end of the Cretaceous that destroyed the dinosaurs. But these have been distant possibilities, not closely related to the modern planet. On the scale of a human lifetime, the Earth seemed a very stable place.

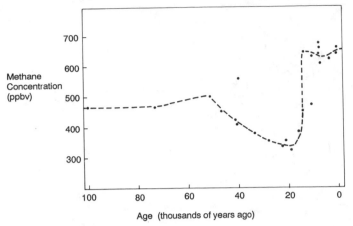

Figure 2.25. Methane variation at the end of the last ice age, from evidence in Greenland and Antarctica. Note the dramatic rise about 13,500 years ago. For recent rise in
methane see Fig. 3.7. (Modified from Stauffer et al. 1988. Reprinted by permission
from *Nature*, vol. 332, pp. 812–14. Copyright © 1988 Macmillan Magazines Ltd.)

This thinking has changed. We now know *how*, though not yet *why* the ice
ages ended. The atmosphere crossed a threshold, and the planet's climate changed
suddenly, not slowly. Sudden change implies thresholds: driving forces accu-
mulate until suddenly there is a runaway and the system moves out of control
until it finds a new stable state.

The change took place about 13,500 years ago. It is well recorded in ice in
Antarctica and Greenland, in peat bogs around the world from Switzerland to
British Columbia, and in sediment in seas such as the South China Sea. The ice
record is the most important. Snow traps bubbles of air as it compacts and turns
to ice. Both in Greenland and in Antarctica it is possible to drill cores into the
ice and extract the bubbles. The air can be analyzed for CO_2 and CH_4. The
isotopes in the core can even be used to give the temperature at the time the snow
fell. The results tell an amazing story.

During the last ice age the world was several degrees centigrade cooler than it
is today. Geologists have long known that in the Northern Hemisphere there
were huge sheets of ice, especially over Canada and northern Europe. At the
peak of the glaciation, virtually all of Canada was covered by ice, as was most
of Britain, Scandinavia, and much of Germany and Switzerland. Even New Zealand and Australia had ice sheets. Sea level was more than 100 meters (350 feet)
lower than today, and Malaya and Indonesia formed a large low-lying connected
land mass. The climate of every part of the world was much different from today.

We know the CO_2 and CH_4 content of the ice age air. The CO_2 level was
between 180 and 200 parts per million by volume (ppmv), around 190 ppmv at
the time just before the change. The CH_4 content was about 350 parts per billion
by volume (ppbv). Suddenly – within a few centuries – the CO_2 level of the air

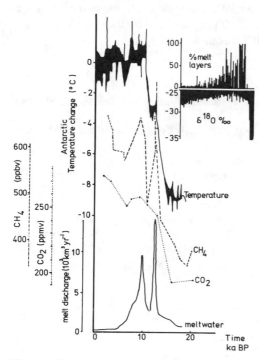

Figure 2.26. Summary of the changes that took place at the end of the last ice age. From bottom: meltwater pulses caused sea level to rise; CO_2 in the atmosphere increased rapidly; CH_4 increased rapidly, in erratic pulses; temperature in the Antarctic increased sharply; (inset) melting of Arctic ice took place in strong but erratic events, while the oxygen isotope content changed suddenly. Time axis is in thousands of years (ka) before present; various vertical axes refer to appropriate curves. (From many sources.)

jumped to 260 ppmv (Fig. 2.24) and the CH_4 level doubled to 650 ppbv (Fig. 2.25). The increase was 70 ppmv of CO_2 and 300 ppbv of CH_4. The greenhouse effect of this jump seems to have been a major factor in knocking the Earth out of the ice age. The ice took many thousands of years to melt, and a further cold period occurred, but the change was permanent. The biosphere recovered rapidly and pollen records from places as far distant as western North America, Europe, and central Africa show a sudden change in vegetation. The ice age was gone; the modern natural world had been created.

There is more than one change like this in the ice-core record. The Antarctic ice cap is deep, and an identical change can be seen in the older record about 140,000 years ago. The climate system has a definite threshold, or switch. It does not slide slowly out of an ice age. It flips over, into a new stable state.

To recapitulate, the change was driven by the addition into the atmosphere of 300 ppbv of methane and 70 ppmv of carbon dioxide. We do not know why the addition occurred, but it pushed the climate over a threshold so dramatic that it created Canada and the Baltic states out of mounds of ice. In the past century we have increased methane by a further 1000 ppbv and carbon dioxide by a further 70 ppmv. We have yet to discover if a threshold awaits us.

Reading list

Barry, R. C., and R. J. Chorley (1987). *Atmosphere, weather and climate*. 5th ed. London: Methuen.

Broecker, W. S., and G. H. Denton (1990). What drives glacial cycles. *Scientific American* (January): 43–50.

Cloud, P. (1988). *Oasis in space*. New York: Norton.

Dickinson, R. E., ed (1987). *The geophysiology of Amazonia*. New York: Wiley.

Lovelock, J. (1988). *The ages of Gaia*. New York: Norton.

Nisbet, E. G. (1991). *Living Earth*. London: Harper Collins.

Warneck, P. (1988). *Chemistry of the natural atmosphere*. San Diego: Academic Press.

Washington, W. M., and C. L. Parkinson (1986). *An introduction to three-dimensional climate modelling*. Oxford: Oxford University Press.

3
The causes of change

I'm truly sorry man's dominion
has broken nature's social union
an' justifies that ill opinion
    which makes thee startle
at me, thy poor, earth-born companion
    an' fellow mortal!
But Mousie, thou art no thy-lane
in proving foresight may be vain.
The best-laid schemes of mice an' men
    gang aft agley,
an' leave us nought but grief an' pain
    for promised joy.
      Robert Burns, "To a Mouse"

3.1 The impact of humanity

People have existed on the planet for several million years. Our own type of humanity has existed perhaps for over 100,000 years: early modern humans inhabited caves in Israel 92,000 years ago. We are the most successful large animal species ever to evolve, if success is measured by impact on the rest of the natural world. This is not simply because of our intelligence, but also because we have endurance enough to run down almost any animal, are equipped with hands to uproot or pick any plant, and are omnivores. Most animals are tied to a very limited range of food sources; in contrast, humanity can eat one species to extinction and then shift to another food source. A characteristic feature of modern humanity is our demand for energy. With energy, we can do anything; without energy we are a species of large chimpanzees. Our modern impact on the planet is caused partly by the use of fossil energy sources and partly by changing the vegetation of the planet. Our actions are transforming the workings and composition of the atmosphere.

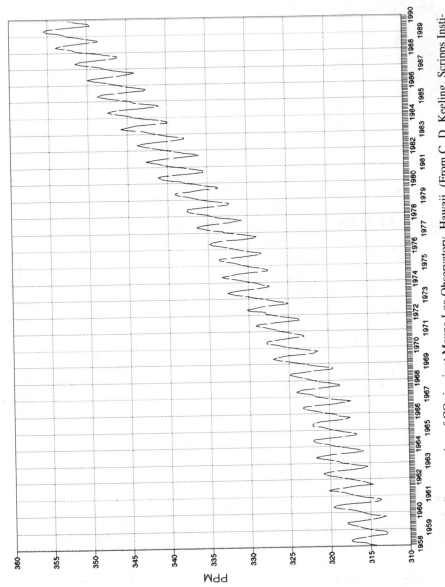

Figure 3.1 (a). Concentration of CO_2 in air at Mauna Loa Observatory, Hawaii. (From C. D. Keeling, Scripps Institute of Oceanography, with thanks.)

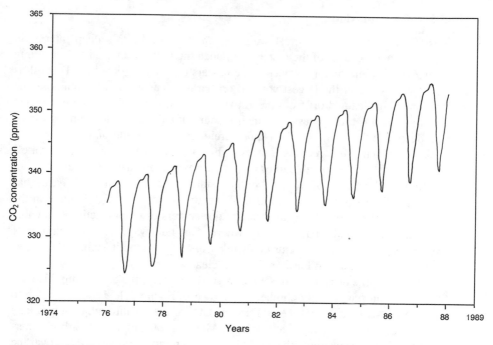

Figure 3.1 (b). Concentration of CO_2 in the air at the Canadian base at Alert, in the far north of the Canadian Arctic Islands. The spring drawdown in CO_2 marks plant growth. The small January ''tooth'' on the annual peak may represent the effect of winter fuel burning, with no process removing CO_2, or may be an accident. (From Atmospheric Environment Services, Canada.)

3.2 The atmospheric results of industrialization

Industrialization began in the green and pleasant hills of England, and almost immediately the more thoughtful minds of that country recognized both the benefits and perils it brought to humanity. The poet William Blake captured this spirit best, in his vision of a New Jerusalem among the dark satanic bymills: industry can free us from the poverty and despair that was the lot of most people in pretechnological society, but at the cost of destroying our souls. Two centuries ago, Earth had nothing to show more fair than the view of London from the Thames; today the summer visitor to England is met by layer after layer of smog as the airplane descends over Windsor Castle to the asphalt of Heathrow. The air has changed, and from London to Moscow, Europe lies under its own eructation, breathing in and exhaling smog. China and many of the cities of North America are the same. The pall stretches around the globe, and the products affect all inhabitants of our planet, to a greater or lesser degree.

3.3 Carbon dioxide

Visitors to the volcanoes of Hawaii can observe the Earth's deep interior in action. From the edge of the crater of Kilauea the tourist can see lava, steaming fumaroles, and in the far distance what appears to be a long low hill. That hill is the distant slope of the largest volcano on Earth, Mauna Loa, and in among the vapors being steamed off from the Hawaiian lavas, most of which are recycled rainwater and natural gases, is a small amount of carbon dioxide coming out of the deep mantle of the Earth. Nature emits carbon dioxide and methane and makes acid rain. More recently, however, humanity too has produced these gases, on a scale that is comparable to or in some cases exceeds natural production.

High on the flanks of Mauna Loa is a station that for the past 32 years has measured the carbon dioxide content of the atmosphere. At high altitude in the mid-Pacific the station is distant from major centers of human activity and able to obtain some of the purest air anywhere. Not far away, near the summit of Mauna Kea volcano, is a cluster of telescopes that exploit some of the best observing conditions on Earth in the high clean air.

Over the three decades of observation at Mauna Loa, the carbon dioxide content of the air has steadily climbed, as shown in Figure 3.1. This rise is due to the use of fossil fuels and the addition of CO_2 to the air through chopping and burning of vegetation, especially forests. Much of the addition of carbon – perhaps half – has been counterbalanced by various removal processes, but over the past 30 years there has been an accelerating increase in the CO_2 content of the air. When measurements began on Mauna Loa, the concentration of carbon dioxide in the air was about 315 parts per million by volume (ppmv). Now the figure is around 350 ppmv. The present rate of increase is about 1.5 ppmv per year. In contrast, during the 1960s the rate of increase was around 0.7 ppmv per year. The growth rate of the increase is close to the economic and population growth rates of humanity during most of the past decades.

Analysis of air trapped in ice cores in Greenland and Antarctica has allowed the reconstruction of the history of carbon dioxide in the air. The threshold that brought the Earth out of the ice age has been mentioned in Chapter 2. In 1750 the CO_2 content of the air was about 275 ± 10 ppmv. In preindustrial time since the Roman Empire, the level fluctuated from as low as 250 ppmv to as high as 310 ppmv, and was normally around 275 ppmv. After 1750, when the industrial revolution began in Britain and when large-scale deforestation began in Eurasia and North America, the level has climbed at an increasing rate (Fig. 3.2).

The increase in carbon release by human activity is shown in Figure 3.3. In the 1920s around a billion tons a year were released; by the 1960s this had increased to 2.5 billion tons;* today the annual release is over 5 billion tons. The most remarkable recent increase comes from Chinese coal mining. In 1984, China produced 772 million tons of coal, which was 22 times the Chinese production in 1949. Much of this spectacular growth has occurred in the past decade, and

---

* Throughout this book, "tons" refers to metric tons.

Table 3.1. *World primary energy consumption in selected years (in million tons of oil equivalent)*

	1979	1986	1987	% Share 1987	% Change 1987 From 1979	% Change 1987 From 1986
By type[a]						
Oil	3124	2899	2941	37.6	−5.9	+1.4
Coal[b]	1968	2318	2386	30.5	+21.2	+2.9
Natural gas	1282	1487	1556	19.9	+21.4	+4.6
Hydropower	424	517	524	6.9	+23.6	+1.4
Nuclear[c]	155	377	404	5.1	+160.6	+7.2
Total	6953	7598	7811	100.0	+12.3	+2.8
By area						
N. America	2136	2033	2081	26.7	−2.6	+2.4
W. Europe	1294	1280	1296	16.5	0	+1.2
Japan	370	372	378	4.8	+2.2	+1.6
Australasia	83	97	100	1.3	+20.5	+3.1
(OECD nations[d]	3883	3782	3855	49.3	−0.7	+1.9)
Centrally planned economies	2215	2658	2740	35.1	+23.7	+3.1
Other countries	856	1158	1215	15.6	+42.0	+4.9
World Total	6953	7598	7811	100.0	+12.3	+2.8

[a]Commercially traded fuels only.
[b]Bituminous, anthracite, and lignite, in terms of oil equivalent.
[c]Amount of oil required to fuel an oil-fired plant needed to produce the same amount of electricity.
[d]OECD: Organization for Economic Cooperation and Development.
Source: British Petroleum Co.

output now is probably close to a billion tons. If economic growth in China follows the pattern of Yugoslavia, an economy that in economic, political, and social terms is a good parallel, by the year 2000 China, which already has some of the worst urban air on Earth, will be producing more CO_2 (along with many noxious pollutants derived from burning coal) than the United States does today. India may well follow the same pattern as China. Its CO_2 emission has increased sharply in recent years, and has now surpassed that of the United Kingdom. Eastern Europe, especially Poland, northeastern Germany, and Czechoslovakia, are massively dependent on coal, although this dependence may decline as the post-Marxist clean-up begins.

The total amount of carbon in the carbon dioxide in the air is about 740 billion tons. This amount is increasing by roughly 3 billion tons per year. At present, each year, the burning of fossil fuels adds about 5–6 billion tons of carbon to the air (in CO_2), and some carbon is also added by biomass burning. These

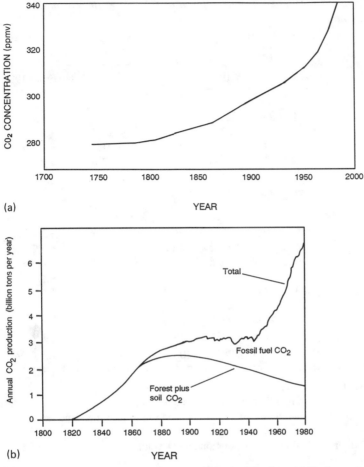

(a)   YEAR

(b)   YEAR

Figure 3.2. (a) Historical record of atmospheric CO_2 since 1750, from glacier ice. (Modified from Neftel et al. 1985. Reprinted by permission from *Nature,* vol. 315, pp. 45–7. Copyright © 1985 Macmillan Magazines Ltd.) (b) Emission of CO_2 from deforestation, changing land use, and the burning of fossil fuels, as estimated by Bolin (1986). Vertical axis is tons C in CO_2.

additions are partly offset by the removal of 2–3 billion tons by biological processes (in plant growth and in the sea), giving the annual increase of 3 billion tons of carbon (11 billion tons of CO_2), or roughly 1.3–1.5 ppm of CO_2 in the air.

Another way of looking at this process is to ask what fraction of all the carbon ever burnt in fossil fuels is still in the air. The answer is probably around 40%, perhaps as much as half. Were we to stop all burning and all deforestation tomorrow, it would take thousands of years for CO_2 to decline to preindustrial levels. This is because of the length of time it would take to mix the CO_2 into the ocean.

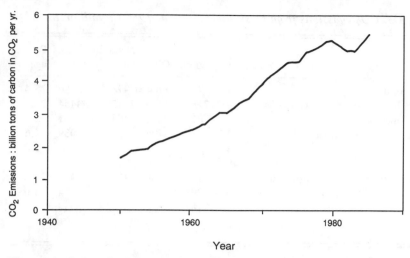

Figure 3.3. Carbon dioxide emission, 1950–85, from fossil fuel combustion and industrial processes. (From World Resources Institute 1987.)

Table 3.2. *Types of energy[a] consumed in 1987 by area (in million tons of oil equivalent)*

	Oil	Coal	Gas	Hydro	Nuclear	Total
N. America	833	486	473	149	140	2081
W. Europe	585	259	207	106	140	1296
Japan	208	68	36	19	46	378
Australasia	33	42	18	7	0	100
(OECD nations	1659	856	734	280	326	3855)
Centrally planned economies	679	1267	632	109	54	2740
Other Countries	603	264	189	135	24	1215
World total	2941	2386	1556	524	404	7811

[a] Commercially traded fuels only.
Source: British Petroleum Co.

The oceans are a major sink for CO_2. There is roughly 50 times as much CO_2 dissolved in the oceans as is present in the air. In the surface layers of the oceans, where light penetrates, plankton use carbon during photosynthesis if enough other nutrients such as phosphorus and nitrogen are available. These plankton supply the base of the food cycle of marine life. The various organisms rain debris – skeletons and shells – down to the sea floor below. These skeletons and shells are dissolved as they fall, carrying carbon down to the deep waters of the oceans. In shallower water, some debris reaches the bottom and is buried, removing carbon entirely from the ocean-air system. Isotopic dating suggests that the turnover rate of the ocean is 500–1000 years, so over the next millennium the seas

Table 3.3. *CO_2 emissions from fossil fuels by area, 1986 (in million tons)*

	Coal	Lignite	Oil	Natural gas[a]	Total	Percentage
Western Europe	807	172	1704	473	3156	15.4
Eastern Europe and Soviet Union	1119	713	1410	1212	4454	21.5
North America	1725	81	2852	1103	5761	28.0
Central and South America	83	0	705	151	939	4.6
Africa	254	0	269	42	565	2.8
Middle East	48	0	394	76	518	2.6
South and East Asia and Australasia	875	117	1306	173	2471	12.0
China[b]	2186	29	426	33	2674	13.0
Total	7097	1112	9066	3263	20538	
Percentage	34.5	5.4	44.1	15.9		

[a] Natural gas losses during transmission are substantial, possibly several percent of total use. These losses, as methane, have an immediate greenhouse impact that is comparable in magnitude to the immediate greenhouse impact of the CO_2 produced when the gas is burnt.
[b] Note the significance of Chinese emissions from coal burning, which are increasing rapidly.
Source: Deutscher Bundestag (1988).

will continue to absorb the carbon added to the air. However, should the oceans warm substantially, an opposite effect may counterbalance this absorption to some extent, because warming water emits CO_2 into the air. Carbon dioxide is also being taken up on land, especially in the remaining forests, which grow vigorously in a CO_2-rich environment and extract carbon from the air, placing it in tree trunks, soil carbon, and so on.

It is difficult to estimate the future increase in atmospheric carbon dioxide levels, because all scenarios depend strongly on assumptions about the rate of emission by human activity and the degree of uptake by forests and oceans. Figure 3.4 illustrates this. It assumes that the growth rate of emissions will range between 0% and 4%, and that the uptake of carbon dioxide by the seas and the land biosphere will be between about one-third and two-thirds of the annual emissions. During the period 1945–73, fossil fuel emissions of CO_2 grew at 4.6% per year. If emissions continue to grow at 4% a year, about equal to economic growth rates in prosperous countries, then by the year 2050 the level of atmospheric CO_2 will be around 900 ppmv, or three times the "natural" levels of 280 ppmv. A 2% growth rate would result in atmospheric CO_2 of roughly double the natural level by 2050.

Historically, the growth in CO_2 emission has roughly matched economic growth rates, so these estimates of CO_2 by the year 2050 are justified by past experience. However, during the energy crises of the 1970s and the economic downturn of the early 1980s, the growth rate of CO_2 emission slowed, and even stopped

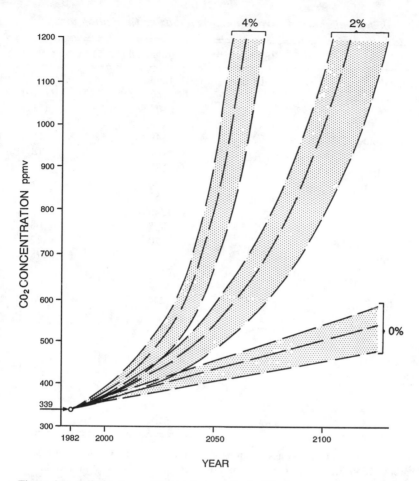

Figure 3.4. Predicted growth rate of atmospheric CO_2 concentration over the next 150 years, assuming growth rates of emissions of 0%, 2%, and 4% per year from a 1982 base. Median line in each curve is plotted assuming 55% of emitted CO_2 remains air-borne; left bounding line assumes 67% and right bounding line assumes 38%. (From Liss and Crane 1983.)

briefly, demonstrating that it is possible to stabilize emissions. If there is 2–2.5% annual growth in emissions, by 2050 the annual output to the air will be 20 billion tons of carbon in CO_2 from fossil fuels (compared to 5.5 billion tons of carbon in CO_2 in 1986), and CO_2 in the atmosphere will increase as shown by the top curve in Figure 3.5. Atmospheric CO_2 will exceed double the natural level by shortly before 2050 in this "no-growth in growth" model. Finally, an optimistic model (lower curve in Fig. 3.5) assumes that emissions are cut by 2050 to less than one-half the present day emissions; in this model, atmospheric CO_2 growth is slowed, reaching a level below 400 ppmv in 2050 AD, compared to 350 ppmv today.

Table 3.4. *Global emissions of carbon in carbon dioxide (in million tons)*

Region	Biotic (1980)	Industrial (1985)	(1988)
North America	19	1293	
United States		1186	(1310)
Canada		107	(119)
Western Europe	−12	779	
United Kingdom		148	(153)
France		107	(87)
West Germany		182	(183)
Italy		92	(98)
Other		251	
Eastern Europe		1346	
USSR	78	958	(1086)
Poland		120	(125)
East Germany		89	(89)
Other		178	
Pacific	43	314	
Japan		244	(270)
Other		70	
China	83	508	(610)
Other centrally planned Asia		44	
South and S.E. Asia	421	166	
Latin America	775	285	
Africa	277	144	
Middle East	7	223	
World	1691	5102	

Note: These figures include many assumptions and estimates. They are stated in million tons of carbon; Table 3.3, which comes from a different source, is stated in million tons of carbon dioxide. One ton of carbon is equal in mass to the carbon in about 3.67 tons of CO_2. Numbers in parentheses are for 1988, from the Carbon Dioxide Information Analysis Center. They are not exactly comparable with 1985 data. Per capita, U.S. output in 1987 was around 5 tons, West German and British output around 3 tons, and French output around 1.7 tons as a result of the French reliance on nuclear power. See also Table 6.14.
Sources: World Resources Institute (1988) and Carbon Dioxide Information Analysis Center (1990).

All this is speculation: which model will come true depends on the setting of policy. At present it appears possible that emissions will be stabilized (the "no-growth in growth" model), implying that the CO_2 level in 2050 will be around 500 ppmv, or roughly double the natural level. However, stabilization is not assured. "No-growth" in this context means only that present emission trends will continue, which will cause rapid growth in atmospheric CO_2. Cessation of emission is, in the present political context, seen as very unlikely. Few econo-

Table 3.5. *Annual emissions of CO_2 and chlorofluorocarbons CFC-11 and CFC-12 in selected years*

	Carbon emitted as CO_2 from fossil fuels and industry (in million tons)[a]	CFC-11 (CCl_3F) emitted (in thousand tons)	CFC-12 (CCl_2F_2) emitted (in thousand tons)
1931	968	0.0	0.1
1941	1337	0.1	3.0
1951	1776	7.6	32.4
1961	2602	52.1	99.7
1971	4267	226.9	321.8
1974	4684	321.4	418.6
1975	4660	310.9	404.1
1976	4924	316.7	390.4
1981	5115	248.2	340.7
1982	5079	239.5	337.4
1983	5068	252.8	343.3
1984	5236	271.1	359.4
1985	5336[b]	280.8	368.4
1986	5548	295.1	376.5
1988	5893		

Note: The impact of initial attempts to control CFC production after the mid-1970s is evident, as is the steady increase again in the mid-1980s. See also Table 6.3.
[a]To convert to mass of CO_2, multiply by about 3⅔.
[b]Difference in 1985 CO_2 figure from that in Table 3.4 is because one is based on production, the other on consumption.
Source: World Resources Institute (1988). Estimates for 1931, 1941, 1988 from Carbon Dioxide Information Analysis Center (1990).

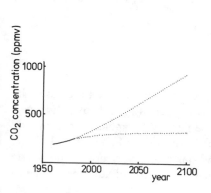

Figure 3.5. Conservative scenarios for future atmospheric CO_2 concentrations. Top curve is calculated assuming that emissions of carbon from fossil fuel consumption rise by 2–2.5% annually to about 20 billion tons per year in 2050 and that land ecosystems (e.g., burning forest) emit 2 billion tons per year. Half of the released CO_2 is assumed to remain in the air. Lower curve assumes that by 2050 only 2 billion tons of carbon as CO_2 are emitted from fossil fuels each year (less than half the present output) and 1 billion tons come from land ecosystems, of which emissions 40% remains in the air. (From Bolin 1989.)

Table 3.6. *Atmospheric concentrations of some important trace gases*

Year	CO_2 (ppmv)	CH_4 (ppmv)	CO (ppbv)
1959	316	c. 1.3	
1978	335	1.52	
1985	346	1.64	50–200

Note: There is a natural cycle in both CO_2 and CH_4 and there are latitudinal variations. Arctic levels of CO_2 in 1991 may peak well above 360 ppmv, and CH_4 well above 1800 ppbv. At Mauna Loa, Hawaii, between 1974 and 1985, the seasonal CO_2 cycle has an amplitude of 6.77 ppmv. This seasonal variation appears to be growing in size, by about 0.05 ppmv per year: the planet is breathing more strongly. During the period of study, CO_2 levels grew by about 1.4 ppmv per year.

During the period 1974–85, 55.62×10^{15} g of carbon were introduced into the atmosphere by fossil fuel and cement production, compared to the total mass of the atmosphere of 5.14×10^{21} g. Of the introduced carbon, 59% remained in the air at the end of the period (Thoning et al. 1989, World Resources Institute 1988).

Methane increased by 11% between 1978 and 1987, with a growth rate of about 0.016 ppmv per year (Blake and Rowland 1988). In autumn 1989 (October–November) a suite of samples collected by us in Prince Albert National Park, Canada, and analyzed by Blake's group contained on average close to 1800 ppbv of methane, on a rising trend, a value now characteristic of northern high latitudes. 1 ppmv = 1000 ppbv.

Source: World Resources Institute (1987), own data.

mists even advocate the option of reducing CO_2 emission to one-half of present-day emissions. For instance, in a prestigious Darwin College Lecture given in 1987 and published in 1989, a distinguished economist quoted, in an effort to get things "approximately right," a prediction that for the next 30–40 years present energy-consumption patterns will continue, followed by several hundred years when society depends mainly on coal. This opinion, which may reflect the orthodox view of economists, would imply very high CO_2 levels by 2050, perhaps as high as three to four times the natural level.

3.4 Methane

Methane (CH_4) is one of the most interesting of the atmospheric trace gases. Its very presence in an atmosphere that contains oxygen is improbable, and proof of life on Earth. It is an important greenhouse gas and is rapidly increasing: the natural methane cycle has been overwhelmed, yet no one knows exactly why. Methane is quickly destroyed in the air, and has a relatively brief atmospheric residence time of 7–10 years. This means that a reduction of atmospheric meth-

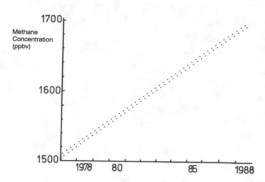

Figure 3.6. Growth in atmospheric concentration of methane (CH_4). (Modified from various sources, including Blake and Rowland 1988.)

ane is perhaps the most obvious starting point if any serious attempt is made to manage the atmospheric greenhouse.

The present concentration of CH_4 in the air is about 1.7–1.75 ppmv (1700–1750 ppbv), in contrast to 350 ppmv for CO_2. This is a very small amount, but added CH_4 has, by weight, nearly 100 times the effect on the greenhouse of added CO_2 (Table 3.7). The methane content of the air depends on location and season. The concentration is climbing rapidly, by 1–1.5% per year, although there is some suggestion that the rate of increase slowed in the late 1980s, to about 1% per year. The level of methane in the air is now more than double the level before the beginnings of Western economic growth several centuries ago (Fig. 3.7).

The direct consequences of CH_4 addition are compounded by the effects that CH_4 has on the abundances of other gases. Methane is destroyed by the hydroxyl radical (OH). Any increase in the methane level of the atmosphere reduces the abundance of OH in the air. OH is the main scavenger chemical of the air, nature's garbage collector. It cleans our environment. As OH is reduced, methane levels climb and overwhelm the cleaning effect, and the concentrations of other chemicals scavenged out of the air by OH also climb. These other chemicals include many unwanted compounds with greenhouse consequences.

Over and above these effects an increase in CH_4 will, in conjunction with nitrogen oxides, lead to enhanced ozone production in the troposphere. Tropospheric ozone (i.e., low-level ozone), unlike stratospheric ozone (which will be discussed later), is a little-desired species, since it is a noxious chemical and it also has a greenhouse effect. The greenhouse effect of enhanced tropospheric ozone adds substantially to the CH_4 greenhouse effect. It should, however, be noted that in the stratosphere, CH_4 can protect ozone by removing chlorine atoms. Methane is thus a very important atmospheric gas, partly because of its own greenhouse properties and partly because of its complicated effects on other important components of the air.

The sources of atmospheric methane can be divided into fossil and nonfossil categories. Fossil sources include coal, natural gas, and gas emitted from warm-

Table 3.7. *The relative effects on atmospheric temperature from a change in concentration of various greenhouse gases, compared to carbon dioxide*

| | Relative effect of change per | |
Greenhouse gas	Molecule	Unit mass
Carbon dioxide (CO_2)	1	1
Methane (CH_4)	21	58
Nitrous Oxide (N_2O)	206	206
CFC-11 (CCl_3F)	12,400	3970
CFC-12 (CCl_2F_2)	15,800	5750

Note: This table is oversimplified because effects vary with wavelength.
Source: Houghton et al. (1990).

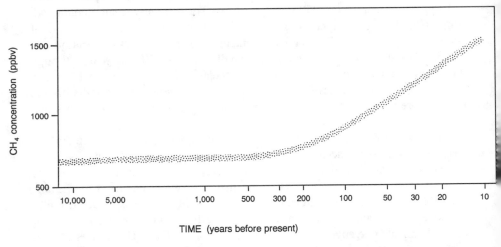

Figure 3.7. Atmospheric methane over the last 10,000 years. (From various sources, including Rasmussen and Khalil 1984.)

ing permafrost, as well as methane formed from carbon in very old swamps and trees. Nonfossil sources include rice paddies, northern wetlands, and peat bogs, termites and cows, and landfills.

3.4.1 Fossil sources of methane

The fossil sources appear to be responsible for a significant fraction (perhaps as much as a quarter to a third) of total methane production. This figure has been determined by a study of the carbon isotopes, ^{13}C and especially ^{14}C, which constitute part of the carbon in atmospheric methane. Carbon 14 dating tells us how long it has been since the carbon in a sample of methane was last circulating in the air. The measurement is difficult, because the amount of carbon as atmospheric methane is minuscule. Huge samples of air (up to one cubic meter) must be taken. Interpretation of the measurement is also difficult, since ^{14}C is released by nuclear reactors and bombs, and this signature must first be accounted for, before the nonnuclear component can be understood. Nevertheless, the likely result is that a quarter or more of the atmospheric methane contains "old" carbon, or carbon that has been removed from the biosphere for thousands of years or more.

Old carbon in methane comes from coal mines, oil and gas fields, asphalt, the seabed, permafrost, and old peat bogs. Coal mines are important sources; it is methane that explodes in mine disasters. Over the last 15 years, since the oil price rises of the 1970s, the rapid worldwide expansion of coal mining must have had effects on CH_4. Many shallow-level mines are gas-poor, but deeper mines often release much gas. In particular, Chinese mines are notoriously gassy, and the way coal is burnt in China, often in small domestic stoves, may also emit substantial amounts of methane. Taking into account all the other effects of coal burning in China, the expansion of Chinese coal mining ranks among the most damaging of the industrial challenges to the natural environment in recent years.

The use of natural gas as an energy source liberates enormous amounts of methane. Natural gas extracted from gas fields is typically transmitted by pipeline to distribution centers or liquefied and shipped. During transmission to the user, gas is lost by leakage or venting. The rate of loss is poorly known, but estimates range from 1% to 5% in North America and Europe and higher, possibly up to 10%, in some centrally planned countries, especially the Soviet Union. The lost methane enters the atmosphere. The amounts may be substantial: they may range up to 50 million tons per year, equal in its immediate greenhouse impact to 60 times that weight of CO_2, though this is probably a high estimate and the methane is rapidly removed by OH. Because of this loss during transmission, the use of natural gas as an energy source can have two to three times the immediate greenhouse impact of an energetically equivalent amount of oil. Methane is also sometimes vented underwater in hydrocarbon production, and it can be given off by asphalt on roads under sunlight.

Another source of fossil methane is gas stored in the seabed and in permafrost. This methane is locked in a structure called a methane gas hydrate, in water or ice. Hydrates are only stable at low temperatures, or under pressures of 10 atmospheres or more, depending on temperature. Enormous amounts of methane are stored this way in the sediments of the seas. In the tropics and temperate zones this storage can only happen under hundreds of meters of seawater, as the relatively warm seafloor sediment needs to be under pressure in order to store methane. In the colder Arctic Ocean, however, much less pressure is required and methane is widely stored under the shallow Arctic seas and also in permafrost. Rising temperatures destabilize the hydrate and cause the emission of methane. The permafrost is currently warming and is probably liberating methane. Mysterious plumes that rose recently from the Soviet Arctic islands, reportedly worrying the Pentagon, which suspected a nuclear test, may have been natural methane bursts from hydrate. One of the nightmares of climatologists is that the liberation of methane from permafrost will enhance the Arctic warming because of the greenhouse effect of the methane, and so induce further release of methane and thus increased warming, in a runaway feedback cycle.

Finally in the list of fossil sources of methane come old swamps and bogs. The peat bogs of the northern part of the Northern Hemisphere store considerable amounts of carbon and are very productive of methane for brief periods in the summer. The methane is newly made, but it exploits a store of carbon that in part may be very old (in terms of Carbon 14) and may have been out of the atmospheric system for thousands of years. Some tropical swamps and bogs may also contain carbon that is centuries or millennia old, from the trunks of old trees.

3.4.2 Nonfossil sources of methane

Methane is not only commonly called "coal gas," it is also called "swamp gas." Much of the methane in the air is produced in swamps and wetlands. This gas contains carbon that has recently been cycled through the atmosphere, and so has young ^{14}C, in contrast to the older sources listed above. A substantial amount of methane is also made by digestion, by methanogenic bacteria in the guts of all animals including humans and especially in ruminants such as cows, and by termites. All of these environmental agents – wetland, swamp, cows, even termites – are now substantially influenced by the actions of human beings. The most important nonfossil sources of atmospheric methane, however, are northern forests, rice fields, cows, and the burning of tropical forests.

The largest prehuman source of methane was probably the wetland, especially in northern forest, in North America and Eurasia. The significance of this source is clear from Figure 3.8, which shows the steep change in atmospheric methane concentration between 30° and 60°N, and a sharp seasonality in the Northern Hemisphere, levels being lowest in early summer and highest in autumn and winter. Conversion of organic matter, such as leaves and wood, produces methane in shallow wetland, forest ponds, and bogs. This production is probably highest in the warmth of summer, but much may be immediately oxidized in the ponds to carbon dioxide; in autumn, evasion may be higher, after leaf fall and during freeze-up. Rice paddies, mostly in southern Asia, are also probably a major wetland source of methane. The methane production rate depends on agricultural practice, including the degree to which the crop is fertilized. The area under water management for rice production is now large and the intensity of that production is high.

Domestic animals, especially cows and water buffaloes, eructate methane in quantity. Each cow emits perhaps 100 grams of methane a day, mostly from the front end. There are about one and a quarter billion cattle on Earth, producing perhaps 45 million tons of methane each year. Other animals also contribute.

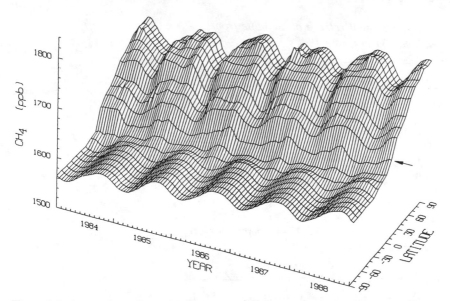

Figure 3.8. Atmospheric methane by latitude. The major sources of CH₄ are in the north and the tropics; the major destruction is in the tropics. Strongly seasonal CH₄ production occurs in the northern forest, north of about 40°N. In the 30°–50°N belt there is industrial production. In the tropical belt, rice fields and termites produce CH₄, but this is also the main region of destruction by OH. The Southern Hemisphere has little production: part of the smooth wave form may reflect northern seasonal CH₄ blown south. (From U.S. National Oceanic and Atmospheric Administration/Climate Monitoring and Diagnostics Laboratory; and P. Steele, unpublished data.)

Tropical forests in the natural state are also producers of methane. The Amazonian forests have large seasonally flooded areas. During the burning of tropical forest, combustion is generally not efficient, and a substantial proportion of the carbon in the trees is converted to methane, not carbon dioxide. This methane enters the air. Similarly, in tropical dry woodland and savanna grassland, frequent fires produce methane. A further important factor is the activity of termites, which recycle dead trees and soil carbon with the aid of methanogenic bacteria in their guts. Termites emit substantial amounts of methane.

3.4.3 The causes of the rise in atmospheric methane

The major sources of atmospheric methane are listed in Table 3.8. The figures are important because of the short atmospheric lifetime of methane – any change in production is soon shown in the atmospheric methane level. However, the figures are also very approximate because the information available about methane is inadequate. Many of the sources seem to be growing in strength. For instance, in 1950 there were around 800 million cattle and water buffalo in the world; by the 1970s there were about 1300 million and the population is rising

Table 3.8. *Annual release of methane from identified sources, late 1980s*

Source	Quantity (million tons)
Animal eructation	60–100
Natural wetland	100–200
Rice paddies	60–170
Biomass burning	50–100
Termites	10–100
Landfills	30–70
Oceans	5–20
Freshwaters	1–25
Methane hydrates	0–50
Coal mining	25–45 (perhaps larger)
Gas drilling	25–50
Total	400–640

Note: Recent results suggest that fossil sources of CH_4 may be greater than estimated here. Hydrate release may increase sharply in future, if the Arctic warms. Total is constrained by atmospheric content.
Source: Cicerone and Oremland (1988), modified.

by 6% a year. The growth in rice cultivation is also an important factor. Over the past few decades, multiple cropping in rice fields has increased, with two or even three crops a year, and the time during which the paddies are flooded has also increased. This has probably favored increased methane production by methanogenic bacteria in the fields.

Natural wetlands too have been heavily influenced by human action. In most of the temperate zones, the area of natural wetland has decreased substantially because human agricultural activity and urban growth have drained swamps and marshes. In the wet tropics forest has been felled, reducing wetland, and erosion has changed drainage patterns, although on the other hand damming of streams and rivers has had the opposite effect.

The most significant change may have been in the north of the Northern Hemisphere, in the forest and tundra of the Soviet Union and Canada. The Canadian wetlands are especially interesting. The population of Canada's national animal, the beaver *(Castor canadensis)*, has exploded recently. Fifty years ago the beaver was almost extinct in North America and in Eurasia *(Castor fiber)*. Today beavers are still rare in Eurasia but in Canada almost every available suitable area is now inhabited by them. Hunting and trapping have reduced predator populations of wolf, lynx, and other carnivores. Beavers are also trapped, but not now at a rate that threatens the population, and survivors reproduce rapidly to dominate the management of streams and ponds.

Beavers used to occupy much of North America, including most of the wetter parts of Canada, the United States, and Mexico, and were once abundant across Eurasia from England to Siberia. In central Canada and northwestern Michigan their population has recently recovered, so that their impact can be gauged. There has been a substantial increase in wetland recently. Beavers have flooded large areas of Canadian forest with their shallow ponds. In some areas between one-seventh and one-quarter of the landscape has been directly affected in some way by beavers. Beavers mine the forests around their ponds for useful vegetation, and one beaver can carry up to a metric ton of wood per year into its pond. Much of this is turned into beaver feces on the bottom of the pond; some rots. The carbon and nutrients of the trees are eventually processed by the bacteria at the end of the pond's ecological chain, producing substantial amounts of trace gases such as methane. In consequence methane production is greatly enhanced in beaver-dominated landscapes. Streams and ponds modified by beavers and floored by fecal mud produce a hundred times more methane per unit of water course length than streams without beavers.

Canadian and Soviet forest management practice takes no apparent notice of the problem of atmospheric CH_4 or CO_2. Any flight across Canada reveals that much of the accessible forest is second- or third-growth scrub, very different from uncut forest in its species distribution and proportion of wetland. Large areas have been flooded by hydroelectric schemes, especially in Quebec. The atmospheric consequences in terms of methane output must be significant, but are as yet unquantified. Canadian scientists with experience of Amazonia have contrasted Canadian forest management unfavorably with that of Brazil. More generally, the lack of management of Canadian and Soviet boreal forest may have had an atmospheric impact that compares with tropical deforestation.

3.5 Other trace gases and atmospheric components

3.5.1 Carbon monoxide

Carbon monoxide (CO) is abundantly produced by humanity. Though not itself radiatively significant, it is important because it scavenges OH from the atmosphere. OH is the species that removes methane. OH is both destroyed and regenerated in reactions involving CO and NO. The more CO there is, typically, the less OH is present. The less OH, the more the methane increases. Part of the recent increase in methane can be attributed to the increase in CO. The main sources are kinds of burning that involve incomplete combustion. Biomass burning, especially in Amazonia, is probably important in producing CO. Another major source is combustion in inefficient devices such as Chinese coal stoves and in automobiles, especially those with gasoline engines not equipped with effective catalytic converters. The gas has an atmospheric lifetime of about two months.

Table 3.9. *Global sources and sinks of carbon monoxide*

	Quantity (million tons/year)
Source	
Technological Sources	640 ± 200
Biomass burning	1000 ± 600
Vegetation	75 ± 25
Ocean	100 ± 90
CH_4 oxidation	600 ± 300
Oxidation of nonmethane hydrocarbons	900 ± 500
Soils	17 ± 15
Total production	3300 ± 1700
Sink	
Oxidation by OH	2000 ± 600
Soil uptake	390 ± 140
Loss to stratosphere	110 ± 30
Total decomposition	2500 ± 750

Source: Seiler and Conrad (1987).

3.5.2 Nitrogen oxides

Nitrogen oxides include nitric oxide (NO), nitrogen dioxide (NO_2), and nitrous oxide (N_2O). All three are important in the atmosphere. NO_x (a term applied to mixtures of NO and NO_2) gases help to control the other trace gases, especially ozone. NO_x in the troposphere comes from a variety of sources, including combustion and biogenic emission on the Earth's surface and downward transport into the troposphere from the stratosphere. High-flying aircraft also contribute to NO_x, as does lightning. The gases are removed by rain, and also by dry deposition after conversion of nitric acid and dinitrogen pentoxide (N_2O_5). NO_x gases are important because they can substantially increase the abundance of tropospheric (i.e., low-level) ozone, also an important greenhouse gas (see Appendix).

Nitrous oxide, which is a greenhouse gas, is also increasing, at a rate of 0.25–0.3% per year. We do not really know why. The sources of N_2O include soil emanations, release from the oceans, polluted rivers and estuaries, biomass burning, fossil fuel combustion, nylon production, and artificial fertilizers. The causes of the increase in N_2O seem to include both a growth in tropical sources, from land disturbance, and also a mid-latitude source from fertilizer use, the popularity of panti-hose, and from burning fossil fuels. Minor sources include anesthesia and so-called green propellants in spray cans. The gas is very stable in the troposphere and is mainly destroyed in the stratosphere. It has a long lifetime in the air, possibly 150 years. N_2O has an indirect effect on ozone concentrations, via

Table 3.10. *Global emissions of nitrogen in NO$_x$ (mixture of NO and NO$_2$) from burning fossil fuels and biomass*

Source	Nitrogen emissions (million tons/year)
Fossil fuels	
Hard coal	2–7
Lignite	1–3
Fuel oil	1–3
Natural gas	2–3
Industrial sources	1
Automobiles	4–8
Total fossil fuels, approximately	13–20
Biomass burning	12

Source: Modified after Warneck (1988).

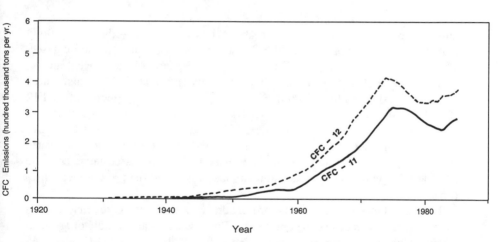

Figure 3.9. Emissions of CFC-11 and CFC-12, showing the drop in the late 1970s after unilateral imposition of controls by the United States, not followed by other nations. See also Table 3.11.

NO and NO$_2$, but the net effect is uncertain because stratospheric NO$_2$ helps to maintain high-level ozone against catalytic destruction by chlorine-bearing compounds. Increased N$_2$O may lead to catalytic ozone destruction by NO$_x$ species.

3.5.3 The chlorofluorocarbons

Apart from methane, the chlorofluorocarbons (Tables 3.5 and 3.11, Figs. 3.9–3.11) are perhaps the most immediately interesting of the atmospheric trace gases

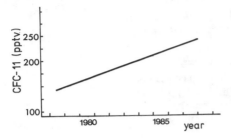

Figure 3.10. Concentration of CFC-11 in the air at Cape Grim observatory, Tasmania, 1978–88, in parts per trillion by volume (pptv). (From Forgan and Ayers 1989.)

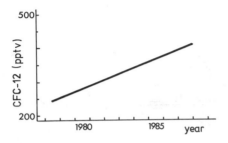

Figure 3.11. Concentration of CFC-12 in the air at Cape Grim observatory, Tasmania, 1978–88, in parts per trillion by volume (pptv). (From Forgan and Ayers 1989.)

although there is no scientific puzzle about their production – they are entirely man-made. Chlorofluorocarbons (CFCs) are extremely potent greenhouse gases. They also are notorious for their impact on stratospheric ozone concentrations. There are several important chlorofluorocarbons and related chemicals. These include CFC-11 or Fluorocarbon F11 (CCl_3F), CFC-12 or Fluorocarbon F12 (CCl_2F_2), carbon tetrachloride (CCl_4), CFC-113 or Fluorocarbon 113 ($CFCl_2CF_2Cl$), methyl chloroform (CH_3CCl_3), Halon-1301 (CF_3Br), Halon-1211 (CF_2ClBr), and CFC-22 or Fluorocarbon 22 (CHF_2Cl). These chemicals are listed in Table 3.11. Their role in ozone chemistry will be discussed later; here the consideration is of their effects as greenhouse gases and their atmospheric lifetimes.

The importance of the chlorofluorocarbons as greenhouse gases may exceed even their significance in depleting ozone. Because CFCs are entirely of artificial origin, their concentrations in the air are still low. In 1990 CFC-12 was 0.45 parts per billion by volume of the air, and was increasing by 0.02 ppbv per year. CFC-11 in 1990 was about 0.27 ppbv. Each molecule of CFC has a very powerful greenhouse effect. Furthermore, when light falls on the natural Earth, it is unlikely to be stopped by a CFC molecule because of the low concentrations. This means that every molecule of CFC-12 in the air that does indeed stop light adds directly to the absorption of infrared radiation, in contrast to the effect of adding an extra molecule of CO_2. To understand this, imagine using strips of adhesive tape to patch up a leaky, drafty wall in a summer cabin. The CO_2 hole is already well patched, so the addition of another strip of tape will not greatly reduce the leak. In contrast, the CFC hole is still wide open, and each new strip of tape has a marked effect. CO_2 is, of course, immensely important because

Table 3.11. *Atmospheric chlorofluorocarbons and related chemicals*

Chemical	Approximate atmospheric lifetime (years)	Estimated 1985 emissions (thousand tons)	Relative depletion efficiency on ozone	Share of total depletion of ozone by CFCs, etc. (%)
CFC-11	76.5	238	1.0	25.8
CFC-12	139	412	1.0	44.7
Carbon tetrachloride	67	66	1.06	7.6
CFC-113	92	138	0.78	11.7
Methyl chloroform	8.3	474	0.1	5.1
Halon-1301	101	3	11.4	3.7
Halon-1211	12.5	3	2.7	0.9
CFC-22	22.0	72	0.05	0.04

Note: Most of these numbers are poorly constrained and have large uncertainties.
Source: Hammitt et al. (1987).

there is so much of it, but each added molecule of CFC can be 10,000 times as effective as a greenhouse gas as CO_2.

The greenhouse potential of CFCs was once dramatically illustrated by J. Lovelock, who calculated that adding a relatively small amount, between 10,000 and 1 million tons, possibly somewhat more, of CFCs to the atmosphere of Mars could set in train events that would warm up Mars to Earthlike conditions. This is well within our present engineering ability. Cooling and hydrating Venus is more of a challenge, but not impossible. It would need the addition of an ocean of water derived from the moons of the outer planets, a shield, possibly of aluminum foil, to intercept some sunlight, and a mirror as a surrogate for rotation. The significance of this digression is, of course, that not only are we changing our own planet but we can change Mars and even Venus for our benefit by the same processes.

CFCs are very stable gases. They can be inhaled by people without ill effect and handled in manufacturing processes without breaking down to produce toxic chemicals. To chemists they seemed ideally safe for human use. However, this stability also means that no tropospheric processes affect them in any major way. They eventually drift up into the stratosphere, where the very energetic light coming in from the Sun finally attacks them, liberating chlorine. Most of the CFCs that have been produced are still in the troposphere. It takes up to 100 years for a molecule of CFC to diffuse up to the stratosphere and then break down. The bulk of the CFCs, in the troposphere, increases the greenhouse effect.

When the molecules finally arrive in the stratosphere they then act to remove the high-level ozone there. The effects of CFCs on ozone are also important in greenhouse terms. The removal by CFCs of ozone from the stratosphere probably cools the high atmosphere but warms the surface of the planet. Although this calculation is not yet fully resolved, it has been suggested that the effect of steady-state emissions of CFCs on stratospheric temperatures is significant.

The reasons for the growth in production of the CFCs have been their safety for humans and their usefulness in industry and the home because of their physical properties. CFC-11 and CFC-12 in particular are made in large quantity. CFC-11 is used in polyurethane foam and in aerosol propellants, while CFC-12 is used in air conditioners, such as those in automobiles. The manufacture and use of these gases is not the problem: it is their emission into the air. For instance, CFC-12 in a car air conditioner is vented to the atmosphere when the air conditioner is recharged or when the vehicle is junked; most countries have no effective policy for retrieving the gas although this would be comparatively easy to do. Similarly, very little of the CFCs in refrigerators or in building air conditioners is retrieved and reused when the equipment is junked. We are also reluctant to give up the use of CFCs in products such as styrofoam. To some extent, control measures are succeeding, and emissions of CFC-11 are now steady at about 200–300 thousand tons a year, but CFC-12 emission has taken longer to stabilize. However, stabilization (as opposed to cessation) of emission does little to improve the problem of growth of atmospheric CFCs, since the lifetimes of CFC-11 and CFC-12 are so long. The problem will now intensify as the CFCs drift up to the stratosphere. They will be with us for centuries, long after production is halted.

3.5.4 Ozone

Ozone is probably the most widely discussed of the trace gases, though African scientists have expressed a certain wry amusement at the contrast between the Western world's concern over skin cancer and the relative lack of concern over the myriad other climatic consequences of changes in radiatively significant trace gases, especially to the poorer nations. Furthermore, in many cities with large numbers of cars low-level ozone is so abundant that the ultraviolet light reaching the ground has decreased, not increased. Nevertheless, the change in stratospheric ozone is a serious and important problem, but for reasons other than skin cancer.

Atmospheric ozone can be divided into two categories, tropospheric or low-level ozone, of which there is often too much, and stratospheric or high-level ozone, of which there is now too little. Stratospheric ozone is made when ultraviolet light from the Sun strikes an oxygen molecule (O_2) in the upper atmosphere. A photon from the light splits the oxygen molecule into two highly reactive oxygen atoms (O), each of which rapidly combines with another oxygen molecule to form ozone (O_3). This gas absorbs ultraviolet light and dissociates into O_2 and O; the O then reforms O_3 with another oxygen molecule and so on, until accidentally two free oxygen atoms collide to make O_2. In the steady flood of ultraviolet light from the Sun the high levels of the atmosphere find a dynamic steady state in which creation of ozone equals destruction.

The small amount of ozone in the atmosphere can be visualized by imagining the whole atmosphere compressed under a pressure of 1000 millibars, about sea-level pressure. If the air were like this, it would only be about 8 km (5 miles) thick. Out of this about 3 millimeters (one-tenth of an inch) would be ozone, in contrast to about 1.5 km (1 mile) of oxygen. Ozone measurements are expressed in this way, first suggested by G. M. B. Dobson who initiated the study of stratospheric ozone. One Dobson unit is a hundredth of a millimeter of ozone in a compressed atmosphere at standard temperature and pressure.

Stratospheric ozone absorbs energy coming from above and from below: on the ozone about 12 watts per square meter (Wm^{-2}) of incident radiation comes from the Sun, and about 10 Wm^{-2} comes from radiation that is emitted upward by the surface and lower atmosphere. The ozone also emits about 4 Wm^{-2} upward and 1.5 Wm^{-2} downward. The net effect is that the ozone absorbs energy, heating the stratosphere. This heating effect is, however, reduced by cooling effects of CO_2 and H_2O. In the natural Earth the net result is a warm stratosphere.

Removal of part of the ozone (turning down the heater) has a cooling effect on the stratosphere. For instance, removing 50% of the ozone cools the middle stratosphere by up to 20°C or more, and also changes the temperature of the boundary, or cold trap, between the stratosphere and the troposphere. This change in temperature of the cold trap has, in turn, an effect on the water content of the stratosphere, so that a cooler stratosphere (with less ozone) means a drier stratosphere, and vice versa. Loss of ozone, by cooling the lower stratosphere, can also make it cold enough to enhance the production of clouds of ice crystals. These clouds promote chemical reactions that in turn destroy ozone, in a vicious circle of feedback.

One of the most dramatic atmospheric changes over the past decade has been the emergence of an Antarctic ozone hole. This hole in the stratospheric ozone layer appears in the Antarctic spring, especially in October, in the region where the stratospheric ice crystal clouds occur. In October 1987 the total abundance of ozone in the air column above part of Antarctica was cut in half, and in the altitude layer from 15 to 20 km the ozone was virtually gone, measuring only 5% of the concentration in the late 1970s.

The discovery of the ozone hole may rank in importance with the disappearance of the peregrine falcons as an "early warning" of environmental catastrophe. The story of the discovery is long and complex. Since 1956 the British Antarctic Survey has monitored ozone levels directly over Halley Bay, Antarctica. The observation team, led by J. C. Farman, has now collected a data set for over three decades. For the first two decades the pattern was of much fluctuation but no real signal (although it is possible with hindsight to see a pattern of decline in the fluctuation). Around 1980 this pattern changed, and a rapid decline set in (Fig. 3.12). The measurements continued; by 1984 the decline in ozone was dramatic. The results were published in 1985 and updated thereafter.

Unfortunately, the data from Antarctica conflicted with satellite results. For many years the Nimbus 7 satellite has been surveying the globe. This important satellite maps total ozone over the entire globe on a daily basis. Since 1982 a

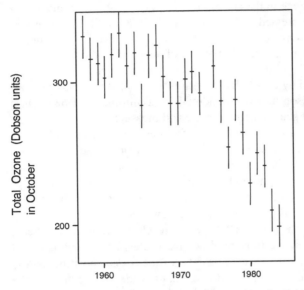

Figure 3.12. The Antarctic ozone data collected by the British Antarctic Survey team led by J. C. Farman. Note the steep decline in total ozone after 1979. This decline has continued: in October 1987 total ozone was about 125 Dobson units. In October 1988 the level recovered somewhat but was extremely low again in October 1989 and 1990. (From Farman et al. 1985. Reprinted by permission from *Nature,* vol. 315, pp. 207–10. Copyright © 1985 Macmillan Magazines Ltd.)

scientist analyzing the data had argued that global ozone was declining rapidly. However, the calibrations of both of the ozone-surveying instruments on board the satellite suffered a shared progressive degradation, so the argument was rejected on the grounds that the data were unreliable. Most scientists are familiar with this type of problem, and unusual results from instruments that are known to be in poor shape are always suspect. One would expect a difference of opinion in any group of competent scientists faced with the problem and the data set. Unfortunately, the Antarctic hole was so deep that it was automatically programmed out as an error and the data were replaced by assumed background values.

   These accidents of misinterpretation are the routine business of scientific measurement. Instruments always have to be calibrated, enormous amounts of data have to be processed, and the scientific message – the signal – is often missed amid the distracting information – the noise – caused by all the other problems. Scientific measurement is difficult. When the British data were published, the raw satellite information was reexamined and the depth and extent of the hole were immediately realized. The first lesson of the story is that long-term ground control is essential in observing the Earth: we cannot rely entirely on satellite measurement. All the scientists involved were competent and ably performing their jobs: Nature simply did the unexpected, and luckily there was ground con-

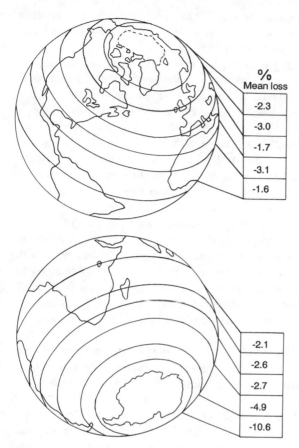

Figure 3.13. Ozone depletion worldwide. These figures are approximate, for the late 1980s, but illustrate the global distribution of the problem. Some recent evidence indicates that depletion is now more substantial than shown. (From Deutscher Bundestag 1988.)

trol to detect it. The second lesson is the opposite: satellite data, once correctly interpreted, are invaluable. Without Nimbus 7 the situation could well have deteriorated for another decade before the British warning was finally heeded. The satellite and the ground data are complementary.

The ozone hole in 1989 equaled in depth and extent the enormous hole of 1987. The stratosphere appears to vary on a rough two-year cycle: in 1988 the lower stratosphere over Antarctica was slightly warmer than in 1987, resulting in a less marked ozone hole in 1988; in 1989 the massive hole of 1987 was duplicated. The hole grew very rapidly in September 1989, and by mid-September covered 19 million sq. km, or 7.5% of the Southern Hemisphere. For comparison, the area of North America (including Panama and the Caribbean islands) is 21 million sq. km. A minimum ozone column of 111 Dobson units occurred on October 7, 1989, about the same as the minimum value in 1987. (To put this number into context, see the values in Fig. 3.12, which first provoked concern about ozone loss.) It is probable that massive springtime Antarctic ozone holes will now occur in many of the years up to 2030 or later.

The Arctic stratosphere, being warmer, has fewer ice clouds and has not developed a massive ozone hole yet. However, on February 6, 1990, a major cooling event occurred over Scandinavia, and the air at the altitude of 22 km reached temperatures as low as $-93°C$. Satellite measurements detected a small hole in the ozone, of 165 Dobson Units, but this hole was not confirmed by balloon measurements, which recorded 285 Dobson units, implying that the satellite was measuring radiation bounced back from cloud tops. Nevertheless, some ozone depletion was observed at about the altitude of 16 km. In general, however, it appears that the Arctic is less likely to develop a massive hole than the Antarctic.

There is still debate about the detailed causes of the Antarctic hole. In the early 1970s it was pointed out that when the chlorofluorocarbons CFC-11 and CFC-12 are carried into the stratosphere, they are ultimately broken down by ultraviolet rays, releasing free chlorine atoms. The chlorine atoms react with ozone and single oxygen atoms, with the net result that ozone is changed to oxygen as O_2 and the chlorine atoms are released again to repeat the process. A single chlorine atom may destroy hundreds of thousands of ozone molecules. The element bromine, which is chemically similar to chlorine, may also be involved.

Normally, most of the chlorine in the lower stratosphere is held in chlorine nitrate and hydrochloric acid. These chemicals do not have an effect on ozone, and act as temporary reservoirs, storing chlorine. In the Antarctic winter, the lower stratospheric wind system isolates the polar air, which cools to $-80°C$ or below, temperatures at which clouds of ice crystals, as well as of hydrated nitric acid, condense even in the dry stratospheric air. The ice crystals in these polar stratospheric clouds provide surfaces in the air on which reactions take place that release chlorine from the temporary reservoirs. This active chlorine very effectively removes the ozone. Arctic air in the lower stratosphere is not usually so cold and reaches $-80°C$ only for short periods. As a result, a long-lived Arctic ozone hole has not occurred yet, although short-lived depletion episodes have occurred.

It is probable that the ozone hole is in part also the result of dynamic processes in the atmosphere. On one day, September 5, 1987, ozone levels fell by 10% over an Antarctic area of 3 million sq. km. This must have been a dynamic effect, just possibly related to the influx of air of complex chemistry from much further north, perhaps in the region of intense biomass burning in Amazonia or Africa. However, the consensus of evidence is that chemical controls play a critical role, with the CFCs dominating. Once the hole is initiated, the ozone loss decreases the absorption of solar energy, because ozone is a greenhouse gas, and the region of the hole cools. This increases the production of ice clouds. The ice particles enhance the reactions that destroy ozone, cooling the hole further. The hole feeds on itself. In the southern spring of 1987 the region of the hole was about 8°C cooler at an altitude of 15 km than it was in 1979.

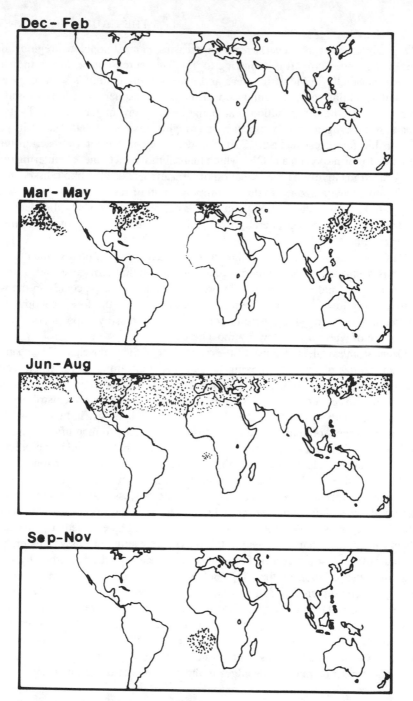

Figure 3.14. Worldwide distribution of tropospheric (low-level) ozone, by season. Regions with tropospheric ozone abundances over 40 Dobson units are shaded. High ozone abundances prevail over the industrial regions of the Northern Hemisphere in summer. The high ozone content of air in the South Atlantic in September to November can be attributed to grass-burning in Africa. In general, low ozone abundances are characteristic over tropical forest, in Southeast Asia and Amazonia. (From Fishman et al. 1990.)

Tropospheric (or low-level) *ozone* is increasing in most populated regions, not decreasing (Fig. 3.14). It is damaging to life. The increase is occurring in many regions especially in industrial areas and over wealthy countries with many automobiles, adding to the greenhouse warming. Tropospheric ozone is a familiar component of the air pollution that hangs over many major industrial cities. Apart from being a greenhouse gas, it is highly toxic to plants and is in part responsible for forest dieback. Ozone is destroyed in the air by reaction with peroxide to form oxygen and OH, which helps remove methane. Countering this, smog reactions involving the oxidation of CH_4 and CO use the photodissociation of NO_2 to produce ozone, so that a complex chain of reactions involving methane, carbon monoxide, and NO can give a net production of O_3. Incidentally, methane may also be peripherally involved, to our benefit, in protecting stratospheric ozone.

Ozone is removed naturally at the Earth's surface by soils, plants, and the sea. Much is destroyed in the southern oceans and in the northern grassland. Forests remove ozone, and the Amazon rainforest is a major sink, especially in the wet season. Very low levels have been measured just above the forest canopy. The destruction of the forest has therefore the effect of increasing tropospheric ozone, and hence of warming the lower atmosphere.

Ozone is an excellent example of the complexity of the atmosphere. The changes that are taking place are removing it from the high atmosphere, where some ozone is useful, and adding it at low level, where it is damaging to the present biosphere. Moreover, many of the other trace gases created or redistributed by humankind, such as methane, the CFCs, and the NO_x gases, influence or compete with ozone, some to restore the status quo but many to magnify the changes taking place. The ozone hole was a surprise – it was not expected or predicted, even by those who had foreseen the global decline in stratospheric ozone.

Surprises such as this are common in the study of climate change. However good we consider to be our understanding of the processes at work in the atmosphere, there is often some overlooked factor that may have major consequences. Most theoretical models of the workings of the atmosphere, whether physical or chemical, predict change by taking the present-day situation, as we understand it, and distorting some aspect. For instance, a model may be perturbed by varying some factor, such as the CO_2 content or the cloudiness, in an attempt to assess the consequences of that variation. The modeling process fails if some major unexpected change occurs, if the model is pushed too far away from present-day conditions. Here is where analogy is important. The small child playfully pushing the glass across the top of a table merely shifts the position of the glass, perhaps with the odd spill of slopped water, until the edge of the table is reached. . . . There must be many table edges in the workings of the atmosphere.

3.5.5 Arctic haze

Arctic haze is a northern phenomenon that illustrates the global scale of pollution. It is discussed here as an example of the photochemical smog that afflicts

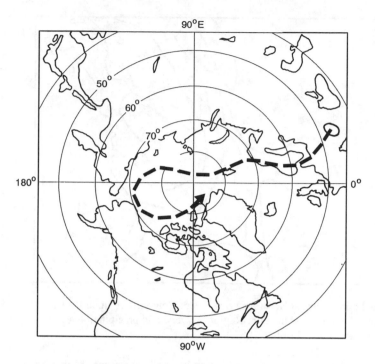

Figure 3.15. The path of one Arctic haze event in spring 1986 from its source in Europe. (From Schnell 1987.)

much of the industrial world. In August afternoons over western Canada the Sun is yellow-white. In polluted London the late summer Sun is yellow, in more polluted Mainz it is yellow-orange, and in the reeking air of Moscow that same Sun may be a dirty red. This pollution of the Northern Hemisphere is becoming a global, not a regional, problem.

The Arctic atmosphere is highly polluted in winter although there is virtually no Arctic industry. Pollution began as early as 1900 and has been increasing markedly in recent decades. Much of the cause of the pollution is the transport of air from Eurasia northward to the pole, coupled with weak removal of the pollution in the cold dry air or by the surface. Many reactions take place in the haze, including destruction of low-level ozone. This destruction may be linked with bromine rather than chlorine as in the stratosphere at the South Pole. The bromine may come from the burning of coal or from marine organisms. Figure 3.15 shows one haze event that was tracked backward to its source in central Europe.

Linked with Arctic haze is Arctic dust. The recent cruise of the West German ship *Polarstern* discovered large areas of dirty ice in the Arctic. The ice contained wind-blown dust from the prairies and steppes of the Northern Hemisphere, possibly in part from deteriorating agricultural regions such as China, Kazakhstan, and the North American prairies. Dust storms are now common in

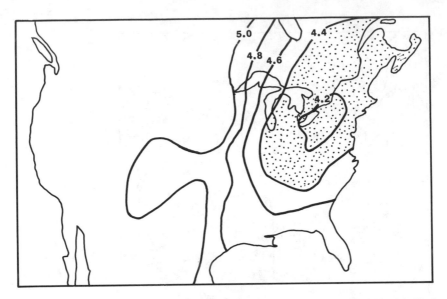

Figure 3.16. Acid rain in central North America, 1985. Contours are in units of pH: the lower the number, the more acid the rain. Regions of very acid precipitation are stippled. (From World Resoruces Institute 1987.)

Soviet central Asia and in Saskatchewan, carrying soil kilometers high into the air. The sea too is contaminated, as the Arctic Ocean is one of the collecting places of the toxic waste of the Northern Hemisphere's industries. The chemicals known as PCBs illustrate this: they have permeated the Arctic food chain. Although they are not in themselves more than mildly hazardous, they are easy to trace and serve as a guide to the much wider and more worrying pollution by other chemicals.

3.5.6 Acid rain

The causes and consequences of acid rain are now so widely discussed that the problem will not be covered in great detail here. Acid rain, a term that has now broadened to include all acid precipitation, was first recognized over a century ago in the British Midlands, the cradle of industrialization. Acidity is measured in pH units, a logarithmic scale where 7 is neutral and lower numbers are acid. North American rain in areas remote from industry has a pH value above 5.0 and naturally contains both weak and strong acids. However, in most areas of industrial activity, the pH of precipitation is much lower – more acid – ranging to pH values close to 4 in the worst-affected regions. Acid rain is widespread in Western Europe, the Soviet Union, and China; damaging acid precipitation also occurs in less publicized places, such as South Africa (which is heavily dependent on coal) and parts of South America.

Much of the acidity comes from sulfur compounds and NO_x in the rain. As

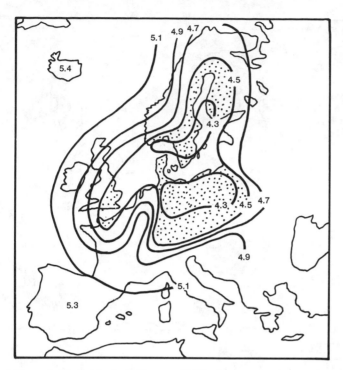

Figure 3.17. Acid rain in Europe, 1985. Contours are in units of pH. Stippling shows regions with very acid precipitation. (From World Resources Institute 1987.)

explained earlier, sulfur is an important natural component of rain, coming from DMS emitted by marine algae, and therefore rain is commonly naturally slightly acid. Human activities have added to this natural acidity in rain, and in regions of dense population up to 90% of the sulfur deposited in the rain is now of human-influenced origin. The causes of this increase are varied. The major sources of acidity added by human beings include the burning of coal, especially for electricity in the industrial countries (strikingly so in the United Kingdom, the eastern United States, South Africa, and Eastern Europe), smelters, automobiles, and sewage and manure (especially in Western Europe) that finds its way via agricultural land into coastal water and enhances DMS production.

Prior to the 1950s much of the pollution from coal burning and smelters was local in nature, as it still is in China and the poorer nations. However, in the richer communities tall smokestacks were built and less coal was used in homes, so transforming a local problem to a long-range regional one as the pollution was inserted above the relatively still layer at the very base of the atmosphere into the winds above. Measures to reduce sulfur emissions have been taken in some jurisdictions but have been minor compared to the rapid increase in coal burning across the entire hemisphere.

The problem is now changing from a local or regional scale to a hemispheric

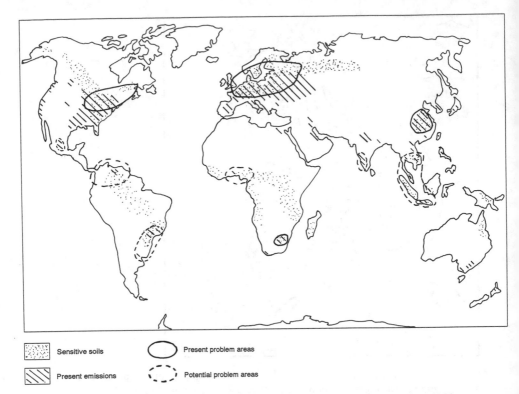

Sensitive soils   ⬭ Present problem areas

Present emissions   ⬭ Potential problem areas

Figure 3.18. The worldwide problem of acid emissions and precipitation, late 1980s. Figure shows regions of acid emissions, regions where soils are likely to be damaged by acid rain, and present and future problem areas. (From Rodhe et al. 1988, modified.)

scale. Polluted air masses have been tracked across the Atlantic and over the North Pole from Eurasia to North America. About 50% of the sulfates falling in eastern Canada probably come from the United States. Much of European acid precipitation comes from emissions within industrial Eastern Europe, especially in northeastern Germany, Czechoslovakia, and Poland. In Silesia, in southern Poland, some regions are now close to environmental collapse. If coal remains the dominant fuel in central Europe the long-term consequences will be profound, not just for Poland which may be dying, both literally and metaphorically, from emphysema, but also for Europe as a whole. Improvements also occur. Over the past two decades, British emissions from coal burning have dropped substantially, to the relief of Sweden. However, as the electricity utility has cut SO_2 emission, the increase in British automobiles, some of them of Swedish manufacture, has sharply increased British NO_x emissions. Other countries in Western Europe have similar histories.

The effects of acid precipitation have been widely publicized. They include the acidification and eventual damage or impoverishment of lakes and rivers, as well as injury to forests, in Europe and also in North America. Assessing the

Table 3.12. *Global atmospheric sulfur fluxes*

Source or sink	Million tons of sulfur/year
Gaseous emissions	
Anthropogenic	103
Biogenic: oceans	36
Biogenic: soils	7
Volcanic gases	7
Particulate sulfate	
Seaspray	150
Mineral dust	?
Anthropogenic SO_4^{2-}	3
Wet and dry deposition	
SO_2 over oceans	15
SO_2 over land	71
Excess SO_4^{2-} over oceans	28
Excess SO_4^{2-} onto continents	42
Seasalt over oceans	135
Seasalt onto continents	15

Note: The transfer onto land of sulfate (SO_4^{2-}), derived in part from DMS emitted by marine organisms, is significant.
Source: Warneck (1988).

effects is very difficult. In Canada, detailed studies of lakes (including experimental acidification) have conclusively shown that the effects are severe. In forests, the analysis is more controversial. Many forests are suffering dieback. The causes of damage may be complex, but acid rain clearly adds to other stresses. Healthy forests have many dying trees, even in normal settings. However, whereas a tree can survive one or more "normal" stresses, such as drought or insect outbreak, adding acid rain may be fatal. Furthermore, the effects are progressive and cumulative. Acid rain is involved in mobilizing soil aluminum. This causes root damage and leaches plant nutrients. Acid fogs and mists can damage leaves directly. Furthermore, the smog that envelops much of Europe in the quiet days of summer reduces photosynthesis by blocking sunlight.

3.5.7 Trace metals and ammonia

Industrial activity and farming practice have injected large amounts of trace metals and ammonia into the air. The sources of emissions are varied. They include processes in nonferrous metal smelters such as roasting, melting, and refining, iron and steel production, the cement industry, urban refuse incinerators (which

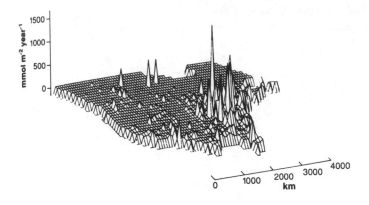

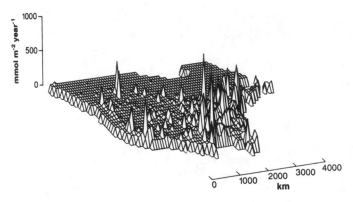

Figure 3.19. Emission of sulfur oxides (top) and nitrogen oxides (bottom) in the United States and Canada south of 60°N, in 1980. Grid cells are 1° latitude by 1° longitude. (From Schwartz 1989. Copyright 1989 by the AAAS.)

are becoming an important source of atmospheric trace metals), and coal burning.

Coal combustion in electric plants and in industrial, commercial, and residential burners is the chief source of airborne mercury, molybdenum, and selenium, and an important source of arsenic, chrome, manganese, antimony, and thallium. Coal-fired electric power stations typically emit large quantities of radon, a rapidly decaying radioactive gas. During normal operation, average coal-fired power stations in consequence emit more radioactivity to the surrounding environment than nuclear power stations of the same size. Oil burning emits vanadium, nickel, and tin. Many pollutants are released by gasoline spills and emitted from additives. Nonferrous metal smelters, especially in copper-nickel and zinc-cadmium production, emit lead, arsenic, cadmium, copper, and zinc. Iron smelting and steel making emit chrome and manganese. Domestic waste also contributes, especially when incinerated or when buried in recharge zones of aquifers.

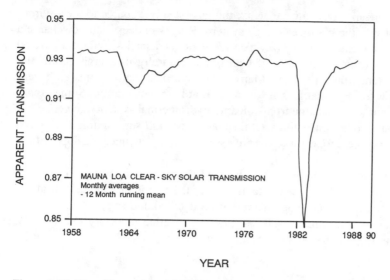

YEAR

Figure 3.20. The effects of the El Chichón eruption in 1982, as measured from Mauna Loa. The "apparent transmission" is a measure of the clarity of the air above the observatory. Note how the sky became less clear after the eruption, then slowly cleared as the aerosol cloud decayed in the years after the eruption. The smaller dip in transmission in 1964–5 may also represent a volcanic explosion, in Southeast Asia. (From U.S. NOAA 1988.)

For instance, long-life and rechargable batteries can contain cadmium or mercury, both highly toxic in certain circumstances and in some ways more dangerous than nuclear waste. Cadmium and mercury do not decay but conveniently disappear from view in the domestic trash can, to go to landfills and from them into groundwater, or even to go to incinerators and the air.

Atmospheric ammonia is another important pollutant. Much is produced by liquid manure, especially in northern Europe, where manure is used in large quantities on farmland. Ammonia is deposited on forest and grassland with serious effects on the vegetation, causing eutrophication and declining of abundance of slow-growing plant species.

These pollutants find their way into the biosphere, either directly via emitted gas or through polluted water. Much of the discharge into rivers and lakes occurs in the rich industrialized world, but pollution in poor nations is also intense, since most poor countries have little or no pollution control. Discharge into soil is also locally severe, especially in central Europe and Japan.

It should be remembered that there are natural analogues for many of these discharges. Major volcanic eruptions belch pollutants (Fig. 3.20). It has been suggested that on occasion these belches have put enough chlorine into the stratosphere to remove the entire ozone layer. Volcanoes can also support hydrothermal systems that discharge quantities of trace elements into the sea and even the air. Mt. Etna, for instance, may be a major source of mercury pollution in the Mediterranean. For many elements, the largest single source to the environment is probably the mid-ocean ridge system. Ridges in deep water discharge a wide variety of elements, many of which are toxic, and shallow sediment-covered ridges, such as that in the Gulf of California, naturally discharge liquids not greatly different from diesel oil. Many ridge hydrothermal systems emit very acid water, with pH sometimes as low as 3.5, at high temperature. Some organisms flourish in the discharge from volcanic hydrothermal systems, in vent communities, but at other times the toxic discharges can kill surrounding life, at sea or on land. Yet these toxic outbursts are essential in the chemical resupply of the biosphere.

Humanity has added to and often dwarfed these natural sources. Trace elements are needed by biological organisms, but the risk is that the industrial addition is overwhelmingly toxic. We require a detailed and sophisticated assessment of natural and artificial sources, and a deliberate planetary management system of all trace metals and toxic emissions.

3.5.8 Hydroxyl

The hydroxyl radical (OH) is the main policeman of the atmosphere. This chemical species removes most of the trace chemicals that we think of as pollution by oxidizing them to other substances that are less harmful to the modern biosphere, and that are eventually rained out or extracted from the air by plants. Many of these oxidation reactions take place in such a way that hydroxyl is returned back to the air, so that a small number of hydroxyl radicals can police and jail a large number of pollutants. Important trace chemicals that are attacked by OH include methane, carbon monoxide, and formaldehyde (HCHO), which are oxidized to CO_2; nitrogen oxides (NO_x), which are oxidized eventually to nitric acid; and sulfur dioxide (SO_2), which is oxidized to SO_3 but eventually dissolves in clouds and is removed as sulfuric acid. Hydroxyl also removes many of the more complex chemicals in urban smog.

Hydroxyl is mostly made by the action of light on ozone in the troposphere, the tropospheric ozone here playing a helpful role in controlling pollution. The ozone breaks down, releasing excited oxygen atoms that combine with water to make OH. The production is greatest in the equatorial lower troposphere. Another source of production is in the urban pollution itself – some of the reactions create OH.

It is very difficult to measure OH in the atmosphere, for a variety of reasons. This means that we do not have a clear idea of the changes that are taking place in the OH distribution in the atmosphere. One fear is that OH numbers may suddenly collapse under the onslaught of methane, carbon monoxide, and so on, leaving the atmosphere unpoliced. The consequences for the biosphere would be drastic.

Our lack of knowledge of OH abundance in the air illustrates our more general ignorance of the workings of the atmosphere. For most of the trace species – CH_4, CO, NO_x, the human output is now comparable to or exceeds natural pro-

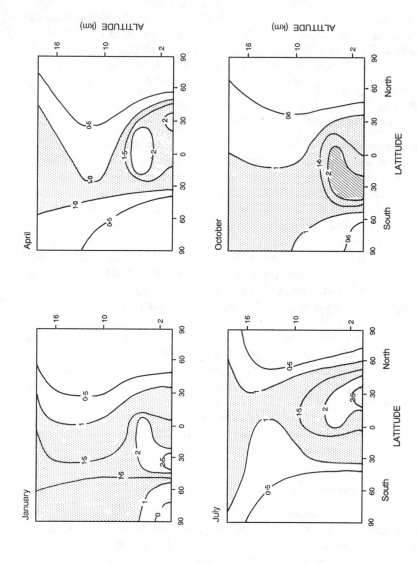

Figure 3.21. Cross section, from North Pole to South Pole, of the OH abundance in the troposphere, not from measurements, which are not possible yet, but calculated from a model. Diagrams show calculated daytime average distributions of OH (10^6 molecules per cubic cm) for January, April, July, and October. (From Crutzen and Gidel 1983; see also Spivakovsky et al. 1990 for a more recent calculation.)

duction, and there is no longer a natural cycle. The levels of these chemicals are now determined by humanity. The policemen, too, are made in large numbers as by-products of human activity, but it is likely that the increase in the numbers of policemen is not keeping up with the increase in numbers of criminals, and the global atmospheric order may therefore be endangered.

3.6 The causes and extent of floral and faunal change

The evolutionary appearance of each species of life alters the entire biosphere, to a greater or lesser extent depending on the success of the species. Human beings, like all other species, have continually modified the biosphere ever since they first appeared. We think of our aboriginal ancestors as living in harmony with the biosphere, but this romantic image is false, a modern version of the seventeenth- and eighteenth-century myth of the noble savage. The first colonists in each continent seem to have drastically changed the environment. The case is not yet proven, but there is strong evidence that the primitive Eurasians and Amerindians extinguished such species as the mastodon and mammoth; within historic times, the Maoris (the last human group to occupy a large virgin territory) extinguished many species and changed the vegetation in the islands of New Zealand.

More recently, Western, technical humanity has had a profound impact on the planet, directly in the occupation of the Americas and Australia and indirectly by imposing European technology and facilitating population growth in Africa and Asia. This impact has been most dramatic since 1950 during the postwar period of economic growth, and 1950 is therefore an excellent arbitrary dividing point. Most humans living today were born since then. Should we ever decide deliberately to manage the planet, the state of affairs in 1950 should become the touchstone against which decisions are made.

3.6.1 Humanity, vegetation, and animals

The direct causes of the change from the "natural" Earth to the modern "managed" Earth are the feeding, clothing, housing, and entertainment of humanity. The largest changes in the biosphere come from the destruction of forest to provide building material and firewood and to make way for domesticated animals, and from the worldwide replacement of natural animal and plant species either by those useful to humanity or by those "weeds and trash fish" that can survive after exploitation.

Humanity has been changing the environment and attempting to come to terms with the consequences since the very beginning of civilization. Bushman paintings in Africa are exquisite depictions of prey animals. At the beginning of Western culture, Mesopotamian kings set aside royal hunting reserves where wild animals could survive for the king's pleasure. Later, the king of England restored the New Forest to wild animals and trees, and among Robin Hood's offenses in Sherwood Forest was the slaughter of the protected deer.

The arrival of humanity coincided with dramatic changes in the animal population and vegetation of the continents. There is still considerable debate about the exact extent to which these changes were directly caused by humanity, and how much of the change was a result of climatic factors, but it seems very likely that human beings were the chief cause.

During the last interglacial period, prior to the spread of modern humans, the continents were much richer in large animals than today. Charles Darwin once wondered why Africa had so many species of big animals, while the other continents were impoverished in fauna. The fossil record gives a different picture. Not long ago, North and South America too had animals similar to elephants, camels, horses, lions: a remarkable fauna of large animals. Europe and northern Asia had abundant elephants, rhinos, hippos, bison, horses, and giant deer. In Australia an extraordinary marsupial fauna had evolved, to parallel the large animal diversity of Africa. These animals included diprotodonts, which were marsupial equivalents of the rhino and hippo, giant wombats, giant kangaroos, and marsupial predators that perhaps took the place of lions. In New Zealand and Madagascar evolution had taken a yet different path, with the development of giant birds.

In each continent outside Africa these diverse fauna disappeared at the same time as humans entered. In Australia, roughly 30,000 years ago, the marsupials were decimated. About 85% of the genera of large animals became extinct, leaving only the kangaroos, emus, and a few other large species. There is still debate about when humans first entered the Americas, but the major wave of colonization took place shortly after the main termination of the last ice age, roughly 11,000 years ago. Suddenly, three-quarters of the genera of large animals disappeared. They seem to have been wiped out by a wave of overkill. Humans, being omnivores, were able to switch from one species to another as extinction proceeded. Other predators, such as the saber-toothed cats and the dire wolves, could not compete and became extinct. Only the large migratory herbivores such as the bison survived, as the kangaroos had in Australia, protected perhaps by their migration to relatively inaccessible summer grazing. In Eurasia the same process took place as the mammoths, rhinos, and hippos became extinct. Only in Africa did the majority of genera survive, perhaps because they had evolved with humanity, and more probably also because in Africa there were parasites and diseases that had evolved to attack humans. To the tsetse fly we may owe the survival of many animals.

The most recently colonized regions of the world are Madagascar (occupied about 1500 years ago) and New Zealand (first occupied about 1000 years ago). On both these islands, colonization was rapidly followed by the extinction of the large flightless birds, such as the rocs or elephant birds in Madagascar and the moa in New Zealand. Fire was a commonly used device in the killing: forests were torched and the birds slaughtered as they emerged.

The notion that so-called aboriginal man lived in harmony with the environment is false, on two grounds. First, humanity is *not* aboriginal ("from the origin") on most continents: we have colonized the world and all humans are alien to Australia and the Americas. Even in Africa, the only truly aboriginal people, still living where they evolved, are probably the Khoisan peoples (Bushmen and Hottentots). Secondly, the case that humanity eradicated the majority of large animals in the Americas, Australasia, and northern Eurasia, while circumstantial, is extremely strong. As our ancestors spread across the planet, wherever we went, we were so powerfully advantaged by evolution that we took what we could and then, when that was gone, settled down to survive with what was left.

The change and disruption of "nature" intensified after the agricultural and industrial revolutions in England, 200 years ago. New varieties of domestic animals and new cropping practices fed a much greater human population; forests were cut down for fuel and to build ships. The prevailing ground cover of Europe changed from a forest to managed grassland and cropland. By 1950 most of the forests of Europe, China, India, Australia, and southern Africa had been attacked, and many had been wholly removed. Over the centuries since civilization began the deep forests of classical Greece have become scrub fit only for goats.

3.6.2 Changes in the Earth's vegetation and animal population

Throughout geological time the plant and animal communities of the Earth have been changing. There is no stability and never has been. Yet for most of the planet's history, ever since the first distinct bacterial groupings evolved from the common ancestor, the different components of the living community have formed an interdependent network. Each species feeds off the bodies or wastes of others, or needs by-products of other organisms to survive, or depends on activities of others. The community of life has always been changing, but each species within the community has adapted to the transient quasi-stability in which it lived.

We human beings do the same in our societies. Those of us who are lucky grow up in so-called stable societies, where there is a set pattern of laws and economic practice. The laws and practice may alter from year to year but the inhabitants of the society shift too, adapting to changes and creating changes. In unstable societies change comes too slowly or too fast. In some, such as Lebanon in the 1980s or Nazi Germany, the fabric of society unravels as interest groups battle or dominate, or the society is destroyed by its interaction with forces that are external to the society.

Humanity's impact on the planet is akin to the situation in such an unstable society. Our short-term demands are overwhelming the long-term dynamic stability of the biosphere. Modern industrial society, in its quest for resources, is behaving like the Norman invaders of England in 1066, exploiting and dominating an established culture. The analogy is carefully chosen for its optimism. The rapacious Norman barons by some miracle became the compilers of the Domesday Book and then somehow over the centuries the amalgam of Norman and Saxon culture attained a longevity unrivaled except by Byzantium. The Western system that emerged from England became the chief cause of our present environmental changes and now dominates the Earth. But there is no planetary Domesday Book yet, no understanding of our treasures.

The cultural expansion of Western technology has been partnered by the biological expansion of the Middle East which has imposed wheat and cereal grains first on northern Europe, and then on North America, Argentina, parts of Australia, and parts of South Africa. The granaries of Rome were in North Africa; they are now desert for the most part. Those granaries of civilization were replaced first by the fields that displaced the forests of northern Europe, then by the grasslands of the Americas, and now by the intensive worldwide monocultures. In the east, in China and Southeast Asia, the water management that was needed to grow wetland rice coevolved with civilization. Here too, monoculture rearranged the environment. Humanity, by cooperating, was able to terrace the land and flood the paddies; cooperative humanity was better fed and more peaceful, and so bred; larger populations needed more rice fields, and so the land was reshaped.

Forest, in particular, has been replaced by grass. Forest grows where it rains, supplies wood, and is vulnerable to fire. As civilization has grown, forest has everywhere retreated. By 1991, the tropical wet forest is substantially demolished in Central America, is restricted in South America and Zaire, is nearly extinct in West Africa, and is under severe attack in the East Indies and parts of Southeast Asia. The integrity of the Madagascan forest is virtually gone, except as relicit islands. On the margins of the rainforest in Africa what survives of the tropical dry forest and parkland has been extensively modified by cutting. In South Africa only a few tiny fragments of true coastal forest remain. In the north, in North America and in the Soviet Union the boreal forest cover is still present, but much of this forest has been cut and replaced by second-growth scrub, very different from the original ecosystem.

Humanity's chief weapons against forest are the chain saw and especially fire. Thirty-eight thousand years ago the first Australians used fire to clear land in the Atherton Tableland, in Queensland. British peasant tribesmen used fire to clear the surface of England. More recently, the Maoris used fire in New Zealand, burning forests and driving the moa birds to extinction. In Amazonia, the same process is occurring during each dry season. In 1987, satellites detected about 6000 fires burning during the peak of the dry season. Dense clouds of smoke covered millions of square kilometers. Air chemistry measurements demonstrated widespread air pollution from the fires. Some of the effects may have had global impact, for instance on the atmospheric content of nitrogen-containing gases.

During the western expansion there has been direct killing of large animals and some game birds and wider but less documented extinction or restriction of many smaller birds, insects, and other life forms. By 1900, many large animal species had already gone or retreated, like the wolves and bears of Europe. The great auk, the North Atlantic's counterpart of the penguins of the Southern Ocean, was amongst the first to go. The passenger pigeon once darkened the skies of America with its flocks of millions; it is gone. The bison and beaver barely survived, a few hundred in Europe and few more in America, though populations there are now expanding again. In Africa, the bluebuck and the quagga were killed, and the only remaining population of black rhino that is large enough to be genetically secure is now gravely threatened in Zimbabwe. The East African race of black rhinos is gone except for a handful of guarded animals and a group introduced by fortunate error some years ago into Addo park in South Africa. There are perhaps 50 northern white rhinos left, many of them in a single camp. The slaughter continues. Plants have been cut, sprayed, or ravaged by introduced diseases, such as those that removed the chestnut from North America's landscape and the elm from much of northern Europe and the United States. What is rarely recorded, though, is the quiet extinction of countless insects and plants that are less lovely or dramatic to humans, and almost wholly unrecorded is the unbalancing of natural communities. Perhaps there now is no living community on earth that is unaffected by humanity. Many marine ecosystems, especially coral reefs, are in trouble. Even in the deep sea the rain of debris from industry and the beheading of the food chain by the hunting of the whales have changed the balance of community.

Life has been replaced by no life. Deserts are common on Earth and have always been here, but never since the Carboniferous period have biologically sterile areas appeared in some of Earth's most fertile spots. This sterility began in the green fields of Middlesex, once among the finest in all England, but now occupied by London. With the exception of a few parks and the odd cricket ground they are gone. The United States has covered a significant proportion of the best land with asphalt. Cities grow without regard to the quality of the land. The modern city building is sealed tightly from nature. Within are climate control systems to remove entirely the inhabitants' link with the commons outside. Outside are parking and asphalt. Urbanization and road building have removed land from the biosphere, not only in Europe and North America, but also more recently in Latin America and East Asia. The story is mixed, life is robust. Butterflies and birds survive on freeway verges, though the neighboring farmland is silent and without color; some cities, like Harare, are forests of tall trees filled with birds. In South Africa the most beautiful of lilies survive beside railway tracks, protected from cows and human pickers.

3.7 A summary of vegetation change

Each region of the Earth is different in its history of vegetation change. Much has been written about the matter. Rather than recount an almost endless litany, it is better briefly to summarize briefly the changes and to look at distinct examples. The causes of the change are outlined briefly here, but in more detail in Chapter 7, which examines the social aspects of human interaction with vegetation.

3.7.1 Amazonia

Amazonia, the vast basin of the Amazon River occupying most of northern Brazil and parts of adjacent countries, is being rapidly deforested, but the causes of

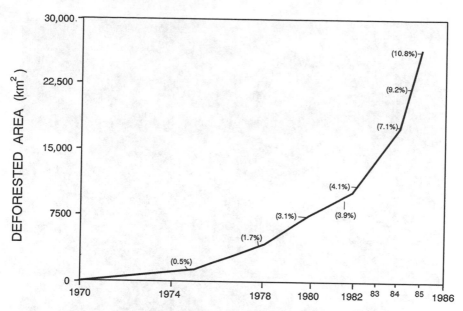

Figure 3.22. Deforestation of Rondônia from 1970 to 1985, showing rapid increase in rate of clearing since 1980. (From Malingreau and Tucker 1988.)

deforestation vary. Much of Amazonia has relatively poor soil. In contrast, the Brazilian region of Rondônia contains better land, some of the richest in Amazonia. Before 1960, the region was home to about 10,000 native Indians and some rubber tappers and prospectors. Around 1974 the suitability of Rondônia's land for agriculture became clear and there was an extraordinary land rush, accompanied by speculation and wild west scenes – shootouts and all. Today the population is probably over a million. In contrast to the failure of other Amazonian colonization projects, the deforestation of Rondônia has had some agricultural success, especially in the cultivation of woody crops.

However, the success is at high cost. Rondônia, with its richer soils, is one of the most important centers of species diversity in Amazonia. It may also have been one of the regions that remained covered by forest during the climate changes that accompanied the Pleistocene ice ages. If so, as a refuge, it may have allowed the survival of many species of forest trees, plants, and birds (including those migrating to North America).

It is difficult to measure the rate of deforestation. Ground-based surveys are of uncertain value, because skill is needed to identify the state of a forest and because of local political problems. Studies using weather satellite data, supplemented by Landsat images, have been reasonably successful, and show an increasing rate of deforestation. In the early 1970s virtually none of Rondônia had been cleared. By 1985 over 20,000 sq. km had been deforested, and the rate was

Figure 3.23. Land clearance in Rondônia, showing the baseroad and field grid. (From Wells 1989. With thanks.)

accelerating. With a million people now in the area, there seems little chance of slowing the destruction unless there are very strict controls on new road building.

The case of Rondônia illustrates the complex problem of deforestation. Some land has been set aside as forest reserve, but effective political action to stop deforestation is difficult because the local population has a strong interest in removing the forest. In Brazil, the incentive to conserve is weak against the local need to exist. Forest is only valuable as it is cut. To halt deforestation there must be more profit from not cutting. Such a financial imperative can only come from external sources – central government and outside aid – linked with the development of nondestructive uses of the forest.

In the less fertile areas of Amazonia the deforestation has been more a matter of tax incentives and corporate activity than of small-scale local settlement. Vast areas of forest have been felled, burned, or treated with herbicide and replaced for a few seasons with cattle pastures that soon lose fertility and pass to poor grassland. Concessions have been granted over large areas. An example of this, quoted by Ghillean T. Prance, is a concession of 346,000 acres (1400 square kilometers) that was under the control of a large West German auto company. This area is comparable to half the size of Rhode Island.

Other areas of Amazonia are being deforested as a result of mineral exploration. Gold and tin production, in particular, has been increasing rapidly, both from small-scale operations by poorly equipped prospectors and from the activities of larger companies including major British and Canadian corporations. Around each gold claim there is an area of deforestation or environmental damage from mines or river dredges. Iron-ore production has also had major impact, directly in the disturbance from mining and from cutting to supply wood-fueled furnaces and indirectly in allowing population spread along the roads built for the iron mines. In conversation with me, a local geologist, carrying out exploration and mapping in the area, described the changes as unimaginable: localities mapped in one season are wholly unrecognizable in the next season. Oil exploration too has had major consequences as settlement advances along lines cut by seismic crews. A substantial portion of the large-scale oil and mineral exploration that has facilitated these changes has been carried out by British, American, and Canadian companies, working in collaboration or in competition with Brazilian corporations. There has been little attempt to develop seismological techniques that do not necessitate line cutting or killing fish in explosions in rivers, although it should be possible to use explosive sound sources and radio-linked geophone arrays in such a way that forests and rivers are not damaged. To service all this activity, roads and dams have been built, frequently with the encouragement and assistance of Western or international aid agencies.

In 1990, satellite surveys showed that 13,818 square kilometers of forest were cleared in Amazonia. This brought total deforestation in the past four decades to 410,421 square kilometers. The area cleared in 1990 was somewhat less than in previous years; this decrease may reflect the unusually heavy rains in the period, but it is hoped that recent changes in government policy, especially the cancellation of tax incentives for cattle ranches, are also reducing clearance.

3.7.2 Central America

In 1950, dense forests still covered a substantial part of Central America, especially in Honduras and Nicaragua. By 1970, this forest cover had been much reduced, but was still mostly continuous. By 1985, the remnants of forests were fragmented, islands of species richness in a landscape dominated by humanity. Central American deforestation is illustrated in Figure 3.24. The causes, as in Rondônia, were varied, but mostly the actions of individual poor farmers and local landowners were responsible. Larger-scale forces also played a role, especially the spread of cattle ranches to provide beef for North American consumers: the hamburger factor.

Recently, Costa Rica has vigorously protected the remnants of its forest. This protection has had unexpected fiscal benefits. Costa Rica has large external debts, some of which have been exchanged for guarantees of forest preservation. Money

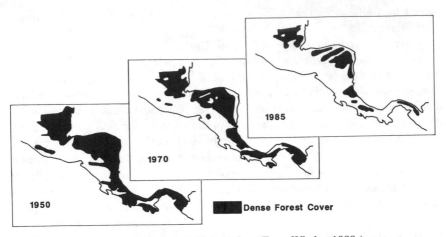

Figure 3.24. Retreat of forest in Central America. (From Whelan 1988.)

has also been raised to protect forest by "selling" acres of forest to the public of North America. Simultaneously, Costa Rica has developed a large tourist industry, based mainly on its good recent record in conservation.

One of the effects of Central American deforestation is beginning to be obvious in the gardens of the United States and Canada. This is the decline in numbers of a variety of migrating birds that used to spend winter in the forest. As the trees are cut, the birds disappear, displaced by cattle: North Americans are eating their songbirds.

3.7.3 The Southeast Asian forest

Rapid deforestation is also taking place in Southeast Asia, especially in the rainforest of Thailand and Borneo. This is a chief source of hardwood for Japan, which takes three-quarters of the exports of timber from Southeast Asia to make items such as disposable chopsticks, food boxes, and scaffolding, as well as houses. As in Amazonia, major fires can occur. In 1982–3, 3.5 million hectares of Borneo were damaged by fire, including 800,000 hectares of lowland rainforest. The Borneo forest is less well studied for its climatological effects than Amazonia, but Borneo is one of the most important fans in the atmosphere, and there are dark risks. Just possibly, cutting Borneo's forest may change India's monsoon or China's rainfall, or change El Niño events – any one of these changes could affect billions and would be irreversible. During the 1980s, according to the World Resources Institute and the United Nations Environment Programme, Malaysia has cut, on average, 255,000 hectares or about 1.2% of its forest area per year. In Thailand, Sri Lanka, and Nepal, the rate has been even higher. China is rapidly stripping the Himalayas. Some reforestation is being attempted, but not restoration of the original ecosystem.

3.7.4 Madagascar

Madagascar has some of the most unusual animal life in the world. Like New Zealand and Australia, it is an ark of life, separated from the main continental masses by continental drift. On this ark a distinct community of plants and animals evolved, notably the lemurs and giant flightless birds. People from Africa and Indonesia colonized the island in the past 1500 years. The subsequent deforestation has been so great that astronauts have nicknamed Madagascar "the bleeding island" because of the intense soil erosion and red sediment in river mouths. Over the past decades, Madagascar has lost four-fifths of its forest, cleared for timber, fuel, cattle, and crops.

3.7.5 Canadian and Soviet forest

The Canadian forest has long been thought to be inexhaustible. Large-scale exploitation began after the American Revolution, when the United Kingdom began to use Canadian timber for its ships, just as Spain transferred much of its naval construction to Havana as the European forests became depleted. During the early part of the nineteenth century eastern Canadian forests were "high-graded": only the best trees were taken and the rest was left as waste. Later in that century and in this one, much more extensive cutting began, for lumber, plywood, chipboard, and pulp. Large areas were clear-cut and then left to regenerate naturally, with little or no attempt at replanting. Today, the fees charged for cutting trees remain low, and felling is extensive and wasteful. "Replanting" is widespread but inadequate in numbers, and seedlings are rapidly installed in the ground, not properly planted. Forest policy varies from province to province, but in general the forest is treated as a resource, to be mined out. This policy contrasts sharply with the Swiss and German policy of managing trees individually in a sustainable regime.

Northern Ontario provides an example of the effects of present forestry policy on the Canadian forest. Huge areas have been clear-cut, with only partial attempts at replanting. One clear-cut covers 2693 sq. km, an area greater than that of some small nation-states. Annually in Ontario, about 1900 sq. km of forest are cut, and about 700 sq. km are planted, with varying degrees of success. A further 3000 sq. km are damaged by fires or insects; some of the fires are natural, but many are deliberate. Overall, it is estimated that about 3100 sq. km of forest regenerates naturally each year, leaving an annual reduction in forest of about 1000 sq. km, and more widespread degradation, out of a total productive forest area of 213,000 sq. km.

Second- and third-growth vegetation now occupies substantial parts of the more productive boreal forest regions, especially in eastern Canada. This regrowth is different in its species balance from the original vegetation. Within the forest, animal populations are probably very different from their precolonial state. Muskrat and beaver numbers in particular have fluctuated wildly, first, declining during periods of extensive trapping and tularemia epidemics, then increasing as natural predators such as lynx were removed. The species balance of the trees in the regrown forest may also have influenced beaver populations. Unfortunately, very little comparative information is available.

Fire patterns have also had a marked impact on the Canadian forest. In some areas, over-mature forests have been protected despite the fact that periodic fires are essential to the forest ecology. Tree species compete in their capacity to burn and reseed; they have evolved by adapting to fires sparked by lightning, which tend to occur during thunderstorms that make the ground suitably wet for reseeding. But arson attacks and human-lit fires, which have been frequent of late, especially in northern Ontario and Manitoba, may be less effective agents of forest renewal. The deliberately or accidentally lit fire tends to occur in hot dry weather and is not necessarily followed by a nurturing rain. The supplanting of natural fire by artificial fire may have resulted in a shift in the species balance of the forest.

The global effects of these changes have not been adequately studied. They are probably large. One of the major sources of methane is the northern forest. The indiscriminate cutting of forest, allowing waste branches to rot, especially in swamps, influences CH_4 production. Beaver activities, encouraged by abundant aspen and poplar in second growth, have a strong effect on CH_4. Significant areas of the northern forest have been deliberately inundated. Large-scale hydroelectric power schemes, such as the James Bay project in northern Quebec, are often carried out without clearing the trees from the land to be flooded. The effect is that of a giant beaver pond. There is little information available on the total output of methane from the James Bay project (the problem does not appear to have been studied adequately), but it must be substantial. In its immediate greenhouse impact it may rival the effect of generating an equivalent amount of power from fossil sources. Canadian deforestation must also have affected global CO_2 and albedo, although these changes are probably less important than the effect on atmospheric CH_4.

Soviet forests are in a similar state. Large-scale utilization of forests there began with the introduction of central planning in 1928, although prior to this a major forest industry existed in the Western European USSR, from the Leningrad oblast to the northwestern Ukraine. After 1928, exploitation shifted to the east, and cutting increased rapidly. More recently, in the past few decades, there has been a decline in timber harvesting in Europe and the Urals, offset by increased cutting in Siberia and the Far East.

Between 1888 and 1908 over 3 million hectares of forested land were cleared in the Russian Empire, and the general forest area had been reduced by 9 million hectares. The total clearance, over historic time until 1914, was about 70 million hectares in European Russia. It is difficult to assess postrevolutionary deforestation, but the available data suggest that the annual allowable cut exceeded annual growth by a factor of 2.5 to 3. Very large areas of forest must have been converted to a poor regrowth of stands of poplar, birch, and alder interspersed with sparse or immature conifers. During the Second World War alone, approximately 20 million hectares were felled or destroyed in areas occupied by the Nazis or subject to military activities.

Timber management in the Soviet Union is very poor. Cutting practice is often extremely wasteful. Replanting is only partial. In 1983 a backlog of 138 million hectares awaited regeneration, an area about equal to the total area of the United Kingdom, France, United Germany, and Italy combined. Annual reforestation is inadequate by 1–1.5 million hectares, and as in Canada the quality of reforestation is often very poor. The Canadian forests have a similarly large reforestation backlog of 22 million hectares. The total combined backlog – cut but inadequately restored forest – in the Soviet Union and Canada, 160 million hectares or more, bears comparison with the total extent of closed forest in Brazil, 357 million hectares. The Soviet-Canadian backlog alone is equal to roughly half the area of uncut Amazonia. It is greater than the combined closed forest area of Indonesia and Malaysia.

In the Soviet Union and in Canada, forest policy has been set by government. About 90% of the 340 million hectares of forest land in Canada is publicly owned. To quote Baskerville (1988), "Like a slum landlord who says he does not paint the walls because the tenants are irresponsible, Canadian society has refused to maintain its forest property, and blames the forest condition on the industrial tenants. In this manner, public ownership of Canadian forests has been a major causative factor in their exploitive degradation. . . ." In general, the management of Canadian and Soviet forests is comparable to the policies in Amazonia, but the degradation of the forest has proceeded further in the northern countries. U.S. forest policy has not been discussed, but is comparable to Canadian and Soviet practice, though in places much more concerned about replanting.

3.7.6 Humanity and vegetation in Zimbabwe: an African example

Zimbabwe has a fascinating pattern of land use that illustrates some of the vegetation problems of Africa. Much of the countryside is broken up into blocks about 100 km square, with blocks of heavily treed woodland juxtaposed against treeless densely populated communal land. The blocks without trees are dry, hot, and bright; the blocks with trees still retain some of their original animal life, and when seen from aircraft or in satellite photographs often seem to have small white clouds over them. The difference in cloud cover has never been properly studied, and may simply be a chance impression gained from one or two photos, but it is interesting.

In south-central Zimbabwe this land-use pattern can be investigated in a region that was once all dry tropical woodland, lying between the small towns of Mberengwa and Zvishavane. The area has relatively low rainfall, around 500 mm (20 inches) per year, mostly concentrated in the summer months but with occasional patches of winter drizzle. Summers are hot, especially in early summer (October to November), when temperatures above 35°C are frequent. Winters are cool, but not cold: 10°–20°C is typical in daytime, with occasional frost.

In 1897 the first geological map of the area was published. This map comments on the state of agriculture. Virtually all of the region was uncultivated and unpopulated. It is probable that some centuries earlier there had been a period of relatively dense population, and archaeological remains go back millennia, but in 1897 the land was for the most part in a nearly natural state. Local oral tradition tells of a time in the last century when there were few people, much warfare, and brave hunters who killed leopards with spears. The vegetation was mainly open woodland.

The impact of Western technology began in the second half of the last century. The population of large animals was first to be changed, as elephant and rhino were hunted. Rinderpest, a disease of cattle introduced from Eurasia, possibly by Italian troops in northeastern Africa, devastated the remaining game population of the area. The hunters were followed by gold miners, who reopened ancient mines and found new ones. With them came wood cutting to supply a railway, the mines, and a wood-fired power station. The geological structure of the area dictated its human occupation. The region consists of a so-called greenstone belt, old igneous and sedimentary rocks, surrounded by old gneissic rocks and large younger granitic intrusions. By 1900 a marked division of land had begun, with gold mining and ranching on the greenstone belt and tribal subsistence farming on the granitoid region. This division was formalized and imposed by laws restricting ownership of land in "European" and "tribal" areas.

By 1929 there had been a dramatic change in vegetation. In the European area there had been massive deforestation to provide timber for the mines, fuel for the power station, and construction material. Certain species of local trees, such as the knobby thorn and the mopane tree, were already scarce. Some farming was being attempted and ranching had begun. A later report, in 1955, repeats the tale. Most of the European farms were by then uninhabited, and apart from cattle ranching the main occupation was timber cutting for the gold mines.

Around this time, a major change took place. With national prosperity came capital investment and rural electrification, but only in the European area. Wood burning declined and there was a reduction in tree cutting. Simultaneously, more sophisticated management of ranch land began. In consequence, by the mid-1970s the vegetation had recovered substantially. The recovery of the wild animals was aided when local farmers set aside much of the territory as a private game reserve. The Zimbabwe civil war of 1975–80 and major drought and climate extremes during much of the period from 1982 to 1987 seriously damaged the flora and fauna (most hippos in the area died in this period), but nevertheless, the ranch land was still in remarkably good condition in 1987. The effects of earlier deforestation still remained, but there had been substantial recovery. In some areas, especially on Belingwe Peak in the south of the greenstone belt, superb virgin stands of forest – amongst the finest in Zimbabwe, with a remnant population of sable antelope – remained, although here cutting has resumed and the forest and antelope are now threatened. In other places, vigorous regrowth had taken place. The larger animals – elephant, rhino, buffalo – remained extinct in the region but smaller game were abundant.

Figure 3.25. Landsat photograph of Mberengwa area, Zimbabwe, in the 1970s. Dark areas are ranch land, light areas are treeless communal land. The Belingwe (Mberengwa) greenstone belt is the sinuous dark region in the lower center. The rectangular dark extension into the light-colored communal land is part of the Union Carbide-owned ranch. Distance across photo is roughly 100 miles (160 km).

The general recovery of the vegetation and wild animals has been greatly to the benefit of cattle-ranching operations. Good management on the largest ranch in the north of the greenstone belt, owned by Union Carbide Corporation, had brought cattle, game, and vegetation through the 1982–7 period with virtually no loss. There is a remarkable contrast between this success during 1982–7 in Zimbabwe and the simultaneous catastrophes in other parts of Africa such as Ethiopia, some of which were less severely affected by drought than southern Zimbabwe. In response to good management, the region has attained a stable and efficient land-use pattern and exports food. Woodland is recovering, game is common. Cattle have replaced the native buffalo, but have not catastrophically unbalanced the ecology of the area. Nevertheless, this is now an anthropogenic ecosystem; it is managed and biased toward the needs and sensibilities of humanity. This is not a natural or primitive ecology. Instead, it is the complex and still evolving product of a mix of climatic, economic, legal, and historical accidents and designs.

Adjoining this region, a few kilometers to the east on the granitoid rocks, a very different story has unfolded. In 1900 this area was lightly populated, with a cover of relatively nutritious bush and savanna grassland. The land became a communal area. Population increased dramatically after 1900, when intertribal warfare ceased, rinderpest faded, food supplies were adequate, and health care improved. Until the 1950s the region served as a refuge for large animal species, hunted to extinction in the neighboring ranch land. By the late 1950s the human population had spread uniformly across the communal land, aided by water boreholes and a rudimentary road system. As the population increased, its impact spread. The animals disappeared.

In 1987 most of the communal land was nearly treeless, a semidesert of scrub and grass. Topsoil was removed by erosion. Amongst the grasses the more nutritious species were disappearing. Poor management of domestic animals was destroying what remained of the resources. Game was virtually absent except for rodents and the smallest antelopes. The best forest in the region, at Mt. Buchwa, was being turned into cropland where slopes permitted. The pressure for wood was so severe that raiding of the heavily wooded ranch land in the adjoining area had begun. Game was extensively poached from the ranch land, where by now it was more abundant. Electricity had not been brought to the area. This meant that all cooking had still to be done on wood fires. Furthermore, people could not read at night or watch TV – mass education was difficult. Roads were very poor and it was difficult to bring in building materials as substitutes for local wood. The only respite from population growth was migration to cities. Not surprisingly, there was also strong feeling that the adjoining well-treed ranch land should be handed over for occupation by people struggling to survive in the communal area. Until 1990, the Zimbabwe government, aware of the dangers and citing constitutional impediments, largely resisted demands for land reform that would rapidly lead to the ecological and economic collapse of the ranch land and a national food shortage. However, the pressure is great, both because of the obvious disparity in population density and because of the racial basis of the historical division of land.

In late 1990, a century after the initial conquest that had divided the land to the advantage of the small group of white settlers, the Zimbabwe government introduced constitutional amendments to the Zimbabwe Declaration of Rights. This legislation empowered the government to punish juvenile offenders by whipping, to carry out capital punishment by hanging, and to buy land compulsorily at prices set without right of appeal to the courts. According to newspaper reports, the declared intention of the government was to buy half the land in the country's commercial farming sector and to settle 5 million peasants on it. The legislation was condemned by Africa's most respected judge, Enoch Dumbutshena, former Chief Justice of Zimbabwe and a man of enormous integrity, who was quoted as commenting that the legislation was the most regressive in the country's history, and that it would have a markedly negative effect on Zimbabwe's economic future. The passage of the legislation was extremely popular and was accompanied by singing and dancing in the aisles of the Zimbabwe parliament.

At about the same time, in Alberta, Canada, the provincial government decided to proceed with the construction of a massive and heavily subsidized tree-processing plant in northern Alberta, overriding the objections of a group of scientists including Prof. David Schindler, one of North America's most respected lake biologists, who had drawn attention to the enormous damage the project would do to vast tracts of forest and to the Athabasca river system. The government was supported in its decision by a variety of local and Japanese interests and by popular enthusiasm among local people such as real estate agents.

From Alberta to Zimbabwe the stresses on the resources of the land are the same. Both decisions were undeniably democratic. In each case a government, perhaps failing at the polls or wishing to please its core electorate, decided to sacrifice good land management in favor of providing land or jobs. In so doing, the governments also sacrifice the interests of the next generation in favor of the demands of this generation. Today's children will grow up to inherit a Zimbabwe where land is poorer and rainfall may be more scarce; young Albertans will lose an important part of the province's forests and will inherit polluted rivers.

In both cases the voices speaking for the natural environment and good land management have been overruled by populist demand. Today's generation places its own interest over the need to safeguard the next, and the trees are cut, the wildlife dispossessed. During the debate in Zimbabwe a leader of the peasant farmers was quoted as saying the wildlife could be moved to poorer land. This is an understandable opinion, and it is obviously much easier and less expensive for a government to expropriate well-managed land and then to mine it to extinction than to revitalize degraded, poorly managed land. Trees, wildlife, and good soil have no votes and no rights.

There is, incidentally, one jurisdiction where trees do have quasi-human rights. This is the Duchy of Normandy, an anachronistic medieval relict belonging to Queen Elizabeth but independent of London, which includes as a bailiwick the island of Jersey. Here, if a tree is to be cut or damaged (for instance in a road scheme), it is entitled to a defense and a court must convene *in the presence of the tree*. By all accounts, these are jolly occasions, but the point is vital: someone must speak on behalf of the tree, and the natural environment does have rights that can override the demands of the present human custodians.

What will the future bring to the Mberengwa area of Zimbabwe? The more likely prospect is for the communal region to become a persistent net importer of food, for gradual encroachment of population into the ranching area, first by poaching and illegal wood cutting, then eventually with some sort of resettlement scheme, followed by a rapid decline of the whole region to semidesert. The less likely alternative is that the communal area will be restored by extensive rural electrification, road building, education, and sophisticated agricultural management. The population in the communal area is well aware of the need for change. In principle, the land can support the population. Poverty constitutes the obstacle. Only a remarkably lucky combination of external factors such as high beef and cereal prices and sound national policy on electrification, education, and road building, together with good local rainfall and intelligent community leadership, can save the region. The problem of communal land in Africa is explored further in Chapters 7 and 8.

3.8 The scope of deforestation

Tropical forests originally covered about 15–16 million sq. km of the Earth's surface. A recent estimate by the British scientist Norman Myers is that 8.6 million sq. km are left, and that between 76,000 and 100,000 sq. km of these forests are being destroyed each year, together with a further 100,000 sq. km that are being disrupted in some way. These are estimates based on old data and rates may have accelerated recently. As a rough estimate, though, about 1% of the forest is removed each year, and rather more than another 1% is being damaged. At this rate, by the end of our lifetimes there will be none left.

The benefits of deforestation include a brief increase in land available for farming, though in many deforested areas this land rapidly becomes infertile as soil degenerates. For the richer nations, the benefits are in hardwood. Consumption of hardwood in North America and Europe has increased sharply since the Second World War, and most homes now have some mahogany, in kitchens or furniture. The remains of the rainforest end up as chairs in the shops of Cambridge, Madison, and Bern or as disposable packaging in Tokyo. Japan has become the dominant importer, taking most of the cut timber from Southeast Asia, although Japan's own forests are heavily conserved.

3.9 The extinction of species

Tropical forests contain the majority of the Earth's living species. An oft-told story in science is that of the researcher who aimed insecticide at the canopy of a forest. A rain of insects fell down, and the analysis of the results doubled the estimate of the number of species on the planet. A few hectares of forest in Southeast Asia or Amazonia have more tree species than North America or Europe; in these trees, especially in the canopy of leaves, are enormous numbers of birds, mammals, reptiles, amphibians, and especially insects. In the flooded forests of Amazonia are equally diverse fish populations. All this diversity is very locally distributed, so that the destruction of a small area of forest may make extinct large numbers of species. For instance, in western Ecuador it has been estimated that 50,000 species have gone in the past 25 years. If this local estimate and other estimates like it are correct, by the end of our lifetimes, at present rates of deforestation, the majority of the species on Earth will have become extinct.

3.10 Inferences

Each country is different, and each region within a country is distinct. In many regions the actions of peasant farmers dominate the management of vegetation.

In others, large corporations play an important role in changing vegetation. In Amazonia, corporations, many of them Western multinationals, have acted to remove or damage forest, as have large companies involved in Indonesia and Malaysia; in Zimbabwe the Union Carbide Company is protecting forest. Each company is subject to local tax laws and other financial and legal incentives. Changes in tax laws and policy, and in the market for timber and beef, will force changes in land use. Corporations respond to their financial environment. Taxation and fiscal incentives are very powerful shapers of vegetation cover. Most governments will consider fiscal changes if the reward is sufficient; thus international financial pressure can become a tool to manage vegetation.

For the mass of tropical humanity struggling to survive, the choices are more limited. The poor do what they do because they have to. Any scheme for managing vegetation that seeks to impose a collective good must be compatible with survival. In protecting vegetation, brute force rarely works; neither do large-scale ''development'' projects. Economic forces, in contrast, are very powerful. In those parts of Amazonia where there are no roads, cutting forest is expensive. If the prices of crops from deforested areas are held down, or if transport is costly, then those crops are not planted. Subtle pressures like these are far more effective than brute force. This topic, introduced here only briefly, is explored in more detail in Chapters 6–9.

The last comment in this discussion of the changing biosphere must be about the oceans. Fishing and whaling are, for the most part, large-scale and capital-intensive industries, especially as carried out by Japan, Taiwan, Spain, the fishing nations of Scandinavia, and the Soviet Union. It is difficult, though perhaps just possible, to discuss deforestation objectively. It is not possible here to discuss dispassionately whaling, and especially drift-netting, which is simply the abolition of life.

3.11 The problem of observation: Landsat

The best tool for studying vegetation is satellite imagery. In 1972 the Landsat system was started. Landsat 5 is capable of a ground resolution of about 30 meters (about 100 feet) and covers the Earth completely every 16 days. It is ideal for study of the vegetation of Earth and could be of immense value to third-world nations, as well as to geologists, farmers, and foresters in North America, Europe, the USSR, and Australia. Early Landsat images were made available at nominal cost, $1.00 per 35 mm slide. Unfortunately, Landsat data are now sold at high prices. A single image costs thousands of dollars. The data are available, but very few organizations and research groups are well enough financed to buy the images. Scientific researchers cannot easily obtain the Landsat images now. Most North American forest camps and mines in the deep bush do indeed have access to satellite imagery, but the pictures that are beamed down to the lumberjacks are of higher commercial value: hard-core pornography, not photographs of vegetation.

The decision to charge high fees for Landsat images has caused much direct damage to North American land management and geological exploration. Worse though, no school or community center in the poor nations can buy the pictures. Limiting access to Landsat may have delayed the warning about the rate of worldwide vegetation change by several years. Had Landsat data been freely distributed to scientists in the poor nations, much might have been avoided. The data are available, but they are locked up.

Scientists have turned to data from weather satellites as a substitute. This information is available at much lower cost, but the resolution is no better than 1.1 kilometer, in contrast to about 30 meters of Landsat and similar satellites. For some uses, however, such as the study of the vegetation of large areas, this poor resolution is an advantage. Weather data were not normally stored, so it was difficult to obtain retrospective information to calculate deforestation. Even this source is inaccessible to scientists from many poor nations where it is diffi-cult or impossible to obtain foreign currency to buy the imagery.

Recently, there has been a further denial of access to satellite data. For some years, scientists have used information from global positioning satellites (GPS) in studies of earthquakes and land and sea levels. This information is very pre-cise, and offered the prospect of providing new methods of studying problems as diverse as predicting volcanic eruptions and measuring the probability of flooding. It is possible that GPS methods could be used to carry out seismic surveys for oil and gas in forested areas without cutting access lines. Unfortu-nately, it has now been decided by the United States that data will be deliberately degraded, to impede the scientific use of GPS data. These measures, taken at some cost, mean that a large variety of environmental applications of GPS data have now become less valuable or have been made impossible. Both Landsat and GPS data continue to be collected, but are only available to those who can pay (in the case of Landsat) or who have the correct security classification (in the case of GPS). We have eyes, but we cannot see; we have ears but we cannot hear. This is Greek tragedy: our society has blinded its own eyes.

Reading list

Barr, B. M., and K. E. Braden (1988). *The disappearing Russian forest*. London: Hutch-inson. Totowa, NJ: Rowman and Littlefield.

Botkin, D. B., J. E. Estes, M. F. Caswell, and A. A. Orio (1989). *Changing the global environment: perspectives on human involvement*. San Diego: Academic Press.

Brimblecombe, P. (1986). *Air: composition and chemistry*. Cambridge: Cambridge Uni-versity Press.

Calder, N. (1991). *Spaceship Earth*. London: Viking Penguin.

Carbon Dioxide Information Analysis Center (1990). *Trends '90: a compendium of data on global change*. Oak Ridge, TN: Carbon Dioxide Information Analysis Center, MS-6335, Oak Ridge National Laboratory, TN 37831-6335.

Gribbin, J. (1988). *The hole in the sky: man's threat to the ozone layer*. New York: Bantam Books.

Prance, G. T., and T. E. Lovejoy (1985). *Amazonia*. Oxford: Pergamon.

World Resources Institute (1989). *World resources 1988–89*. New York: Basic Books.

4
The consequences of change

Our global climate system may be on the verge of crossing a threshold. Worse, it is possible that we have already crossed over and that the change to a new regime has already begun; perhaps some catastrophe, which we have yet to recognize, is now inevitable. The fire may be alight. It may be too late to put it out. We simply do not know. We can only hope that we can predict change. The example of the unpredicted ozone hole is not reassuring. We can expect surprises.

Models of the future climate of the planet are based on our ability to understand the present climate. First, it is necessary to determine the factors that dominate the working of the modern atmosphere, controlling the broad global patterns of distribution of rainfall and temperature. Computer models can be constructed to simulate the broad patterns of atmospheric circulation. Once the modern climate is understood, those computer models that can successfully simulate today's conditions and the past record can be perturbed – for example, by increasing the greenhouse warming effect of the atmosphere – in an attempt to predict the future.

However, climate is not a linear system. It is immensely complex, and it is quite likely that the models may be subtly incorrect or fail to predict a major

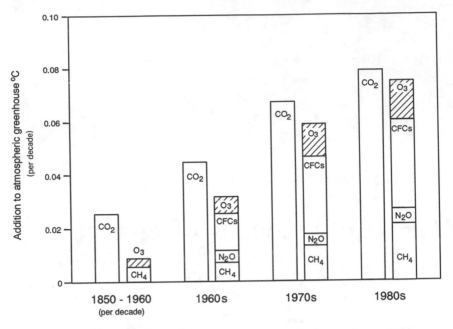

Figure 4.1. Relative importance of additions to the atmospheric greenhouse, 1850–present. Each pair of columns shows the estimated addition to the greenhouse effect in each decade from CO_2 (left column) and the trace gases (right column). Note that the trace gases have increased in importance recently. Increments marked with diagonal lines are very speculative. (From Ramanathan et al. 1987.)

change. Imagine, again, the small child pushing a glass across a table. Each time the child pushes, the glass moves a little. This we can model, and say that the next perturbation will move the glass another inch. But eventually at the edge of the table the response is nonlinear: something different occurs. The atmosphere too has edges, like the change at the end of the last ice age, when major events take place. Unfortunately, we do not know where these edges are.

   In searching for edges, it is possible to use the geological record of past climates as analogue systems. For instance, there was a period several thousand years ago, known as the Altithermal or Hypsithermal. This period may have been warmer than today, and study of the weather during this time may provide some clue to the future, giving us an idea of whether an edge is awaiting us. Nevertheless, prediction is uncertain: models are simply well-educated guesses.

4.1 Forcing the climate: feedbacks from atmospheric change

The various recent changes in atmospheric composition have, in effect, placed an extra blanket around the biosphere. Geophysicists call this a forcing effect. We cannot yet state accurately what the result will be, but it is a sound general principle that if at night one throws an extra blanket on a sleeping child, the child will become warmer, all other things being equal. Unfortunately, we do not know if all other things are equal.

Figure 4.1 illustrates the extra blanket. Humanity has added warming components to the atmosphere, including carbon dioxide (CO_2), methane (CH_4), and the chlorofluorocarbons (CFCs). The preindustrial atmosphere had about 280 parts per million by volume (ppmv) of CO_2. A sample of air collected by AES Canada for me in May 1989, at Alert in northern Canada, had 359 ppmv of CO_2. If the CO_2 content of the air is doubled over preindustrial levels, the direct warming effect can be calculated to raise global temperature by about 1.2°C. At current rates the doubling of CO_2 may take place by the year 2100. The prehuman natural abundance of CH_4 was about 650 parts per billion by volume (ppbv). The same sample of air collected at Alert had about 1760 ppbv of CH_4. Other samples collected in autumn 1989 in the Canadian forest had 1800 ppbv of CH_4. When the effects of all the minor trace gases are considered, it is clear that a blanketing effect or greenhouse increment that is equivalent to a doubling of CO_2 will be reached more quickly – in the first half of the next century, perhaps by the year 2030 or even earlier.

Until about 1960 the main addition to the atmosphere was about 4.2 ppmv of CO_2 per decade, producing a direct warming of about 0.025°C per decade. There was also a lesser warming from added CH_4. This warming rate of 0.025°C per decade calculated prior to 1960 increased strikingly in more recent years. In the 1960s the calculated rate of increase of direct warming doubled, and the influence of trace gases such as CH_4 became more marked. By the 1980s the rate of increase in the warming from CO_2 alone was three or four times that of the century prior to 1960. The minor trace gases had increased even more, to the extent that the annual increment added to the greenhouse by trace gases now equaled or exceeded that from CO_2. To calculate the warming effect of the addition of trace gases is difficult because of the chemical complexity of the atmosphere. For instance, adding CH_4 to the air can either increase or decrease ozone (O_3), depending on the concentration of other gases such as nitric oxide and nitrogen dioxide (together called NO_x). In places where NO_x is abundant, such as cities filled with automobiles and industry, low-level ozone may increase as CH_4 is oxidized to CO_2; in other regions where the air is poor in NO_x, the oxidation of CH_4 to CO_2 may actually remove ozone.

This warming may seem trivial, but there is substantial evidence that the direct warming effect is multiplied by various feedback effects, so that the actual warming that takes place much exceeds that calculated directly from the greenhouse effect. The feedbacks are complex and extremely difficult to estimate. They also interact with each other. For instance, snow and ice may become less common, so the planet as a whole will become darker – a positive feedback, making the world warmer. However, if the world is warmer, there will be more moisture in the air so it may snow more, making the Earth whiter – a negative feedback.

There is little agreement among climatologists on the total feedback effect. A commonly cited estimate is that the feedback factor is positive and somewhere between 1.2 times and 3.6 times the direct effect of adding CO_2, CH_4, CFCs, and so on, to the air. This estimate is derived backward from climate models that suggest a global warming of $1.5° - 5°C$ if CO_2 doubles. However, the feedbacks are very poorly known and urgently need further research.

To conclude, the net effect of adding chemicals to the air is difficult to predict. Most models suggest a strong warming, but the conclusion is not secure.

4.1.1 Cloud feedback

The effect of cloud cover is very great, but poorly understood. If the Earth warms, more water will evaporate and be held in the atmosphere and cloud cover may increase. Depending on the types of cloud and their properties, the feedback effect could be either positive or negative. Furthermore, the extra water in the air may allow the production of extra hydroxyl (OH), which will have significant chemical effects on the various trace gases in the air, especially in removing methane.

The cloud problem is exceptionally important, because it may hide a potential threshold mechanism that could drive the climate system into a wholly new state. At present, tropical clouds are roughly in radiative balance. They reflect sunlight back out into space, cooling the system, but they work equally as greenhouse components, warming the system. Clouds in the middle to high latitudes of the Northern Hemisphere act differently, and have a major net cooling effect.

Both the cooling and heating effects of clouds are huge, comparable to the effect of a massive, manyfold increase in CO_2. The changes in CO_2 and CH_4 may set off small shifts in cloud patterns that could have massive implications for the climate system. If clouds above the North Atlantic started to behave like tropical clouds, the planet would warm dramatically. Equally, if the tropical clouds were to become more like the northern clouds, the Earth would cool, possibly even to an extent that compensated for the greenhouse effect of added CO_2, CH_4, and CFCs. We cannot yet tell.

Changes in cloud regimes would mean significant modifications to rainfall patterns even if the Earth's temperature as a whole alters little. This is the optimistic outlook. The nightmare is that a major change in cloud patterns could bring the atmosphere across a threshold of organization. It may cross over into a new stable pattern, just as it did at the end of the last ice age. We cannot really guess if this is likely, but the possibility exists. If a threshold is crossed, it is probable that human beings, who have arranged themselves to live under skies where it now rains, may have to rearrange themselves elsewhere.

4.2 Is the Earth warming?

The obvious way to answer this question is to examine the climate record over the past century and ask if there is evidence for planetary warming. We know that atmospheric CO_2 has increased by about one-quarter above prehuman concentration, and that other gases have added to the forcing. Doubling CO_2 is, in principle, roughly equivalent to turning up the heat from the Sun by 2%. The effect of the past increase in CO_2 – much less than doubling, so far – is hidden in natural fluctuations and is as yet difficult to detect. It is hard to be sure if the Earth as a whole really is warming although the evidence is beginning to become stronger. The historical record of temperature comes from a variety of sources. The data come both from the land and from the sea.

Data from the land come from the meticulous observation of temperature by the many weather stations scattered across the planet. During the later period of European colonization there were detailed temperature measurements from most parts of the world, although recently in some places the data sequence has been interrupted by local political events or financial problems. Generally, since about 1861 there is a rich worldwide temperature record. Marine data have also been collected in abundance since then, from the daily records of ships at sea.

The data sets include tens of millions of observations over the past century or more. Unfortunately, there are large problems in interpreting the data. For example, many of the sea temperatures, including all the early data, were measured in an uninsulated canvas bucket. As the bucket was lifted out of the sea, the water evaporated and cooled. Later, measurements on many ships were taken from the water intake to the engines. These temperatures were roughly half a degree centigrade higher than the bucket measurements: the sea did not warm (as the raw data implied), it was simply that the collection method changed. More recently, satellite measurements of sea surface temperatures have been available that are more consistent. On land, similar problems occur. For instance, many early North American weather stations were on the edges of small towns or hamlets. Over the years these hamlets have grown to towns filled with trees, buildings, and people. In winter, especially, these towns act as heat islands and are warmer than the surrounding countryside by several degrees. Given such complexity, the only resort is to estimate corrections to the data.

The general conclusion, after all this processing, is that the world is indeed getting warmer. Figures 4.2 and 4.3 illustrate two compilations of global temperature variations. In general, the results show little warming in the nineteenth century, marked warming in the 1940s, relatively steady conditions from then to the mid 1970s and then a rapid increase with the warmest conditions recorded in the 1980s. It is tempting to conclude that the warm years of the 1980s were indeed the result of the addition to the greenhouse gases in the atmosphere.

This conclusion, however, is not certain. Weather is a fairly chaotic affair, as far as farmers and sailors are concerned, and we expect it to do odd things. A crude way of considering the variation is to guess that weather fluctuates in much the same way as many other natural processes fluctuate, and to ask what is the chance that we are simply sampling a very odd set of years. The answer is that the present unusual weather differs from normal by somewhere between two and three standard deviations; in other words, we would expect a year like 1981 to occur at a rate of between five times a century and once in about four centuries. To get several such years in the 1980s is possible but unusual. These arguments, however, are statistical, and can be challenged on the statistical grounds that we

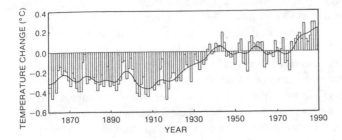

Figure 4.2. Average temperature changes on the Earth, 1861 to 1989. The solid curve shows 10-year smoothed values. The changes are shown relative to the average temperature for 1951–80 as 0.0. Preliminary data for 1990 suggest that it was the warmest year ever, roughly 0.4°C above the 1951–80 average. (From Jones et al. 1986 and Houghton et al., 1990).

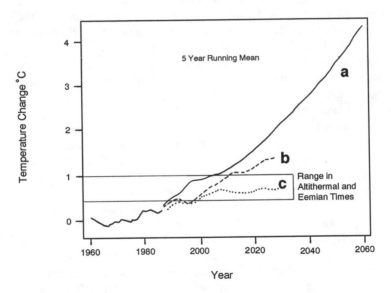

Figure 4.3. An estimate of the temperature history of the Earth, since 1960, with possible future trends **a, b,** and **c.** For more details, see section 4.3 of the text. The Altithermal and Eemian periods are discussed in section 4.6. (From Hansen et al. 1988.)

do not know what normality is, or what deviations from normality to expect. It is a good guess but not certain knowledge that the planet is getting warmer. Complex special events, such as the El Chichón volcanic eruption in southern Mexico, probably played a role in the climatic variations in the 1980s, but there were also special events in the 1920s. We cannot yet be sure, but it *seems* that we have warmed up the Earth. If so, we can predict that matters will now accelerate. Weather is weather: one volcano can erupt and cool the entire plant for a year, and fluctuation is great. Nevertheless, the prediction is that the next century will be substantially hotter than, say, the 1960s.

4.2.1 The Arctic geotherms

The most convincing single line of evidence for warming comes from the rocks and permafrost of the North American and Soviet Arctic. The temperature gradient in the outer layer of the Earth depends on several factors: the ability of rocks to conduct heat, the radioactivity of rocks, the supply of heat from the interior, and the surface temperature. If surface temperature is increased, the signal from that increase moves downward at a rate that can easily be calculated. Imagine pouring hot coffee into an earthenware coffee mug. The temperature increase inside the mug takes a little time to be conducted to the outside, depending on how thick the mug is. Similarly, if the surface of the Earth is heated, the warmth slowly travels down into the ground. The way in which temperature varies with depth into the Earth is known as a geotherm. Each point at the Earth's surface has a geotherm that depends on local surface temperature, the nature of the rocks, and heat flow from depth. If surface temperature changes, so does the geotherm.

Drill holes in cold permafrost land or ice in Alaska, Arctic Canada, the northern Soviet Union, and possibly also Antarctica, clearly show the effects of surface warming in the rocks, as atmospheric heat moves downward into the ground. The amount of warming and the time of warming can be calculated. The whole Arctic seems to have warmed by several degrees centigrade over the last half century. This is the best method for directly measuring global warming. The actual heat is measured, so this is not a proxy method. Meteorological data do not show Arctic warming: nevertheless, the geotherms are convincing evidence that in the Arctic at least, the climate has become warmer.

This Arctic evidence does not, however, mean that the planet as a whole is warmer. It may simply mean that the efficiency of transfer of heat from the equator to the poles has improved. More long-term monitoring is needed. Permafrost is ideal for this, because there is no fluid flow through the ground to disrupt the heating record. A high priority of research into global change should be the establishment of a worldwide chain of stations monitoring permafrost drill holes in the Arctic and Antarctic and also on high mountains in temperate and tropical regions. It is ominous that the only region where there is certain evidence of warming is the Arctic: many climate models predict that Arctic warming will be the first detectable sign of planetary warming.

4.3 Predictions of the future

Figure 4.3 illustrates the implications of several scenarios of change. Scenario **a** was developed by J. Hansen and associates of the Goddard Institute for Space Studies. Perhaps on the high side of ultimate reality but on the low side of current political expectations about the state of the economy, it assumes that the growth rates of the emission of trace gases that were typical of the 1970s and 1980s will

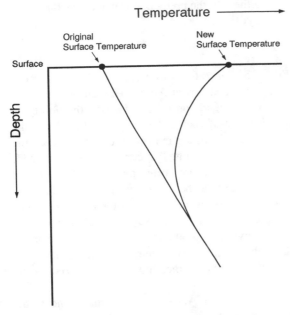

Figure 4.4. Trends in Arctic geotherms. Under steady circumstances, with a stable mean annual surface temperature, the geotherm near the surface (top few km) plots almost as a straight line with a steady increase in temperature with depth. If the surface is warmed, the heat pulse takes some years to travel downward through the ground, and so the geotherm is perturbed and the ground is warmer at the surface than at shallow depth. Most Arctic geotherms, as recorded in drill holes in permafrost, now show this perturbation resulting from surface warming. (Modified from Lachenbruch and Marshall 1986.)

continue indefinitely, or at least in the next few decades. Given the current behavior of the industrial economies, this is, in political terms, a conservative model: the implications of most political hopes and promises are that growth will be faster, more typical of the late 1980s than of the more depressed periods in the past 15 years. Furthermore, industrial growth even at this rate, around 1.5% per year, is much less than in the past century, around 4% per year.

Scenarios **b** and **c** are also shown in Figure 4.3. These models embody optimism: they assume that some degree of control is imposed on emissions (model **b**) or that strict control is imposed (model **c**) and that several planetary cooling events caused by huge volcanic eruptions occur. Scenario **b** is perhaps at the limit of optimism; scenario **c** assumes, hope triumphing over experience, that CFC emissions are eliminated by 2000 and that by the same year the net emission of CO_2 and other trace gases will stop. In this case emissions would just balance sinks: in other words, oil and coal burning would be drastically reduced and forest growth drastically enhanced.

The most likely near-future trend, even with some degree of international control, is **a,** or hotter than **a.** By 2030 to 2050 the atmospheric changes will probably be equivalent to a doubling of CO_2, or 1.25°C in direct temperature increase across the planet. When the feedback effects are considered, this leads to a global warming of somewhere between 1.5° and 5.5°C. Some climatologists say the effect may be as small as the bottom end of this range, others suspect the global warming will be 5°C or more. An evaluation by the Intergovernmental Panel on Climate Change, published in 1990, put the warming as 0.3° to 0.6°C in the century prior to 1990 and predicted an increase above present temperatures of 1° by 2025, or about 0.3°C per decade, with an uncertainty range of 0.2° to 0.5°C. This estimate may be conservative. Most likely, by 2050 warming will be around 2°–4°C. This sounds small, but it is comparable to or greater than the difference between temperatures in the ice ages and today. There is agreement that substantial change is on the way, and this change should become unmistakable sometime after the year 2000. By this date the warming effect will probably be distinct from the annual and decade-by-decade fluctuation of weather. Whether the global warming will be one and a half or five and a half degrees, we do not yet know.

4.4 The problems of prediction

It is hard to predict the change in the behavior of the atmosphere and biosphere as given quantities of CO_2, CH_4, and other gases are added. As discussed above it is also difficult to predict what these quantities will be, how much of these gases will be added in the next few decades. In 1970 many economists attempted to predict the future price of oil, but none successfully prophesied the extraordinary series of discontinuities that was to follow. Similarly, we cannot now predict the political factors that may affect the future consumption of oil and coal; we can only guess. More unsettling, though, is our inability to predict the consequences of that oil burning and coal mining. In the previous section, the best available rough estimate – a warming by the middle of the next century of between 1.5° and 5.5°C, most probably 2°–4°C – was quoted. The warming that actually occurs may be even less than 1.5°C, but there exists the gloomy possibility that it may be far greater than 5.5°C, along with massive changes in weather patterns.

Much of the uncertainty lies in assessing the feedbacks. An example is the effect of warming on polar ice. If the feedback effect is strong, the planet will obviously warm more than if feedback is weak. Probably there will be a substantial feedback, but it is very difficult to estimate, in part because the two poles are so different from each other. An example of a nightmare, mentioned already, is that the methane trapped in Arctic permafrost will begin to escape in large quantities. The probability of this is small, but it is possible and if it took place would strongly enhance the positive feedback.

The most important uncertainty is in the behavior of the clouds, which outweighs all other feedback factors. Any significant change in cloud patterns could rapidly and substantially change the planetary climate. We do not yet understand cloud radiation balances well enough to predict how change will occur, but we can recognize the danger of sudden, massive change.

Many other feedback mechanisms exist, some positive, some negative. Changes induced by humanity may also be classed as feedback. For instance, if Arctic rivers in the Soviet Union and in Canada are diverted southward to supply water for dry continental interiors, there may be an effect on Arctic temperatures and hence on the ice-snow feedback. The list is long; some major feedback effects may not yet be known.

4.5 The physical effects on the oceans

One of the problems of climate modeling is estimating how much heat will pass into the oceans. To date, global climate models have not been very successful in modeling oceans. As warming occurs, first the surface layer of the ocean warms, then the deeper water responds. This very large heat store delays the warming; equally it would retard recovery if the atmosphere were eventually controlled to reduce the concentrations of greenhouse gases. Changes may also occur in the way the oceans transfer heat around the globe. Small variations caused by changes in wind patterns, say, in the Gulf Stream or the North Atlantic deep-water circulation would have powerful effects on European weather. On a slightly longer time scale, small fluctuations in global deep-water movements could radically alter the world climate.

The Scottish islands that are warmed by the oceans so that they support winter golf and palm trees also show the effects of changes in sea level. Most have fine examples of raised beaches from postglacial rebound of the land and from episodes of higher sea level. If the planet warms and the oceans warm, the water will expand thermally and raise sea level. Expansion in the next 50 years will probably add 10 cm to sea level. Secondly, as warming occurs the polar ice may melt. This is not certain to happen: it may be that the net effect will be to thicken the Antarctic ice cap when more snow falls on it from a warmer, moister atmosphere, which would reduce sea level. However, the experience over the past century is that sea level has risen at measuring stations in some areas, such as Western Europe and eastern North America, by about 15 cm; the rise has apparently been faster in the past half century, perhaps over 2 mm per year.

The actual change in sea volume is difficult to assess. Part of the apparent rise in sea level in the southern North Sea may be a result of changes in land height near the recording stations that are consequences of the last ice age. During glaciation, the ice cap depressed Scandinavia and raised the land further south; now the reverse is taking place and the land that once was raised is now sinking, as the Earth responds to the removal of the load. However, there is probably a small true global rise in sea level. It is very difficult to extrapolate the trend to the future, but some estimates suggest that if the apparent rise in sea level continues, with the same rate of geological sinking and the same relationship to global temperature change as seen in the past century, then by 2030 (when a greenhouse effect equal to the doubling of CO_2 concentration may have been reached), perceived sea level should rise by 25–165 cm relative to population centers on the coasts of Europe and eastern North America.

Much depends on effects that are difficult to predict. For instance, if nearly all the world's small glaciers melted (a probability if global temperature rose 6°–7°C), then the sea level would rise about 33 cm. On the other hand, by 2030 there could be so much more snowfall in Antarctica that accumulation over Antarctica would increase by over 10%, dropping sea level by 30 cm. Greenland, in contrast, may lose ice. It is extremely difficult to estimate the net effect. The problem is complicated further by such factors as the dust storms of the northern continental interiors, which have recently been depositing much dust on Arctic ice, darkening it and enhancing melting.

Some evidence comes from the last interglacial period. About 125,000 years ago the sea level in equatorial regions was probably about 7 meters higher than today, and the surface of the Greenland and Antarctic ice caps may have been 2° to 3°C warmer than today, or 200–300 meters lower than today, or some combination of both factors.

There is a lurking danger in the Antarctic, quite different from the Arctic terror that methane will be released in vast quantities from the permafrost. The problem is in West Antarctica, where ocean warming may lead to the collapse of marine ice sheets, as bottom melting removes the restraint on grounded ice. Estimates of the time that the ice sheets would take to collapse range from 100 years to many centuries. It has been suggested that this collapse has already started, but the evidence is not convincing. If the sheet did collapse, sea level might rise by 5 meters. This is thought unlikely.

It is difficult to reach a conclusion about sea level. Perhaps an eventual rise of very roughly a meter is the most likely consequence of the warming from a greenhouse increment equivalent to doubling CO_2. The Intergovernmental Panel on Climate Change, in their 1990 report, came to a slightly lower conclusion: they predicted a rise of about 6 cm per decade, or 18 cm by 2030, or 66 cm by late in the next century. One meter or less seems, at first thought, to be a small amount. But for the inhabitants of coastal cities a meter would be expensive. For instance, the defense of London by the Thames barrier and the Netherlands by the barriers across the Rhine delta have been extremely costly, many billions of dollars. The problem is not in the daily tides but in the increased reach of storm surges and in managing water flows on land below sea level. Moreover, industrial nations can afford the billions of dollars needed to raise dikes or move sewage outfalls. For Bangladesh, where a single storm surge in the Ganges delta has killed hundreds of thousands of people, the consequences of a rise in sea level by a meter may be more severe. For some island nations such a rise would mean extinction.

The circulation of the sea, like the circulation of the air, has the general effect of transferring heat from the equator toward the poles. The circulation is strongly dependent on the density contrast between cold, denser water in the Arctic and off Antarctica, and the warmer, less dense surface water. Ocean circulation patterns may therefore change quickly, reacting to changes in the weather and heating regime above. Global warming would probably intensify the Gulf Stream and increase the oceanic transfer of heat to the Arctic in the short term. The long-term changes are not obviously predictable.

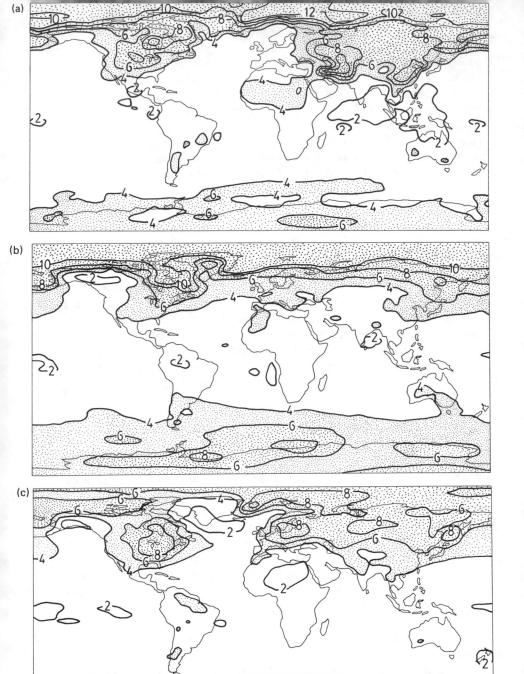

Figure 4.5. The change in the air temperature at the Earth's surface that would result from a change in the greenhouse effect equivalent to the doubling of CO_2, expected by about A.D. 2030. These are equilibrium models: they contrast the state of the Earth under the present atmosphere with its state under a $2 \times CO_2$ atmosphere, but do not consider the effects during the transition. Three high-resolution models are illustrated, first for the Northern Hemisphere winter (a,b,c), then for the Northern Hemisphere summer (d,e,f). All three models give similar estimates of the overall temperature change, of

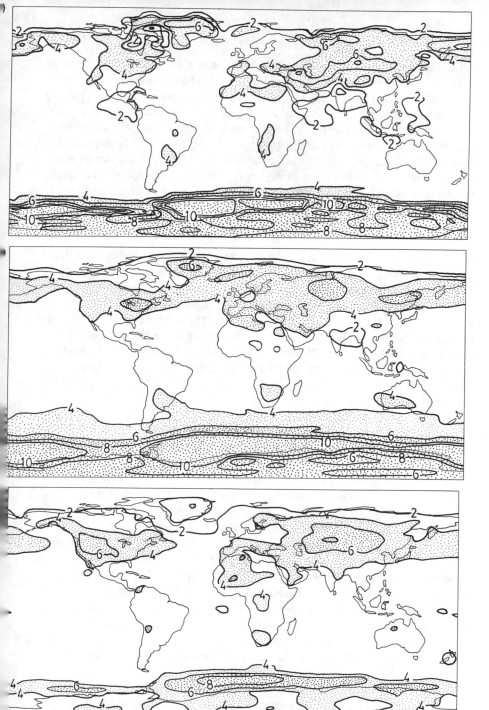

Caption to Figure 4.5 *(cont.)* 3.5°–4°C. In the diagrams, regions where temperature changes by more than 4°C are stippled. **(a)** Canadian Climate model, Dec.–Feb.; **(b)** Geophysical Fluid Dynamics Laboratory model, Dec.–Feb.; **(c)** United Kingdom Meteorological Office model, Dec.–Feb.; **(d)** Canadian Climate model, June–Aug.; **(e)** Geophysical Fluid Dynamics Laboratory model, June–Aug.; **(f)** United Kingdom Meteorological Office model, June–Aug. (Redrawn from Intergovernmental Panel on Climate Change report, World Meteorological Organization [Houghton et al. 1990].)

Because fresh water is less dense than salt, major river flows are important in controlling the density of the sea (see 2.2.5). Any change in the flow of a large river could, conceivably, change the planetary circulation, and it is possible that this did happen in the past, when water was diverted naturally from the Mississippi system to the St. Lawrence. Further back in time, about 40 million years ago, the Arctic was warmer and supported forests, not ice. It is possible that ocean circulation then ran backward compared to today, with the denser water being the more salty warm water in the tropics. As in the atmosphere, the oceans may have thresholds that mark the switch from one circulation pattern to another.

4.6 The regional effects of change

If more energy is transferred from the equator to the poles, weather is likely to get more violent. Storms may be more powerful, winds stronger, lows deeper. But this will not necessarily occur. The loss of Arctic ice may have the opposite effect, allowing the pole to warm and reducing the violence of the heat transfer. It is extremely difficult to model accurately what the regional effects of climate change will be; as was shown above, even to estimate the feedbacks and average global warming is not easy. Nevertheless, several attempts have been made to assess the way in which the global change will affect the various regions of the Earth. Most attempts adopt the climate models that were used in estimating the overall global warming. They remain educated guesses only, because we understand clouds so poorly.

A secure conclusion from the global models is that climatic change *will* occur. The models also show that the change will not be uniform. The planet will not warm up evenly. Change will be locally distributed: some regions will warm markedly, others only slightly, and a few may cool. Figure 4.5 illustrates the conclusions of three climate models; they differ in detail but imply that in general there will be a significant global warming by about 2030, when the increase in the greenhouse effect will roughly be equivalent to a doubling of CO_2

In recent years, attempts have been made to model the way in which climate responds, over time, to atmospheric changes. Equilibrium models, such as those illustrated in Figure 4.5, calculate the climate as it would be with a different atmosphere (typically with a greenhouse effect equal to that given by doubling CO_2) and contrast that with the present climate. In contrast, some newer models attempt to model transient changes, so that the year-by-year alteration in climate can be estimated. These models typically include the oceanic circulation and allow for the time it takes to heat the ocean. The Earth's climate has not yet fully responded to the changes in the past century.

Models that allow for the transient effects tend to show that it takes longer than a decade for the climate to warm by a significant fraction of the final equilibrium warming, for models that predict a warming of about 4°C for doubling of CO_2. By about 2030, the climate may have warmed by 2°–3°C, not by 4°C. Significant regional differences also occur between the transient and equilibrium models. Figures 4.6 and 4.7 illustrate predictions from one such model. The

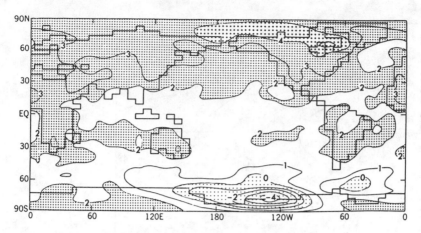

Figure 4.6. Results from a coupled ocean-atmosphere model by Stouffer, Manabe, and Bryan (1989). Panel shows the geographical distribution of the change in surface air temperature 60–70 years after the beginning of a gradual increase in atmospheric CO_2, at a rate of 1% per year, compounded (equal to a doubling of the effect of atmospheric greenhouse gases other than water by 2030). This model is markedly different from the earlier models (Fig. 4.5) but similarly shows strong Northern Hemisphere warming, especially in Canada and Northern Eurasia. Interestingly, the model predicts some cooling in Antarctica. (Copyright 1989, Macmillan Magazines Ltd.)

Southern Hemisphere, which is mostly ocean, tends to warm more slowly than the north. The transient models also tend to show less warming in the northern North Atlantic, but strong warming in the Arctic. These transient models, being more realistic than the equilibrium models, offer the hope that warming and climate change will be slower than popularly imagined some years ago; however, the heat storage of the oceans is so great that once warming has set in, it will take many years to reverse. These models are in accord with the experience of the past decades.

Rainfall change is linked with temperature change: as the climate warms, the air can carry more water and so global rainfall should increase. Again, however, this intuitive conclusion needs to be tested by detailed modeling. The effects may be localized, and in many regions rainfall may decrease. Furthermore, annual rainfall is not so important as the water balance that results. Cambridge in England has the same rainfall as the fringes of the Kalahari Desert in Southern Africa, but Cambridge is usually moist and green, not dry semidesert. This is because Cambridge is cooler and there is less evaporation. Most, but not all, climate models predict severe changes in moisture availability, especially in continental interiors.

A guide to the future is the past, which can be used as a key to the changing present. Rapid climatic changes have taken place many times in the relatively well-recorded recent past (up to a million years ago). Warm episodes in the recent past can be used as analogues of the future.

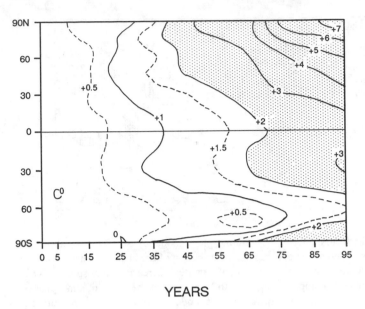

YEARS

Figure 4.7. The gradual warming of the planet in the model of Stouffer, Manabe, and Bryan (1989). Diagram shows how warming may occur over the next 95 years, zonally averaged for latitudes from the North Pole to the South Pole. Initially warming is slow and distributed worldwide. Later, the marked warming of the northern continents begins, but warming is slower in the south because of the thermal inertia of the Southern Ocean. (Copyright 1989, Macmillan Magazines Ltd.)

A commonly used example is the so-called Altithermal episode, from 9000 B.C. to around 6000 B.C. The climate may then have been several degrees warmer than it is today (the range is shown in Fig. 4.3), at least in parts of North America and Eurasia. Furthermore, some scientists have suggested that the rainforest of Amazonia was probably confined to much smaller areas than at present, although this is not proven. However, there are important differences between the Altithermal and the conditions expected in the early part of the next century. These include the presence of large residual ice sheets in the early Altithermal in the Northern Hemisphere, as well as differences in atmospheric circulation and in the distribution of solar radiation received by the Earth. Nevertheless, the analogy is interesting and instructive: it would imply warm dry conditions on the North American prairies. The results from several climate models support this inference.

Another analogue may lie in the Eemian interglacial, a relatively warm period that took place about 100,000 years ago (see range in Fig. 4.3). This period is, however, less well known geologically. It is possible that the Antarctic ice may have begun to melt in this time. Longer ago, continental drift reduces the value of direct analogues, but much can be learned from studying vegetation. For instance, forests and alligators used to flourish during the Eocene in the Canadian Arctic Islands, which even then were close to the pole. The implication is that forests can grow even in the polar regions, if the weather is warm enough, despite the months-long dark period in winter.

A final instructive analogue is the medieval warm epoch, from A.D. 800 to 1200, in which the Vikings settled in Greenland and Newfoundland and the monks of England grew excellent wine grapes. It is interesting that several fine vineyards now exist in England again, not least among which are those in the valley of John Milton's cold and miserable River Cam.

4.6.1 The Arctic

Many climate models suggest that the Arctic will warm markedly, especially in winter, and there is strong evidence that this is already taking place (see section 4.2.1). Some models imply that the polar surface temperature will increase by as much as 10°C when the greenhouse increment reaches a level equivalent to a doubling of CO_2. Figure 4.5 illustrates this: in several models, winter warming of 8°C occurs in the Arctic, and summer warming of 6°C occurs in methane-hydrate-rich western Siberia. Antarctic effects are also strong, but influenced by the massive ice cap. The various models differ in detail but agree in generality. Summer temperatures in the Arctic will probably be substantially warmer by around 2030 if they are correct.

The response of the Arctic pack-ice to warming is interesting. Some research suggests that even a 4°C rise would remove the bulk of the ice. Figure 4.9 illustrates one model of the behavior of sea ice. The top panel shows the extent in summer of Arctic ice, calculated from present temperatures, and the bottom panel shows the calculation repeated after a 5°C warming. Models such as these imply that there will be little summer ice in the Arctic by 2030. The change, if it takes place, will have important implications for Arctic communities, for sea temperatures in the North Atlantic, and possibly for the methane content of the air if permafrost methane is liberated.

4.6.2 Soil moisture

Predictions of change in soil moisture can be extracted from global climate models, representing the balance between precipitation and the combined effect of evaporation and runoff. The results are not reliable and differ between models, but there is some agreement that the interiors of the northern continents, in mid-latitudes, will become much drier in summer. This is not a secure conclusion – we cannot be sure of what will happen – but for the farming districts of North America and Eurasia it is a matter of economic concern. In the North American prairies, for example, a study of the economic consequences of the changes predicted by one of the climatic models shows that the extra costs will be high, with reduced grain yields.

More important for the bulk of humanity are the changes expected in India, China, and Africa. Some models (see Fig. 4.10) predict major changes in soil moisture in Asia, especially in China. This is where change could not easily be

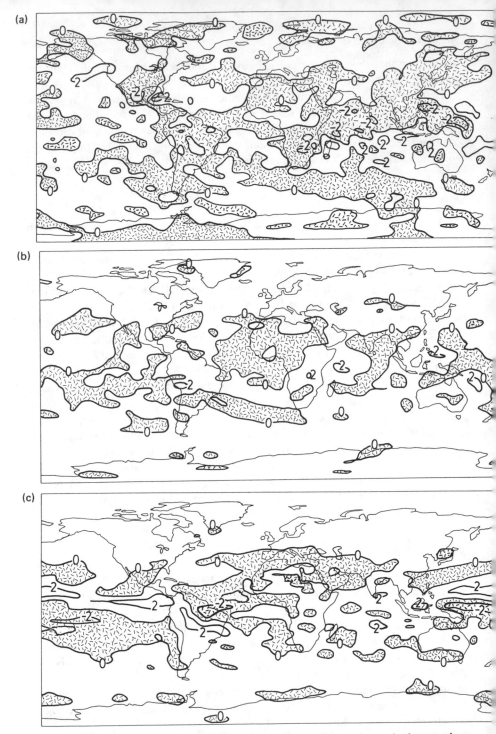

Figure 4.8. Changes in precipitation that would result from a change in the greenhouse effect equivalent to a doubling of CO_2. The models here are the same equilibrium models as those in Fig. 4.5. Regions where precipitation decreases are stippled. Contours show changes in precipitation in mm per day. (a) Canadian Climate model, Dec.–Feb.; (b) Geophysical Fluid Dynamics Laboratory model, Dec.–Feb.; (c) United Kingdom Meteorological Office model, Dec.–Feb.

Caption to Figure 4.8 *(cont.)* **(d)** Canadian Climate model, June–Aug.; **(e)** Geophysical Fluid Dynamics Laboratory model, June–Aug.; **(f)** United Kingdom Meteorological Office model, June–Aug. (Redrawn from Intergovernmental Panel on Climate Change report, World Meteorological Organization [Houghton et al. 1990].)

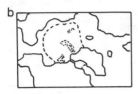

Figure 4.9. The extent of Arctic sea ice in summer. Top panel (a) shows ice limits (dashed lines) and extent of perennial sea ice (stipple) as calculated for the present climate. Lower panel (b) shows ice cover if climate warms by 5°C. (From Washington and Parkinson 1986. Copyright 1986 Oxford University Press, by permission.)

tolerated – billions of people would be affected. For many other countries, major changes in soil moisture are also predicted. Alternative models are similar in their conclusions, though the details differ. Incidentally, many models predict increased precipitation over the British Isles in winter. All these models are simply models, though, and reality may be very different, especially if somehow the dynamics of the monsoon circulation are changed, ocean currents change, or cloud patterns alter.

What these models show is not that we can securely predict changes in soil moisture – we cannot – but that soil moisture *will* change, and change markedly. This will have profound implications both for natural vegetation and human agriculture across the planet. Millions of people will find themselves failing to grow food in land that has become too dry, or attempting to produce traditional crops that are unsuited to new, wetter conditions. Either the people will have to migrate or they will have to adopt new farming methods. Forests and other natural vegetation will be stressed, either from drought or from conditions that are too warm and wet for the former ecosystem. Drainage systems in regions of increased rainfall will take centuries to adapt, and flooding may become common, especially in deltas of rivers draining catchment basins whose deforested upper reaches cannot retain water.

The greatest concern is that something will occur to change the rainfall patterns over the Indian peninsula, Southeast Asia, or China. Any global shift in the climate is likely to affect the monsoon in a way that is difficult to predict. Rainfall and soil moisture changes in India or China may cause massive famine or large numbers of environmental refugees or both.

4.7 The response of vegetation

As climate changes, the natural vegetation and the agricultural patterns that human societies have created over the past centuries must also change. Climate is always shifting and this is not, therefore, a new process; what is new is that much of the Earth is now a garden, with its vegetation controlled not by natural processes but by the design or accidental impact of humanity. The response to change of both the natural vegetation and agricultural practice is thus not only determined by the shift in climate. Human needs and behavior play a major role.

Two examples are the recent southward spread of desert and semidesert in the Sahel region of West Africa and the eastward encroachment of the Kalahari Desert and Karoo scrub of southern Africa. In both cases the effects of climate fluctuation have been amplified or even caused by overgrazing and wood cutting. Marginal lands frequently experience a string of years of poor rainfall, followed by wetter episodes. During a sequence of poor years the natural animal population normally dies off. Human beings, however, have access to stored and underground water from boreholes and can maintain stocks of animals such as cattle and goats that are well above the carrying capacity of the land in poor years. In peasant societies the animals represent wealth: there is a social incentive against destocking. After a sequence of poor years the vegetation is devastated, perennial species have gone, and in the grassland few seeds remain to initiate the recovery. Woodland is destroyed by people for fuel and building material or overbrowsed by animals. After a return to wetter conditions the annual vegetation responds rapidly, but perennial soil-binding growth is lacking. With little vegetation to protect the soil, heavy rain brings soil erosion and flooding. After removal of the woodland, water is less efficiently recycled, cloud cover may be poorer, and the surface color of the region changes. Typically the local conditions may become hotter and dryer. The illustrations here are from Africa, but similar processes have occurred on every inhabited continent.

4.7.1 The impact on vegetation of increases in CO_2

Within limits, many plants (those known as C3 plants) flourish in air that is much richer in CO_2 than the present atmosphere. At first thought, a rise in atmospheric CO_2 should favor these plants and increase biological productivity across the planet. Yields of most crops should be higher, trees should grow faster. However, in many regions the productivity of plants is limited by other factors, such as rainfall and the availability of nutrients. Furthermore, many plants mature or age too rapidly in high CO_2 conditions. The overall worldwide effect of increased atmospheric CO_2 is thus difficult to judge. On balance, a doubling of CO_2 in the atmosphere may cause an increase in growth and yield of common staple crops such as wheat, soybeans, and rice. These may show gains of perhaps 10–50% in productivity, depending on conditions. Other crops (known as C4 plants), such as maize, sorghum, and millet (staples in Africa) and sugarcane are less helped by high CO_2 levels. They may show little or no gain in productivity.

Little is known about the long-term effects of increased CO_2 on forests, but it is possible that trees, which are C3 plants, in general will behave in the same way as the first class of crops discussed, with increased growth. However, no long-term experiments have been done. Possibly the geological record, especially from the Cretaceous when atmospheric CO_2 concentration appears to have been high, may provide a good analogue. Furthermore, in many forests the limitations of nutrients or rainfall may be significant. What follows is simply crystal

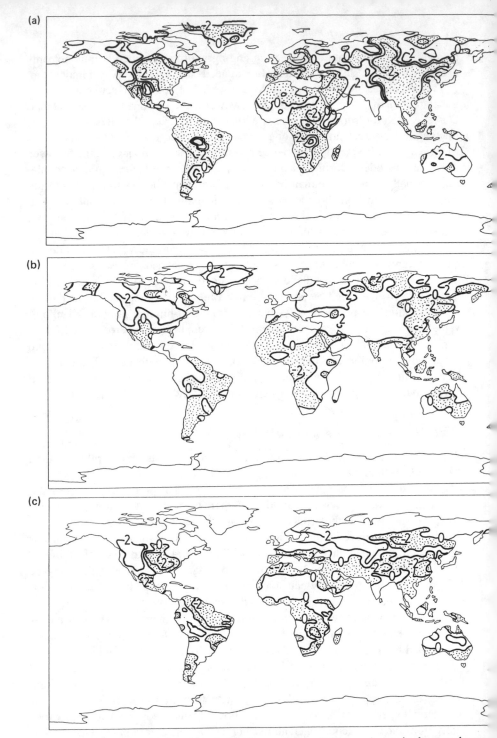

Figure 4.10. Changes in soil moisture that would result from a change in the greenhouse effect equivalent to a doubling of CO_2. The models here are the same equilibrium models as those in Figs. 4.5 and 4.8. Regions where soil moisture levels decline are stippled. Contours show changes in soil moisture levels, in cm. (a) Canadian Climate model, Dec.–Feb.; (b) Geophysical Fluid Dynamics Laboratory model, Dec.–Feb.; (c) United Kingdom Meteorological Office model, Dec.–Feb.

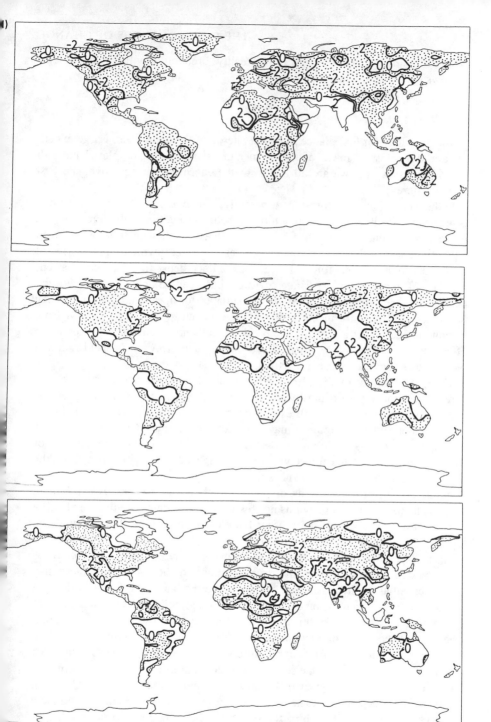

Caption to Figure 4.10 *(cont.)* (**d**) Canadian Climate model, June–Aug.; (**e**) Geophysical Fluid Dynamics Laboratory model, June–Aug.; (**f**) United Kingdom Meteorological Office model, June–Aug. (Redrawn from Intergovernmental Panel on Climate Change report, World Meteorological Organization [Houghton et al. 1990].) Canadian model includes variable soil mixture capacity, other models do not.

gazing. Possibly it may approach future reality, probably it will not, but at least it shows the *types* of change that may occur.

4.7.2 North America

North America is at first glance the least altered of the continents. Large expanses are still forested; much of the former grassland is now cropland, but still growing grasses. Yet we are all aware of the enormous changes over the past three centuries.

In the North, Canadian forest is, apparently, largely intact. But this impression is incorrect. As already stated, much of the southern margin of the forest, which is the most productive part, has been cut at least once, with little attention to replanting, so that the proportions of species in the regrowth are quite different from those in the original forest. Huge areas of once-dense forest are now barren, or scrub. The southernmost fringe of forest has been completely removed, especially in Ontario, wherever the soil is suitable for agriculture. Wetland may have increased recently in the forest. These wetlands are fed by carbon from deciduous trees in the regrowth, and produce methane, adding to the greenhouse effect. In the richer, more varied forests of southeastern Canada, acid rain and other stresses are causing severe damage.

Computer models of the effects of climate change indicate that the boreal forest will probably shift northward. Should significant polar warming occur, especially in winter, forest will extend over much of the present tundra. In the east, the moister temperate forest may also move northward. However, the models only consider changes in climate. It is difficult to imagine how the advance of the northern edge of the forest will be accomplished in regions (much of the Arctic) where there is little or no soil. Probably it will be a slow process, a sporadic advance of a few tens or hundreds of meters a year. Over a human lifetime there will be little obvious northward advance unless deliberate planting is undertaken. The size of the present backlog in replanting after logging (see Chapter 3) suggests that this is unlikely.

In the south of Canada, in the prairies, change may be more dramatic. If, as some models predict, warmer and dryer summer conditions prevail, there may be a substantial northward retreat of the southern edge of the forest in drought years, caused by dieback or catastrophic burning, followed by grass growth. There would probably be political pressure to clear and farm once-forested land. The northern boundary of the prairies would be likely to move rapidly northward as far as plowable soil permits agriculture to advance (which is not far in much of Canada). As this occurs, the southern Canadian prairie might pass to semidesert. Major dust storms appear to be becoming more frequent in the Canadian prairies. These deplete the topsoil of the farming regions, though they do have the interesting effect of enriching the soil in the forest regions further north.

Further south, in the great plains of the United States, vegetation patterns are already determined by policy decisions, on which the financial rewards of various types of farming depend, and by water resources. The dry prairie ranch land may be greatly extended, agriculture retreating if regions west of the Mississippi become substantially warmer and drier. Many of the Rocky Mountain states face a water crisis in the near future as groundwater reserves that date back to the ice ages are depleted by irrigation. Unless Canadian water is diverted south, much of the land now irrigated with groundwater will be abandoned. However, the diversion of water from the Arctic and the St. Lawrence may in itself have global consequences for ocean circulation, by changing the salinity of the North Atlantic and Arctic Oceans. In the eastern United States the present patterns of forest and farmland may not change greatly, except that there may be substantial die-back where forest is stressed by temperature change, drought, and by acid rain. If so, local and regional microclimate regimes may alter.

Models of the economic impact of agricultural changes caused by shifts in U.S. weather patterns in a high-CO_2 world vary greatly. One model suggests a net gain of $10 billion (in 1982 dollars) a year; another model suggests a loss of a similar amount. The Texas wheat crop may be much poorer, but the Iowa corn crop may even improve. These effects are in addition to the $5 billion annual cost of changes in atmospheric ozone. In general, however, equilibrium climate models suggest that the eastern and central United States will become much less productive agriculturally.

4.7.3 South America

In South America the dominant vegetation changes will come less from the controls imposed by the global climate system than from local human behavior, which if it continues as at present, will reduce the Amazon rainforest to fragments by the end of the first quarter of the next century. The forest is being replaced, for the most part, by scrub or by relatively unproductive grassland, which may receive substantially less rainfall than is typical of the region today. Deforested soils in these areas rapidly turn to laterite, a form of natural concrete, and become permanently unproductive. Rain will run off to the rivers, rather than being recycled via the forest. River flow may become more variable, and flooding more frequent and irregular. By changing the input of latent heat into the equatorial air, deforestation may affect the jet-stream circulation over North America, altering rainfall patterns there too.

In northern, central, and northeastern Brazil, the impact of the reduction in rainfall may have severe effects on agriculture. In southern Brazil, virtually all of the coastal forest is gone, locally replaced by eucalyptus plantations. It is possible that this region and parts of Argentina may also become drier, especially in summer. No doubt the Falkland Islands will become less cold and more like present-day Britain.

4.7.4 Europe

Many models have been made in attempts to predict the future climate of Europe, and the answers are almost as diverse as the models. Many suggest, perversely, that northwestern Europe will get colder and wetter, especially in the winter; other regions, especially in summer and in the south, may become hotter or drier. Europe, of course, is deeply urbanized, and all the land between the cities that is not paved is intensively and expensively managed. The changes in vegetation that occur will be entirely directed by humanity. Vegetation is controlled by political events. To take an extreme example, if, for instance, the Soviet Union and Canada were to join the European Community, under the common agricultural policy, the production of wheat to feed Europe would shift to the natural grasslands of the steppes and prairies, and the intensive cereal monoculture of much of Western Europe would probably be replaced by a more varied landscape of mixed farming and trees.

The stresses from coal-based industrialization are producing a collapse of forest vegetation across much of Eastern Europe. The land under the heavily polluted economies of Poland, eastern Germany, and Czechoslovakia suffers some of the most acid rain on Earth. Table 4.1 lists acid rain damage that has already occurred. To some extent this loss of forest may be offset by minor reforestation of coastal mountains in Mediterranean Europe, as sheep and goat farming declines.

4.7.5 Asia

For human beings, the most important consequences of global change will occur in Asia, simply because it is here that the bulk of humanity lives and where most of the annual increase in human population occurs (though the growth rate is higher in Africa and South America). Furthermore, it is in the countries of eastern and southern Asia where the most rapid industrialization is likely in the next few decades.

In Soviet Asia the effects of climate change will probably parallel those in Canada. The marked warming in the Soviet Arctic as recorded in permafrost thermal profiles has already begun. The southern boundary of the forest may retreat, the northern boundary may advance, slowly if it moves naturally, faster if it is planted. Several climatic models predict a severe decline in summer soil moisture in important grain-growing regions. Within many agricultural zones, massive changes have already taken place. For example, in the Aral basin there has been an ecological catastrophe caused by ill-thought-out irrigation projects. A large area of the sea is already dry salt flat, and much of it will soon be gone unless water is diverted from the north, which will in itself have consequences for Arctic climates.

For southern Asia, the potential for unpleasantness is high. Even small changes in climate may force the movement of tens of millions of people within countries or against the barriers of national boundaries. We cannot guess the change – for instance, one model predicts much drier conditions in Indochina, while another suggests wetter conditions for the same area – but all models agree that change will come. The deforested regions of Indochina, Thailand, and Central China may fare worst.

Table 4.1. *Acid rain damage to forests in Europe, up to August 1986*

Country	Forest expanse (sq. km)	% damaged
Finland	194,000	35.0
Norway	83,330	4.9
Sweden	265,000	4.0
West Germany	73,230	51.9
Netherlands	3,090	50.1
Belgium	6,160	18.0
Luxembourg	820	51.6
France	150,750	1.85
Switzerland	12,000	36.0
Italy	63,630	5.0
East Germany	29,000	12.0
Poland	86,770	26.2
Czechoslovakia	46,000	26.1
Hungary	16,700	11.0
Yugoslavia	95,000	10.9

Note: Figures for percentage damaged are not reliable or strictly comparable from country to country. The West German figure, for example, of almost 52% includes only slightly damaged trees that may be stressed in other ways; only a few are actually dying. In contrast, the East German and Polish figures may be underestimates. More recent estimates from Poland are around 50%.
Source: Myers (1989), using various sources.

Important urban changes will probably also occur. If Chinese coal production rises at the rate of the past decade, urban China will be the most polluted inhabited region of the globe. The local environmental quality will become very poor, almost inhuman. There will also be a global impact from the methane and carbon dioxide emitted. If Chinese economic growth continues at the 1980s rate (3–15%, varying from year to year), the consequences for global air pollution will be extremely severe.

Deforestation will probably continue in the more remote parts of China, affecting microclimates and perhaps increasing the variability of stream flow and the probability of flooding. Chinese colonization of Tibet has been accompanied by devastating exploitation of that country's forests, with unassessed climatic and biological consequences. China is so large that the local decreases in rainfall may be matched by increases elsewhere, so that poorer harvests in certain areas may be matched by better crops in other areas. This will cause major internal migrations of population. In Vietnam, with a smaller area, any adverse climatic change will produce famine, especially in the north of the country, or massive emigration.

India is heavily dependent on the monsoon. The geological record suggests that the monsoon has varied greatly in the past and may have been weak or absent during the ice ages. Some global climate models suggest that with warming the monsoon will strengthen. If so, rainfall will increase. On the other hand, the monsoon may also become more variable. Should the monsoon fail for several consecutive years, the demands on world food supply may be beyond the available stocks. The most precarious part of the subcontinent is Bangladesh, which may be seriously affected either by change in sea level or in the flow regimes of the Ganges and Brahmaputra rivers – too little in the dry season, flooding in the wet season. Deforestation and irrigation practice in the upper parts of the river basins are out of the control of Bangladesh, which may become a victim of the misdeeds of others and its own fecundity.

A rapid increase in the wealth of India is highly desirable, but if it is partly based on coal the regional and worldwide implications will be serious. If India's coal production expands to equal that of China today, atmospheric CO_2 and CH_4 will rise substantially.

Southeast Asia, especially Indonesia, Malaysia, Burma, Thailand, and Papua New Guinea, possesses the remnants of the Asian rainforest. As mentioned before, the Thai forest is almost gone, and Indonesia and Malaysia are rapidly cutting down their trees, chiefly for use in Japan. As destruction continues, the latent heat transfer into the atmosphere above Borneo and the other islands may be modified. It is not possible to predict the long-term consequences, but there is a modest but significant chance that phenomena such as the Indian monsoon and the El Niño–Southern Oscillation will be drastically influenced by this deforestation, which should be nearly complete within the next half-century. The global effect of removing the Borneo forest probably equals that of clearing Amazonia. For humanity, which mostly lives in Asia, it may have an even deeper effect because of the extraordinary climatic importance of the Southeast Asian archipelago.

The memory of most misdeeds fades. Five hundred years from now, Hitler will be remembered, like Nero, as a monster, but there will be no physical memento of his evil. In contrast, five centuries from now the surface of Sarawak will still bear the scars of deforestation. One cannot blame a peasant for wood cutting, but the legislators, corporations, and planners cutting the forests of Southeast Asia, Brazil, the Soviet Union, and Canada are aware of what they are doing, as are their customers. Half a millennium from now their actions will still be obvious to those who look on the face of the Earth.

Perhaps more than any other rich nation, Japan lives by its wits alone. Japan has over 120 million people in a land area a little smaller than Montana. Much of the land is mountainous and the area under cultivation is small. The forested area is vigorously protected and trees are carefully conserved. With few resources, Japan is vulnerable to external change. Its need for resources and its dependence on fisheries are among the principal causes of global change. The list of effects induced by Japan includes deforestation in Southeast Asia, the general degradation of the oceans, and the support of coal mining in Australia, South Africa, and western Canada. Furthermore, Japanese foreign aid to the poor countries has had some serious environmental consequences. Because of its wealth, Japan now ranks very high among the destroyers of the biosphere. This may begin to have economic consequences, such as campaigns by environmental groups in the other industrialized nations to boycott Japanese goods.

4.7.6 Africa

The saddest of continents is Africa, and it will probably be Africa that suffers most in global change. Other continents have either wealth, or space, or large-scale human organization. Africa has none. Its countries are poor and small in comparison to India and China, and its people cannot easily move across boundaries to regions that are agriculturally more fortunate. Furthermore, the climate is complex and diverse. Regions of high rainfall are closely juxtaposed to deserts. Small changes in weather patterns can produce large changes in the habitability of present population centers.

Africa is divided into a series of climatic belts. In the west equatorial zone, there is a region of high rainfall and generally high biological productivity. Flanking it are the seasonal woodlands and savannas, passing to semiarid and arid belts, the Sahel in the north and the Kalahari in the south. Each year the Intertropical Convergence Zone sweeps south to Zimbabwe, then returns northward to the Sudan. Years of strong convergence are years of high rainfall; times of weak convergence bring drought, especially to southern Africa and the Sahel. El Niño events in particular can cause major droughts.

In Africa the effects of global climate change are amplified by local stresses on the biosphere caused by agricultural practice. The advance of the Sahara in the north and the Kalahari in the south has taken place over the past few decades. This advance has been caused or intensified by destruction of vegetation, including wood cutting and overgrazing, especially in the Sahel. Any weakening of the monsoon in the north equatorial belt across Africa will accelerate the ecological decline, and may lead to biological and economic collapse in some nations. For example, Egypt is wholly dependent on the Nile. In 1960 Egypt had 26 million inhabitants; by the year 2000 it will have over 60 million. Much of the Nile water comes from Ethiopia via the seasonal floods of the Blue Nile; the rest is from the steadier flow of the White Nile. The Aswan Dam has been criticized for its cost and its many harmful effects on the ecology of the Nile delta. It did, however, enable Egypt to survive the droughts in Ethiopia in the 1982–7 period. The floods of 1988 in Sudan and Ethiopia came just in time to retrieve a critical situation in Egypt. However, prolonged drought in Ethiopia may return; if so, there may be famine not only in Ethiopia but also in the much larger population of the delta. Once collapse of food supply occurs in any degree, political institutions tend to suffer. The result is a weakened ability to withstand other crises.

In southern Africa, which also experienced severe droughts in many years of the 1982–7 period, the population and ecology of Zimbabwe and Botswana survived fairly well. In contrast, Mozambique collapsed. Zimbabwe and Botswana, with good management, were able to feed their population and by destocking to protect the land to some extent from overgrazing. In many parts of Mozambique there was mass starvation, exacerbated by civil strife. Rebuilding the economy there will be extremely difficult, even if peace returns. Furthermore, as a perceptive Mozambique government officer has pointed out, the long-term effect of severe malnutrition on the children of the 1980s will be to produce a generation of mentally damaged people. The scale of Mozambique's disaster, in contrast to the successful drought management in Botswana, demonstrates the strong link between political factors and the effects of global change. If Africa in the early years of the next century is managed as well as Botswana, the prospects are moderately favorable. If, however, it is managed as have been Uganda and Ethiopia, or unfortunate Mozambique (many of whose problems are external), the famines may exceed the capacity of Western nations to help.

A further factor in the prospects for Africa is the emergence of AIDS. In Zimbabwe, 10% or more of the population may be carriers of the virus by the time this book is published; in many parts of central Africa the infection is even more widespread and demographic projections suggest that in some regions the population may eventually decline despite a very high birth rate. AIDS is spreading rapidly in the poor nations. It is a measure of the misery of Africa and of the lack of sympathy in the West that this horrific prospect is quietly welcomed by many both inside and outside the continent: not only the bodies of Africa are in peril but also the soul of the West.

What are the prospects for Africa in 2030, when the greenhouse increment may be equivalent to a doubling of CO_2? With much luck, some countries – Botswana and Zimbabwe, for example – may be fairly secure, with well-educated populations, reasonable ecological management, and declining population growth rates. For the rest, the prospects are of rapid reduction in per capita incomes and episodic famines. Millions or even tens of millions of people may die or try to migrate, emigrate, or burst the locked doors of the West. The land will be in collapse, its ecosystems devastated and many of its animal and plant species eradicated. There will be strife or revolution in many countries. In short, the Malthusian correctives – war, disease, famine – will be in full power, the continent a surging sea of misery.

Reading list

Bolin, B., B. R. Doos, J. Jager, and R. A. Warrick, eds. (1986). *The greenhouse effect, climatic change and ecosystems*, SCOPE 29. Chichester: Wiley.

Friday, L., and R. Laskey, eds. (1989). *The fragile environment*. Cambridge: Cambridge University Press.

Gribbin, J. (1990). *Hothouse earth – the greenhouse effect and Gaia*. London: Bantam.

Mitchell, J. F. B. (1989). The greenhouse effect and climate change. *Reviews of Geophysics, 27*: 115 – 40.

Pearce, F. (1989). *Climate and man: from the ice ages to the global greenhouse.* London: Vision Books.

5
To manage the planet

Do not muzzle the ox while it is treading out the grain.
Deuteronomy 25:4; I Timothy 5:18.

5.1 Judging the problem

"The Common Law of England," A. P Herbert once observed, echoing the judicial logic of generations, "has been laboriously built about a mythical figure – the figure of 'The Reasonable Man.' " The evidence detailed in the previous chapters is that the Earth has been damaged and that the damage will become worse. The accusation is that the past and present behavior of humanity is responsible for this damage, which is increasing to the point where the well-being of all people, animals, and plants is threatened. What would a jury of reasonable people decide? Would a reasonable society take action?

The evidence for the prosecution is circumstantial. We can analyze the planetary atmosphere. We know that we are changing its composition and that the rate of change is accelerating. The evidence is firm that we are rapidly adding to the atmospheric greenhouse around the planet. When we look at the past record, we can see that rapid and great changes in the climate have occurred. The most worrying discovery is that the change that marked the end of the last ice age and the beginning of the modern climate seems to have been very sudden, and linked to a rapid increase in atmospheric CO_2 and CH_4. This addition, which may have caused the end of the ice age, was less in quantity of gas than the recent extra amount added by humanity.

We can, as yet, only barely detect a signal of modern planetary warming through the noise of the normal fluctuation of weather and volcanic eruption, but the thickening greenhouse is there and computer modeling indicates that the warming is likely to come. The ice cores from the Arctic and Antarctic show conclusively that polar heat input has already risen. The geological record and the changes at the end of the ice ages teach us that there are thresholds in the climate system. If the system is forced, for instance by an increase in the greenhouse effect, it does not respond smoothly, except within small ranges. Instead, if pushed too far, it flips rapidly from one stable state (such as the ice age) into another stable state (such as the present climate). We cannot easily model this, and we do not know whether we are safely in the middle of the stability range of the present state or if we are close to a threshold, a surprise. Whether or not we are close to an edge, we can expect change of some sort. All models are simply educated guesses, but we expect the changes to be significant to the life of humanity, and that they will come soon.

The second piece of evidence for the prosecution is the measurable but poorly studied destruction of the vegetation and animal life of the planet, especially in the rainforest. We know that the vegetation of the Earth interacts with the atmosphere, and we suspect that it provides many of the fans and chemical processing systems of the global air conditioner. But we do not yet understand the system fully. We can measure the energy transfers and the extent of some of the chemical reactions; we know that they are on a scale that must be of planetary significance; however, the system is complex and so we cannot yet say exactly how it operates.

Beyond the evidence for major changes in the air and the biosphere, there are innumerable smaller pieces of evidence – pollution of rivers, contamination of soil, and so on. And then there are the species that have disappeared. For instance, the rhino is surely more than an ornament to the planet, a disposable trinket. It is important, a part of the function of the biosphere. The rhino is like the miner's canary: if it is healthy, so is the general state of the environment in which it lives.

The prosecution, then, argues that the Reasonable Man or Woman must accept that the atmosphere is changing and that the animals and vegetation, especially in the forest, are critically important. If so, the argument continues, the reasonable society must act to call a halt until the problem is better understood. There must be, at least, a restraining order.

The defense is quite different. It is not based on detailed evidence. It is simply an almost silent presumption of innocence. Put another way, humanity has always acted as it is acting today, why should what we do now be less acceptable than what we did a century ago? Victorian England burned coal and cut trees and there was no harm seen to be done, why cannot modern America, Russia, and China burn coal, and Canada, Brazil, and Indonesia cut trees? "The air," to quote Aeschylus, "is heaven's protectorate." It is not humanity's business.

There is, so far, no unambiguous proof of climate change – if this be murder, then where is the corpse of the planet? In any civilized court the defendant has the right to remain silent, and is innocent until proven guilty. We may deplore the loss of the rhino, but is it really important? Need it stand in the way of development? The biosphere is not dead, yet. The proof is not there.

The prosecution replies that this is not a case of murder; the planet, we hope, is not yet dying. This is a court of humanity, not of law. No conviction is called for, but a restraining order is. The evidence is extremely strong that, without urgent action, something may happen that is deeply damaging to the welfare of the planet. The defense counters, again, with the presumption of innocence. It has always been thus, and it will be very expensive to change now. Mistakes have been made in the past with overprecipitate environmental action. Take, for instance, the example of asbestos. One kind, blue asbestos, is deadly. The other, white asbestos, is a completely different mineral for which there is no good evidence that it is any more hazardous than other building materials. Yet, because of the verbal association with blue asbestos, billions of dollars are being wasted in the removal of white asbestos. Too-hasty action over the environment could lead to much greater waste.

What is the verdict of a jury of reasonable people?

The evidence is wide-ranging and covers almost all aspects of human life. If the prosecution is right, expensive corrective action is needed. Humanity needs to be set on a new path. The implications of inaction, of finding for the defense, appear at first glance not to be expensive. Yet if the court finds for the defense and is wrong, the consequences are unthinkable: grave difficulties for the richer nations, chaos and the Malthusian correctives of war, famine, and pestilence in the poorer countries, the extinction of species everywhere, and for our children, the loss of beauty from the planet.

The reasonable jury finds for the prosecution. The evidence is strong enough. There may be some mistakes, but the case is overwhelming. Prosecution and defense are ordered to go out, the prosecution to seek more evidence, the defense to consider how to halt the damage. And since prosecutor, defense, and judge are all one person, humanity, they go out to find the simplest, most effective, least difficult, and most acceptable future.

Being human, we are optimists.

For generations, lawyers have pointed out that an act of God is what no reasonable person could have expected. The future of our atmosphere is something that we *can* expect, and it is something that is within our power to control. But politics is the art of the possible. For humanity deliberately to manage the planet seems impossible. The story of the attempts to limit CFC production is instructive. In response to scientific evidence, all that humanity initially managed to produce was the 1987 Montreal protocol, an international agreement that had all the vigor and hope for the future of an elderly dodo. Since then, individual nations have taken action and in 1990 the London agreement was signed, a much more rigorous document than the 1987 attempt. But as far as the more general problem of climate change goes, we have done nothing. The nations argue, each looking to its own economy and competitive advantage. We gather here as on a polluted and darkling plain, while the bright sea of nature recedes and in its place ignorant nations clash by night.

Yet our ramshackle systems of government can manage the world's monetary economy with remarkable success. The four decades since the Bretton Woods Conference that created the modern international financial regime have been the most prosperous and the best managed in human history. This may ring hollow to the peasants of Cambodia or Uganda, but the generality of humanity has been well looked after. The guiding principles have been those of enlightened self-interest, mingled with the smallest leaven of altruism. A world society that in the past decades has been capable of managing the finances of the planet is fully able to take on the further burden of true stewardship – the deeper meaning of "economy." Our decision making is disseminated, but not the worse for its diversity. The complex network of national interest and regional and international cooperation – that tangled, acronymic matrix of G-7 meetings, OECD reports, GATT, meetings of ministers, informal agreements, and market forces – is quite capable of shaping the planet, provided there is a clear demonstration of self-interest. This self-interest may in the democracies be the survival instinct of politicians in the face of strong environmental lobbies, or in less democratic nations the bribes of money or stability proffered by aid agencies, but it can be created if the environmental voice is loud and there is profit to be had.

5.2 The cost of the commons

When building an industrial plant, businesses and planners have regarded environmental impact simply as a cost, to be used in profitability analysis just like any other factor. Traditionally, certain assets have been treated as belonging to the *commons,* owned by all humanity, and thus given zero cost. For example, 25 years ago when a power station was built, the builders recognized a direct expense of building a chimney so that the smoke could be removed from the immediate locality and dispersed into the global atmosphere. In contrast, the more general public cost of a polluted global atmosphere was disregarded. The assumption was that the commons belong to no one, have infinite assimilative capacity, and so can be freely used.

More recently, the environment has become regarded as a system that can receive waste, but with a finite capacity or a finite ability to assimilate the waste. Assimilative capacity is regarded as an open-access resource, to be exploited on a first come, first served basis, or, in some jurisdictions, to be sold and traded. A senior law professor at my own university was once asked to define ethics; his reply was that if it is legal, it is ethical. Because the laws tend to postdate the recognition of an environmental problem, the first businesses or societies to exploit any part of the environment are set up without legal control and, at the time, are ethical by contemporary standards. Later, when the damage to the environment is recognized, strong vested interests have grown up that depend on the activity causing the damage, and these forces resist new legal controls. In a culture that equates ethics with laws, the damaging activity continues until eventually it is constrained, after a change in legal controls, by the economic costs (e.g., fines or taxes) or the personal costs (e.g., imprisonment) of complying with new laws. In the worst case, the laws of chemistry prevail: the community is penalized directly by the destruction caused.

The assumption that the environment has an assimilative capacity carries with it the implicit consequence that the environment will be changed by the activity that needs to be assimilated. Any injection of waste into the environment will alter it, so the use of the concept of assimilative capacity by economists is a misstatement of the notion of acceptable change. The idea of assimilative capacity is attractive to planners and analysts because it can be costed, like any other resource. When applied to the environment, however, it is an invalid concept: there is no such thing as environmental assimilative capacity, only degrees of environmental change. Similarly, the idea of waste is attractive to analysts, because it can be measured. To the environment, not only waste but also resources and product are changes – all three need to be assessed.

Some economists and legislators have responded to global change by attempting to evaluate the costs to the commons of human activities. This approach is partly successful, but suffers from the criticism that it can be compared with the ultimately unsuccessful approach to nuclear power station design: if a safety system fails, then put another one on top, and then again, until the inappropriately designed reactor is covered with a Christmas tree of restorative devices. "Costing the commons," as it is colloquially known, can be attempted by taxing CO_2 emissions, placing duty on wood imports: making the polluter pay. In part, this approach must be followed. However, it cannot easily be successful because the environment is global but human behavior is nationally divided. Some nations may refuse to act cooperatively.

It is typically extremely difficult to measure either the costs or even the extent of environmental change. Furthermore, it is almost impossible for independent competing nations to agree upon national property rights as they apply to the global environment. Beyond the political problems involved, most scientists would deny the validity of any such attempt: the global system cannot be divided. "The atmosphere knows no boundaries, and the winds carry no passports," said the British diplomat Sir Crispin Tickell, sometime president of the UN Security Council. If property rights could be defined, then – in a well-behaved world – the social costs of pollution and environmental damage could be calculated. But they cannot. The biosphere is too interconnected and too complex to divide.

The recourse is to ethical systems, if it is accepted that ethics come before, not after, laws and costs. The United Nations Stockholm Conference declared that countries have "the responsibility to ensure that activities within their jurisdiction or control do not cause damage to the environment of other States or of areas beyond the limits of national jurisdiction." This declaration does not consider costs or compensation. It rejects them in favor of a direct prohibition on any national action that will harm other nations or the commons.

This is excellent, but immediately falls to the criticism that it is unrealistic in an aggressive and uncooperative political world of nation states. The countries that are harmed cannot prove the harm while the action occurs, because the harm is in the future, not the present. The countries that do the harming can either appeal to contrary scientific opinion or simply deny liability. The problem is exacerbated by the division between the large-scale atmospheric emitters, which are mostly wealthy and skilled, and those nations that are most vulnerable, which are typically poor, unskilled, and unable either to prove their case or to enforce their opinions.

5.3 Attempting to calculate social cost

One approach to the problem is indeed to attempt the almost impossible: to try to calculate the social costs of pollution, in the hope that numbers, in dollars, will convict the polluters of error and convince them of the need to change. This approach is based on the assumptions that the environment's response to waste can be estimated, and that change will occur in a predictable, costable way. Most scientists would be unhappy with these assumptions, but nevertheless they need to be considered because they are the foundation beneath some attempts to model the economic consequences of global change.

As an example, consider CO_2 emissions. Each year, over 20 billion tons of CO_2 are emitted into the air from vehicles and power plants. A single recent storm in England caused \$5 billion in damage, according to early estimates. In an imaginary argument (there are as yet no data), one could, perhaps, argue that a single year's CO_2 emission will cause, somewhere in the world, one extra such hurricane or cyclone in each future year, producing \$5 billion in annual extra damage, and that this extra annual damage will continue for the next 2,000 years until most of the CO_2 is extracted to the seafloor. In other words, to compensate for the damage or for protective measures, the emitters would need to set up a fund that generated, in real terms, \$5 billion of income each year for 2,000 years. The capital cost of this would be perhaps \$100 billion, say \$10–\$15 for each ton of coal and oil burnt. This calculation only considers storms. The incremental annual cost of all other aspects of climate change – drought, starvation, collapse of government – outweighs storms a hundredfold or more: to a dying Africa, the cost is infinite. Thus the real cost of burning one ton of fuel may be \$1000 or more, over and above the cost of extracting the fuel and disposing of the solid waste.

This type of calculation can be challenged on several grounds. First, it is too poorly constrained – the numbers simply are not known. Instead of \$1000, they may be as little as \$50 per ton, though more likely they are closer to the higher figure. We simply do not know, and moreover, we have no way of finding out. It is impossible to predict future storms and droughts, however sophisticated our computers. Furthermore, it would be impossible to transfer the money. It is easy enough to decide who should pay – power stations, car owners, and so on – but it is not easy to decide who should benefit. Nations could make claims, but it is not obvious whether the claims should be according to population or to area. In

Table 5.1. *Commercial energy consumption by region and fuel (in petajoules[a]), 1986*

Region	Oil	Natural gas	Coal	Nuclear	Hydropower
North America	34,078	19,239	19,725	5,431	6,582
Latin America	9,019	2,981	925	63	3,442
Western Europe	24,582	8,060	10,468	5,803	4,233
Middle East	4,535	2,010	92	—	109
Africa	3,421	1,193	2,772	42	674
Asia, Australia[b]	17,137	3,772	10,957	2,290	2,378
USSR	18,632	21,157	15,751	1,474	2,198
China	4,154	507	22,241	—	1,181
Other centrally planned countries[c]	5,070	4,183	13,746	502	950
World	120,627	63,102	96,682	15,605	21,747
% of CO_2 emissions[d]	42.9	15.6	41.5	0	? decaying forest
% of CH_4 emissions	low	high	significant	0	probably significant
% of energy use	38.0	19.9	30.4	4.9	6.8

Note: Immediate greenhouse effects of fossil fuel use result not only from CO_2 emissions, but also from increased CH_4 and O_3. Natural gas transmission losses are substantial, and the consequent CH_4 greenhouse contribution may outweigh the CO_2 contribution from gas burning by a factor of 2–5. This may make natural-gas use a greenhouse contributor of the same magnitude as oil and coal. Coal mining and inefficient burning, too, produce CH_4, increasing the greenhouse effect from coal. Oil use in motor vehicles produces tropospheric O_3, also a significant greenhouse gas. Large hydroelectric schemes, especially in Canada, are probably major net emitters of CH_4. Industrial processes, such as cement manufacture, emit CO_2.
[a] A petajoule is 10^{15} joules, equivalent to about 239 billion kilocalories. One million tons of oil is equivalent in energy to about 42 petajoules.
[b] Omits China and minor centrally planned Asian countries.
[c] Includes Eastern Europe and minor centrally planned Asian countries.
[d] For conversion between relative CO_2 emissions per unit of energy: 0.75 for coal; 0.62 for oil; and 0.43 for gas.
Sources: World Resources Institute (1988); Wallis (1990), citing Smith (1988), *CO_2 and climate change, IEA Coal Research*, London.

the real world, some major emitting nations such as Poland or China simply cannot pay out foreign exchange to external victims of pollution. They are too poor. The system would be impossible to enforce. Finally, and perhaps most cogently, we have the scientific objection. We *cannot* cost change, because it is too risky to allow change to occur. The biosphere is not a computer model or a perfect factory. It is a complex, fast-changing reality. If we continue to assume that we can cost change, to misquote a famous economist, "in the long run, we are all dead." The argument thus reaches no clear conclusion except that humanity should attempt to minimize its impact on the biosphere.

Despite their weaknesses, all the approaches outlined above probably have some merit. Our future may be like our past, a mixture of empirical attempts to solve problems, some of which work, some of which do not. In the next sections, the practical problems are examined and some empirical and unideological escape routes are offered. However, one broad remedy is a global bargain: defining the commons, and each nation's share in the global responsibility. In practice, such a bargain would take much time to develop. It would probably begin with outline agreements among the major nations to behave better; over the decades such agreements could be expanded and made more powerful in their encroachment upon national sovereignty. Given such a global bargain, the management of the commons can begin, using not costs, but tools: taxation, subsidy, and regulation. The long-term effect may be a more cooperative, more peaceful planet.

5.4 The industrial atmosphere

Ever since the industrial revolution began in Britain two centuries ago, the atmosphere has been regarded as an infinity, locally polluted perhaps, but so enormous that it could not be damaged. We have now reached its limits. How can the air be managed?

The first task is to differentiate between the types of atmospheric pollution. Most of the chemicals that we place in the air do not need to go there at all. With some thought and some cost they could be disposed of elsewhere, or, by changing packaging habits, they need not be generated. It is tempting for a city like Detroit to decide to incinerate its garbage into the atmosphere, but not essential. A few federal laws or taxes can stop the temptation immediately, transforming a global pollution of the air into a local problem again. We have used CFC gases in refrigerators, in air conditioners, and in expanded plastics, simply because these gases are cheap and not toxic to humans, but with some disruption – say five to ten years of transition – we are learning to extract and rescue them or use substitutes so that little is vented into the air. The core of the problem, though, is more difficult – CO_2, CH_4, and other gases released or created during the burning of hydrocarbons. We can only eliminate the industrial emission of these gases by major rearrangement of our society.

Each gas has its own social background. Methane has a short lifetime and reducing its abundance is the obvious immediate mechanism of managing the greenhouse and a major target for control once CFC emission has halted. It is produced naturally, by industry, in coal mining and burning, and in forestry and agriculture. Managing CH_4 will involve significant changes in forestry and farming practice. Carbon dioxide is longer-lived, though it cycles rapidly, and is an unavoidable product of burning fossil fuel. To stop the growth of atmospheric CO_2, we will have to abandon fossil fuels or capture CO_2 in forests or both. The NO_x gases are strongly associated with automobiles. Immediate improvements can be made by emission control (at a price in increased CO_2 emissions), but more lasting change will require a shift away from the gasoline engine. This is reductionist thinking – solving individual pieces of the problem. A planetary solution also needs a more general approach, in tandem with a detailed piece-by-piece attack.

5.5 Energy

For humanity there is only one basic resource, and that is energy. With energy, we can do anything; without energy we are feudal.

Our continued habitation of this planet and our exploration of other habitats depend entirely on how we obtain and manage energy. If growth is defined as the improvement of the welfare of humanity or as the improvement of the global economy (in the real sense of the word), there is no limit to growth except stupidity. Growth does not necessarily entail CO_2 emissions, or even more use of energy; it is something other than ever-increasing production of cars, refrigerators, and VCRs.

All nations need energy. Historically, there has been a direct correlation between energy use and wealth. In rich nations, the relationship is becoming less direct and the increase of wealth is not necessarily linked to an increase in energy use. The richer Western nations in 1984 accounted for over half the primary energy consumed by humanity. A citizen of a Western nation consumes, with great waste, about 100 times as much energy as a citizen of sub-Saharan Africa. High prices are the best incentive for conservation. If energy were more expensive in the West, consumption would fall but wealth would fall less.

Poor people, however, whether the old and the poor in the West or the generality of humanity in the poor countries, desperately need cheap energy. The only escape from poverty for Africa and the poorer nations of Asia is access to abundant energy.

The problem is illustrated by the position of the African peasant family. At present, most African peasants gain their energy from fuelwood, cut from a supply of trees and bushes that is rapidly being depleted. Rural electrification would free them (mostly the women) from the daily struggle to gather fuel, and would even allow some reading at night. It would also save the remaining trees. If energy were cheap, the benefits of electrification would outweigh the costs (which is one way in which Western economies have grown). But if energy is expensive, no African peasant can hope for electricity; indeed, even the fuelwood will soon disappear and life will depend on burning a declining supply of cow dung. The poor nations can escape from poverty only if they have abundant cheap energy.

During a drought in central Africa some years ago, I once gave a ride to an elderly man standing by the road. Knowing the seriousness of the food shortage, I asked him about it. All he could say was, "the people are crying." I had energy and could travel to a place that had food. He did not, and came from a district so short of energy that some people were pulling plows themselves, not even possessing ox power. In better years, however, the people of the region can afford to import goods, listen to radios, and light their huts with oil lamps. A few even have old tractors. To them, any rise in energy costs will be catastrophic, forcing them to abandon all hope of escaping poverty: all years will become bad years. The political leaders of the poor nations know this, and will strongly resist any Western-led move to reduce coal and oil consumption by raising prices.

These needs of the poor countries for cheap energy contradict the environmental impulse in the rich states to conserve energy. If energy is too expensive in a conservation-minded West, but remains cheap in poor countries, then the industries in the West will point out that they cannot compete with products of manufacturers in poor countries. Political lobbies demanding cheap energy will mount in the rich nations as factories close down and production is shifted offshore to the poor countries where the energy costs are less because local governments have refused to raise them. Thus conservation-by-price in the rich nations will rapidly lead to trade conflict with poor countries, as Western industry accuses the poor states of dumping goods made with cheap energy, and the poor countries counter with accusations that the conservation-minded rich countries are condemning them to perpetual poverty.

The consequences of such policies would link energy conservation in the West with forced energy deficiency in poor nations. Forced conservation would be vigorously resisted by the poor nations. China would refuse to stop mining coal; India would refuse to give up its struggle to improve the lot of its people; Brazil would refuse to halt the burning and hydroelectric flooding of the forest. The rich nations, ultimately, would have to force submission by military means.

There is a parallel to this in the unselfish unilateral attempt by the United States a decade ago to reduce the emission of CFCs. It was a brave attempt, and successful for a year or two, until other nations in Europe and elsewhere simply took American restraint as an excuse to produce more CFCs themselves. The United States government was unable to persuade these other nations to restrain themselves. On their irresponsibility we can blame some of our present problems. For energy, which is more important than CFCs, it would be impossible to enforce conservation worldwide.

Consider now the selfish option. The rich nations continue their economic growth just as they have for the past 40 years. The planet warms. China disappears beneath a self-imposed smog of coal smoke. The Maldives and part of Bangladesh vanish beneath the sea. In the West climate change occurs, but the West is used to change: it is fundamental to the dynamic of our culture. Without change, we suffocate; more important to politicians, periods of low growth mean lost elections. So we welcome change. The economic problems are overcome, although in North America, Manitoba may become a cattle ranch, southern Saskatchewan a desert, Louisiana and Florida marine parks, and Alaska a source of excellent white wine. In contrast, the poor countries are not used to change. Their economies are too close to the edge of human existence, and their political systems too fragile to survive rapid climatic change. Disaster strikes. Flood, drought, and famine afflict Africa and Asia; country after country is battered by change and with it hunger and social unrest. Governments, stressed beyond endurance, collapse. The tropics become a chaotic landscape of misery, refugees flooding across nations and threatening borders as the wealthy nations watch and close their doors. But no door will hold against refugees – some millions will get through, the remnants of the elite of the poor lands, with enough money, luck, or deceit to escape.

Doors locked against some are doors locked against all. The bloc of rich nations, if stressed from outside, will itself disintegrate into antagonistic national or regional units, each a fortress walled by dikes to repel the tidal wave seeking refuge or help. Shall we see tanks patrolling the Mexican border, Europe's navies sinking innumerable dinghies crowded with Egyptians, Australians fearfully repelling Indonesian peasants, or the Soviet Union building a latter-day Great Wall along the Chinese border? Before that happens, there is a need to look carefully at the pragmatic alternatives available to us.

Reading list

Allaby, M. (1990). *Living in the greenhouse: a global warning.* Wellingborough NN8 2RQ, England: Thorsons.

Leonard, H. J. (1988). *Pollution and the struggle for the world product.* Cambridge: Cambridge University Press.

Mannion, A. M. (1991). *Golbal environmental change.* Harlow: Longmans; New York: J. Wiley & Sons.

National Academy of Sciences (1991). *Policy implications of greenhouse warming.* Washington D.C.: N.A.S. Press.

Redclift, M. (1987). *Sustainable development: exploring the contradictions.* London: Routledge.

Repetto, R., ed. (1985). *The global possible: resources, development and the new century.* New Haven: Yale University Press.

Schneider, S. H. (1989). *Global warming.* San Francisco: Sierra Club.

Tickell, Sir C. (1986). *Climatic change and world affairs.* Lanham, MD: University Press of America.

Weiner, J. (1990). *The next one hundred years: shaping the fate of our living Earth.* New York: Bantam.

6
Reducing our impact: the means to generate and conserve energy

> . . . had he been more in the habit of examining his own motives, and of reflecting to what the indulgence of his idle vanity was tending: but, thoughtless and selfish from prosperity and bad example, he would not look beyond the present moment.
>
> Jane Austen, *Mansfield Park*

Power, in real and not metaphorical terms, comes from either the Sun or the Earth's interior. From sunlight stored or unstored, we obtain coal, oil, and natural gas; hydroelectric, and wind power; wood, dung, and other combustible biological products; and solar energy. From the interior we can extract uranium for use in generating nuclear energy or we can use stored nuclear and accretionary heat by tapping geothermal sources.

Ultimately, no energy source is renewable – even the Sun will, aeons hence, run down – but solar, hydroelectric, wind, and biomass power are renewable from year to year, because they are all ways of harvesting the steady input of sunlight. Geothermal energy is not renewable in human lifetimes, and is best described as a form of mining, because heat from the Earth takes thousands or even millions of years to recover once it has been extracted, depending on location.

Unfortunately, most renewable sources of energy will, if exploited on a large scale, disrupt the environment horrendously. For instance: the large hydroelectric installations of northern Quebec have widely inundated precious ecosystems and generate methane; each power windmill is an eyesore, and can be dangerous; and tidal stations destroy the local ecology – damming the Bay of Fundy would

151

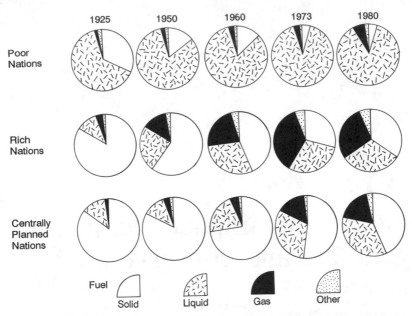

Figure 6.1. Regional energy production, by type, in selected years, 1925–80. "Other" includes hydroelectric and nuclear power; this category does not allow fully for wood, cow dung, etc., which are difficult to quantify. (From Darmstadter 1986.)

disrupt marine life and change the tides as far away as Boston. To depend on biomass we would have to dedicate vast tracts of the most productive land to monocultures of crops such as sugar. And the use of geothermal power, as in New Zealand or Iceland, creates thermal and chemical pollution. Direct solar power is probably the renewable source that is least damaging to exploit, but it requires factories to make the collectors and ground to site them on.

Renewable energy, in general, is only renewable when seen in the very limited local context of power supply. When the larger commons are considered, the air or the ecosystem, so-called renewable energy is almost always exploitive or disrupting in some way. Some installations, especially some dams, cause far more irreversible damage to the commons than nonrenewable sources. The partial exception to this is solar energy. Unfortunately, it remains very expensive. New research and new technology may change this.

The conclusion, then, is that no present energy source is unpolluting. It is true, however, that some sources of energy pollute less than others. The most obvious of these relatively clean sources is the absence of consumption: conservation.

6.1 Controlling greenhouse emissions

It is extremely difficult to set scientifically based targets for the concentrations of greenhouse gases in the atmosphere. The implication of the UN Stockholm

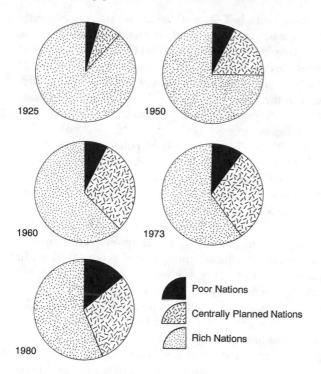

Figure 6.2. Percentage share by regions of worldwide energy consumption in selected years, 1925–80. (From Darmstadter 1986.)

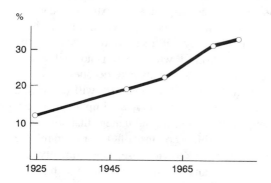

Figure 6.3. Share of worldwide primary energy consumption that is used to generate electricity, 1925–80. (From Darmstadter 1986.)

formula, that there should be *no* additions, seems intuitively the safest policy. However, this position is based on scientific opinion, not on political reality. Most political discussion of the greenhouse effect centers on proposals that are similar to the 1987 Montreal protocol on CFC emission. These proposals would constrain the nations to slow growth rates in emission or even to reduce emissions, but not to *stop* that emission.

Even the proposal for zero growth in CO_2 worldwide emissions is generally regarded as overoptimistic by much of the political community. Zero growth is a deceiving misnomer. The meaning is that annual global CO_2 emissions should remain constant at the present level of over 5 billion tons of carbon or about 20 billion tons of CO_2 annually, while atmospheric content steadily climbs, with presumably some reduction in the emissions of the rich nations to compensate for growth in emissions by poor nations (particularly China and India). Even this modest target for CO_2 control goes strongly against much political policy. For instance, it sharply conflicts with 1990 British transport policy, which is based on an estimated increase in the number of automobiles in that country of up to 140% by 2025 – nearly two and a half times as many cars as today. Chinese growth plans, to take another example, are coal-based in part, and strongly conflict with the zero-growth proposal.

The CO_2 content of the air is at present increasing by about 10 ppmv every seven or eight years. If this continues, according to the zero-growth proposal, our children in their old age would breathe an atmosphere of about 450 ppmv of CO_2, compared to the present level of 350 ppmv and the natural level of 280 ppmv. Methane is at present increasing by about 1% a year, or 15–20 ppbv. Political goals omit mention of methane. The air in 75 years' time may have roughly 3000 ppbv of CH_4, compared to the natural level of 650 ppbv. This assumes that the chemical controlling action of OH will remain as today. More likely, OH will not respond linearly, and the atmospheric lifetime of CH_4 will grow longer and longer, with the concentration rising rapidly, say to 10,000 ppbv, as OH is overwhelmed. Simultaneously, other pollutants such as CO that are controlled by OH will increase equally rapidly, or even show runaway growth.

The zero-growth option, because of these consequences, is extremely dangerous. "Zero growth" sounds comforting, but it refers to growth in rate of damage, not to stopping that damage. It is certainly not a safe option, because the atmosphere would be moving to a composition for which there is no analogue in the geologically recent past. Indeed, because the oceans take decades to warm, we do not yet even know what our present (1990) atmosphere will do. In the case of the zero-growth atmosphere of the next century, we simply cannot accurately predict what would happen. There is a profound danger that some sort of climatic runaway would develop, either from a collapse of OH or from release of Arctic methane in hydrate or from some unknown cause. Consequently, the zero-growth option is deeply disturbing to many scientists concerned with global change, despite the favor in which it is held by political leaders and economists.

A much more radical but safe choice would be to adopt the reduction path outlined in Figure 3.5, which assumes (lower curve) that emissions of carbon from fossil fuels will drop to 2 billion tons per year and release from ecosystems will drop to 1 billion tons per year. This is to reduce all greenhouse emissions – CO_2, CH_4, CFCs, and trace gases – to half or less than half of present emissions. This can be done within the next 10 to 20 years as the capital equipment of the industrial world (power stations, cars) depreciates and is replaced. There is, however, no way of quantitatively justifying this path, which would lead to annual net carbon output of about two billion tons added to the air or about 37% of present CO_2 emission. It would lead to a CO_2 level of about 400 ppmv at the end of the next generation, compared to the present level of 350 ppmv and the natural level of 280 ppmv. Even 400 ppmv may prove to have fearful climatic implications when the ocean finally warms, but it is certainly much safer than taking the world to 450 ppmv of CO_2 and the danger of a methane runaway. But it is *still* probably too much. The only safe future is one in which there is *no* net increase in atmospheric greenhouse gases.

In 1980, the World Coal Study at the Massachusetts Institute of Technology came to this conclusion on the greenhouse effect: "We also believe that the present state of knowledge about CO_2 effects on climate does not justify action to limit or reduce the global use of fossil fuels or delay the expansion of coal use even if a mechanism for such concerted action by all nations were available."

This statement makes two points: that the evidence for constraining fossil fuel use remains inadequate and that even if the evidence were good, nations would not agree. Both points can still be made today, a decade later. Nevertheless, it is now good scientific intuition, though not yet proven, that not only must we constrain fossil fuels, we must also reduce *net* greenhouse emissions to very low levels. Politically, this opinion is nowhere held, except by fringe groups.

6.1.1 Setting practicable targets for emissions

To summarize the preceding discussion, we can choose: (1) to wait until further research proves unequivocally that greenhouse emissions should be controlled, and in the meantime allow unrestrained growth, as advocated, for example, in the quotation from the 1980 World Coal Study; (2) to stop the *growth* in emissions, so that they remain at the current level of about 5.5 billion tons of carbon in CO_2 a year and around 500 million tons of CH_4; (3) to reduce emissions to the Figure 3.5 target of 2 billion tons of carbon, or 7.3 billion tons of CO_2, within a few years; or (4) to adopt the U.N. Stockholm position and stop *all* net emissions.

Of these four options, the second is the most likely, but the third is probably the most immediately beneficial to the broad welfare of humanity if then followed by a move toward total cessation of emissions as soon as possible. It is, of course, arbitrary to choose as an interim target a level of 2 billion tons of carbon emissions per year, not 1 billion or 2.5 billion, but it is an option that is compatible with scientific intuition about a safer path. The fourth goal, total elimination of all net emissions, is intuitively necessary, but recognizably impracticable in the near future. It must, surely, be attained as soon as possible. The first option, unrestrained growth, is extremely dangerous, as is also the second, "zero-growth in growth" option.

In 1988 the Toronto conference on the changing atmosphere called for a 20% reduction in CO_2 emissions by the year 2005. The oil price fluctuations associated with the Kuwait crisis of 1990–1, and the simultaneous economic downturn, may help in initiating a decline in emissions. The challenges of sustaining reductions in emissions would be considerable, however. Such reductions would require substantial changes in the industry and transport of humanity and would imply that, after an initial planning stage, most rich nations would have to spend the decade 1995–2005 in changing the ways they use energy. The target, a 20% reduction, is modest but attainable. It is most probably much too modest; more acceptably, it should be seen as a first step, to be followed by the requirement that in the next decade, 2005–2015, all greenhouse emissions be reduced to a level equivalent to 2 billion tons a year of carbon as CO_2, worldwide. If the Toronto and 2-billion-ton targets are to be met in successive decades, we should begin to rebuild our industry and transport so that by 2005 CO_2 emissions are down 20% and, more important, major new capital projects are then under way to reduce net emissions by roughly two-thirds in the following decade. In the decade after that, net emissions can be brought down to zero.

The discussion at Toronto and most other discussion elsewhere has concentrated on CO_2 emissions. Not explicit in this, though needing to be stated, is the assumption that other greenhouse gases – CH_4, CFCs, tropospheric ozone, and so on – are included by proxy of equivalence in the CO_2 figure. In other words, the "2 billion tons of carbon as CO_2" target includes all other gas emissions, according to their greenhouse impact as stated in terms of CO_2 – for instance, one ton of carbon as CH_4 would equal about 25 tons of carbon as CO_2.

To conclude, all greenhouse emissions should be rapidly reduced at least by the Toronto target of 20% (but preferably more) between 1988 and 2005, and to the equivalent of 2 billion tons of emissions a year of carbon as CO_2 in the following decade. As soon as possible a further cut should ensue, to equilibrium (no net emission), and eventually to a managed atmosphere in which CO_2 and CH_4 are controlled by international agreement, added or subtracted to maintain a stable climate. By then, we should understand the workings of the climate much better. *We should aim for a greenhouse-neutral economy as soon as possible.*

This conclusion is subject to the caveat that CH_4 and CFC emissions should not simply be regarded as more powerful analogues of CO_2. Methane emission carries the great risk that atmospheric OH may be overwhelmed. Thus all use of CH_4 for energy, especially in natural gas, must be carefully controlled. CFC emissions, because of their effect on ozone, need immediate sharp reduction.

6.2 Conservation of energy

Energy conservation is the most obvious way to reduce greenhouse emissions. In the initial stages of greenhouse control, it will be by far the cheapest option. Much experience has shown that, megawatt for megawatt, it is initially substantially cheaper to invest in saving energy than to convert from fossil fuels to other sources of power. Thus, in the rich nations, the obvious way to begin to reduce CO_2 emissions is to conserve. However, all resources eventually become more expensive as they are exploited and this applies also to conservation: the deeper that conservation bites into energy use, the more difficult and more expensive it will become to conserve further. To use a human analogy, for most people it is easy to diet and lose a few pounds, but much less easy and perhaps even dangerous to the constitution to lose 50 pounds. Eventually, major nonfossil energy sources will be needed to supplement the initial impact of conservation.

6.2.1 Conserving by price

The tools to achieve conservation are price, legislation, and research. Price is the most obvious starting point. In the 1970s the oil price shocks caused substantial changes in oil consumption in the rich nations. Per capita consumption fell, but this fall caused only local and limited disruption to the economies of the rich nations. The initial gas station lines in the United States were followed by the manufacture of progressively more fuel-efficient cars.

It can be argued that historically, the price of energy has been artificially low, because a realistic price would include the cost of waste disposal into the atmosphere. As discussed above, this argument subtly contradicts the U.N. Stockholm position and the view of many scientists that the disposal of waste into the global atmosphere *cannot* be costed. The atmosphere does not have absorptive capacity; instead, it changes according to forcing. Nevertheless, the cost to other human beings in terms of the changes they will suffer, as argued in section 5.3, can be estimated, and the very rough estimate presented there was, per ton of coal, $10–$15 for direct weather damage and $1000 or more for the total cost per ton, in all its climatic effects. A median figure would be roughly $500 per ton of coal, $350 per ton of oil and $250 per ton of natural gas. These numbers are obviously arbitrary, but they approximate the real cost of energy.

A real-cost price of this sort would rapidly end the coal industry, being substantially higher than the present price of substitutes for fossil fuels. The imposition of a pollution tax to establish a real-cost energy price is politically improbable, in the short term, but might be attempted gradually, increasing annually in impact. To maintain fiscal neutrality, the revenue gained would have to be spent elsewhere, probably on compensating for the effects of the tax by subsidizing alternative energy sources and public transport, and aiding poor and elderly people who would be especially hard hit.

The opposition to taxes of this sort that make energy prices realistic is not only from interest groups of producers but also from consumers in each nation. Some interest groups, especially the coal industry, will inevitably suffer in attempts to control CO_2 emission, but the problem of consumer opposition to taxes is more complex. The argument, already explored in section 5.5, is that environmentally conscious nations are at a competitive disadvantage: a state with expensive, realistically priced energy cannot compete with the products of a state with cheap energy and few pollution controls.

A recent example of this occurred in Ireland in the 1970s. The Irish government allowed its Industrial Development Authority to offer industrial sites to foreign companies with the implicit permission to pollute. The purpose was to attract foreign capital and industry away from other more strictly controlled regions of Europe. Spain, too, became attractive to industry because of lax pollution controls. These examples illustrate the fear that cheap energy will be used as a carrot to attract manufacturers away from nations with realistically costed fossil fuels.

The solutions are twofold. The better one is international agreement, for instance within and between large trading blocs such as the European Community and the United States–Canada free trade area. Where there is no agreement, selective compensatory sanctions, especially import duties, can be applied against nations that refuse to cooperate in setting realistic energy prices. Such use of trade sanctions to protect conservation attempts would conflict with some existing international trade agreements, especially the General Agreement on Tariffs and Trade. The problem is examined in a later chapter.

For most of the poor nations, however, price rises in energy will simply be politically unacceptable. This, ultimately, limits the extent of conservation by price.

6.2.2 Conserving by legislation

A wide variety of legislative devices has been proposed to enforce conservation. These devices range from simple rules, such as ordering lights to be turned off, to complex mixes of legislation and fiscal policy that merge legislated conservation with market forces. Some of these devices are as much conservation by price as by legislation but are more subtle than straightforward price rises.

Perhaps the most effective legislative actions to reduce fossil fuel emissions were those taken in the 1970s as a consequence of the Arab oil embargo of 1973. The primary reason for the legislation was a perception of supply shortage, not a concern for the environment. Consequently, the United States enacted in June 1974 the Energy Supply and Environmental Coordination Act, which unfortunately suspended vehicle emission standards for some years in order to allow vehicle manufacturers more time to improve fuel economy. Over the past 15 years, the results of this and related legislation have included a significant improvement in the fuel economy of U.S. automobiles, but a rather less successful improvement in the control of exhaust pollutants such as carbon monoxide and NO_x. The experience illustrates the general problem of legislative tinkering: improving one thing (fuel economy) can be offset by political bargaining to relax controls on another (exhaust emissions). The problem is general and comes from the laws of thermodynamics – catalytic converters or lead-free fuel can cost energy and increase fuel consumption; scrubbing the exhaust gases from coal-fired power stations also costs energy and increases the total coal burning.

Despite these problems, the experience of the last 15 years in the U.S. motor industry does demonstrate that significant reductions in CO_2 emissions can be ordained by legislation. During the period 1973–84, fuel consumption by the average U.S. automobile fell sharply. The present average fuel economy of cars, worldwide, is about 13 liters per 100 km or 18 miles per U.S. gallon (mpg). Many modern cars, designed under the pressure of fuel economy legislation, can attain fuel usage of 4.4 liters per 100 km (about 55 mpg) on the highway, and it should be possible to attain fuel economies as efficient as 2 liters per 100 km (about 120 mpg). Californian legislation, promoting "zero-emission vehicles," should promote a significant switch to electric vehicles by the year 2003. World-wide, improvements in fuel efficiency in the rich nations affect the designs of the same automobile models sold in poor nations; the continuing legislative pressure in the United States and Japan may therefore globally improve fuel efficiency at a rate that compensates for the rapid increase in car numbers. With luck, increased legislative pressure in the United States and Europe may result not only in a sharp reduction in per-mile gasoline consumption and CO_2 output in the rich nations but also in the poorer countries.

Legislation can also be used to reduce the energy consumed in heating buildings, both residential and commercial. This is probably best achieved by the use of building codes, in programs that are nationally legislated but locally implemented. An analogy can be drawn with the improvements in many nations over the past two decades in fire safety standards, in access standards for disabled people, and in earthquake-proofing. Typically, codes of practice are drawn up, either as national policy or in influential regions such as populous states, provinces, or large towns. Gradually, these codes are copied by other regions or neighboring countries. They have had significant impact in earthquake-proofing, as a comparison of the 1989 Loma Prieta (San Francisco) earthquake with the recent Armenian earthquake demonstrated: few people died in California, tens of thousands in Armenia, the difference being mostly a result of building codes. The codes were imperfectly implemented in California, but nevertheless they made an enormous difference to the casualty list.

The potential for improvement in residential energy use is illustrated in Table 6.1. It should be possible to reduce fuel usage in the average residential housing stock as much as 50% by retrofitting existing houses and perhaps by up to two-thirds in new houses. There are problems, of course, especially ones involving the buildup of radon and dangerous chemicals in tightly sealed energy-efficient houses. These can be overcome by the use of heat exchangers to replace air. An important point that is often ignored is that good insulation is as important in housing in hot regions where air-conditioning is used as in cold areas where heating is necessary. Most housing in the southern United States is extremely poorly insulated, despite the enormous amount of energy needed for air-conditioning.

Domestic appliances and lighting of domestic and commercial buildings are also wasteful of energy, and here too legislation can be used to reduce consumption. For instance, refrigerators account for about a quarter of residential electricity consumption in Sweden and the United States. Refrigerators in industrialized countries now require 30–70% less electricity than models made 15 years ago, but there remains room for improvement. Lighting too is very wasteful. A Brazilian study showed that shifting to more efficient refrigerators could save, by the year 2000, 3.5 gigawatts (GW) of baseload installed power (total present installed power in Brazil is about 45 GW), and that switching from incandescent

Table 6.1. *Space-heating requirements in single family dwellings (in kilojoules per square meter per degree day)*

United States	
Average housing stock	160
New (1980) construction	100
Energy-efficient housing, Minnesota	51
Super-insulated home, New York City area (calculated)	15
Sweden	
Average housing stock	135
Houses built under 1975 Swedish building code	65
Energy-efficient housing	36

Source: Williams (1987).

to more efficient fluorescent lights was 2.5 to 5 times cheaper than building new hydroelectric power stations. Since the saved power represents fewer dams, especially in Amazonia, the environmental impact is substantial.

Legislation is needed to bring this about. A recent comparison between energy in Texas and California demonstrates this. Between 1977 and 1984, California developed strict conservation controls, while Texas allowed market forces to rule without legislated incentives or regulations to conserve. In the year 1984, the average Texan used 1424 kilowatt-hours more energy than in 1977, while a Californian used 267 kilowatt-hours less. Consequently, during the period 1977–84 California added new access to electricity equivalent to three standard power plants, while Texas in contrast added eleven. The cost to California of conservation proved to be three to five times less than the eight new power plants that were not built would have cost.

Most domestic devices are made or sold in the regulated industrial states – the European Community, North America, Japan. Legislated or regulated economy of energy can be enforced relatively easily by controlling the relatively few manufacturers of domestic appliances, just as automakers are being forced to produce fuel-efficient vehicles. Market forces can be used to supplement direct efficiency regulation. For instance, in Europe there is only a small number of manufacturers of refrigerators, and most appliances fall into clearly defined categories, in terms of cubic capacity, shape, and so on. One way of using market forces would be to tax or reward manufacturers according to the deviation of their products from average energy consumption in that category. This would be fiscally neutral, but a manufacturer who produced a more-wasteful-than-average washing machine would bear a taxation burden, while a manufacturer with an efficient machine would be rewarded annually, according to sales. If the penalties and rewards were sufficient, the effect of competition among manufacturers would be a rapid reduction in the average energy consumption of their products, as each manufacturer attempted to undercut the average. The rewards would allow lower prices and thus larger sales for energy-efficient appliances, further reducing the average. Legislative policy of this sort uses market forces as an amplifier and is intrinsically more powerful than simple control or dictate.

6.2.3 Conserving by town planning

For most of this century, the cities of Europe and North America have been deliberately planned. Planning decisions, generally made by local authorities, have governed the distribution of homes, industrial facilities, and shops. In consequence, these planning decisions have also controlled the need for transport between places of residence, work, distribution, and leisure. In the long term, the reduction in need for transport must therefore be closely linked with local and regional planning decisions.

Town planning texts in the early part of this century envisioned idealized garden cities, in which houses were laid out in leafy suburbs upwind of the city center and industrial areas were located downwind, so that pollution and smoke could blow away to the countryside. This style of planning is now deeply set in the bricks and concrete of the rich nations, with general careful segregation between homes and industrial areas. White-collar work has also been segregated from homes. In Europe, historic city cores have been preserved, but newer residential suburbs and industrial areas have been separated from each other. More recently, distribution has also been segregated, as shopping has become concentrated around large supermarkets and malls, surrounded by car parks. In North America, with less historical background, urban sprawl has developed, poorer people living close to industrial areas or old city cores while the middle class commute between sealed, air-conditioned or heated houses and offices and shopping malls. We have grown accustomed to working in buildings with no windows, only transparent walls, cut off from the outside environment by the climate-control systems in each building. The layout of cities and towns, by controlling distances that need to be traveled, dictates much use of energy; the sealing off of the inhabitants from the environment contributes to a lack of perception of the impact of human activities on that environment.

Four examples of town design are worth considering: Saskatoon, Canada; Cambridge, England; Zurich, Switzerland; and Harare, Zimbabwe. All are university cities. Saskatoon is a creation of the twentieth century. The mean annual temperature is slightly above freezing. It has roughly 200,000 inhabitants. The city center is an open mesh of office buildings and large parking lots; each new building has to provide adequate parking. Aerial photographs show a landscape that is half given over to automobiles. Suburban shopping is mostly at malls, with large parking sites. Nevertheless, Saskatoon does have a good, dependable public transport system, all by bus. No part of the city is more than a few hundred meters from a bus stop; service is typically every quarter hour and timetables are strictly kept, with buses connecting so that passengers can transfer from route to route. Schedules match university terms and lecture hours. The city formerly had streetcars (trams), but abandoned them. It has, however, had the good sense to reserve corridors for future public transport routes. Pollution is beginning, with photochemical smog on still days, despite the clean air around the city. In city design, no thought has been given to atmospheric pollution problems.

The inhabitants of cities on the Canadian prairies arguably emit more CO_2 and CH_4 per capita than any other citizens of the industrial world, in part because of the climate, but also because of the heavily car-dependent North American lifestyle (see Fig. 6.11). Saskatoon is typical. Although the houses are well-insulated, the city is otherwise almost aggressively uncompromising in its lack of adaptation to the climate. Houses are typically detached single buildings, and many of the apartment blocks are sited at the edges of the city, with long commuting distances. With a very few exceptions there are none of the row houses so typical of Europe or parts of eastern North America, despite the efficiency with which row housing utilizes space and conserves heat, while still providing gardens.

Universities should lead their societies. In Saskatoon the university has an efficient and beautiful core layout, with a central grass bowl surrounded by a ring of dense buildings. This is a pedestrian area, with services and vehicles banished to the outside. However, the rest of the university has the general layout of a slightly upmarket airport industrial area, with scattered buildings surrounded by large car parks and roads. To the north is a science park built over remnant prairie. Walking across campus is difficult, especially to the science park in winter, because of the long distances and lack of paths. The university occupies a lovely riverbank site. One of the best spots used to be a fragment of prairie and riverbank in front of the university hospital, a small paradise of birds and rodents, with ducks, pelicans, and beavers below. To quote Joni Mitchell, a Saskatoon-bred singer and poet, they paved paradise and put up a parking lot. They then built a five-lane highway to it.

The street layout of the city in general has little to distinguish itself from many cities in much warmer climates, say, Canberra or Bloemfontein. Indeed, the older part of Regina, the capital of Saskatchewan, has a town plan that is almost identical (even down to the layout of the trees in the main square) to Harare, the capital of Zimbabwe, which was founded at the same time. The younger parts of Saskatoon and Regina would not look out of place in Albuquerque or Houston. Industrial activity and warehousing are mostly confined to one sector of the city, where standards of beauty are lower. This means that some workers in this sector have to commute long distances, despite the small size of the city.

Virtually all of Saskatoon's city design, like that of most middle-sized cities in North America, has been shaped by the needs of the automobile, coupled with the requirements for basic services. In recent years, with the popularity of mountain bikes, cycling has become possible all year, and many people attempt to commute this way even in January. However, with the exception of a few designated road shoulders, there are virtually no bicycle paths for commuters. There are no cross-city bicycle routes, and only a rudimentary network of routes leading to the university, one of the main commuter foci. Typically, cyclists are mixed with cars, which is dangerous for cyclists, or with pedestrians, which is dangerous for the pedestrians. Commuting on foot is very difficult, except in the immediate neighborhood of the university, because of the long distances and the danger of crossing urban highways in snow. There are no winter cross-city ski routes. One Canadian city, Ottawa, has developed winter commuting routes for skaters, using a canal. Saskatoon, being flat and very cold, could experiment with skiing and skating tracks for commuters, but this has not been attempted.

More generally, however, the low density of the city makes it very difficult to avoid using automobiles. Distances are simply too great to walk, cycling is too dangerous. The long distances reduce the effectiveness and increase the cost of public transport. Residents must use cars to go to school, to shop, to visit friends. Each year the city expands outward into the surrounding farmland, rather than inward by building on its parking lots or reducing the space allotted for roads. The environmental movement at the university is strong, but probably the chief topic of student politics is the cost and shortage of parking on campus.

A serious attempt to reduce the greenhouse emissions of the city would involve a large-scale rearrangement of commuting patterns, carried out over many years as buildings age and are replaced. Commuting distances could be shortened by mixing places of work and housing more closely, eliminating the industrial ghetto and the shopping strips and malls that are beyond walking distance of their consumers. Cross-city cycle paths, with priority at crossings, would coax commuters away from cars, especially if parking lots were made more expensive by taxation, and prohibited in new construction. Urban rail and streetcar lines would add to the already good public transport, making it more pleasant to commute by transit than by car. The population density would increase if parking lots in the downtown core and along strip malls and straggles of offices were replaced by condominium or row housing. The net effect over about two decades would be a denser city that used far less fuel, huddled together in the cold climate, with far less need to commute and perhaps a more closely-knit texture to society. Saskatoon is in many ways typical of North American small cities. Its challenge is cold, but heat places similar constraints on cities in the southern United States and demands similar solutions if greenhouse gas emissions are to be reduced substantially.

The second example is less encouraging. Cambridge, England, was once a lovely city, filled with old gardens over which the spires dreamed no less than in Oxford. At its center was the Cavendish Laboratory, that fountain of modern science which gave us the electron, nuclear physics, and DNA. Cambridge is still lovely, but the best view of the city spires is now over a large freeway. The most impressive spire today is a hospital incinerator flue (see Fig. 6.12). Some years ago the Cavendish Laboratory was moved to a "green field," and is now housed in reductionist coops suitable for a chicken farm. The modern heart of the city, close to the old site of the laboratory, is the large Lion Yard car park. On most afternoons, long lines of cars form with engines running; Fridays and Saturdays over 100 cars wait to enter at rates of one to two per minute. The city experiences gridlock each afternoon on rainy days. The university has no transport policy to regulate commuting. Public transport is rudimentary and partial. It is not integrated. The rail station and bus depots are far apart and bus services are irregular, run by competing companies, and serve outlying villages very poorly.

Scheduling is not coordinated between bus companies and rail services. There is no monitoring of air pollutants, though the smog is often visibly thick. There are many bicycles, mainly used by students but also by commuters, but the city has an extremely high accident rate for cyclists, one of the highest in England.

Former generations of academics used to stroll through the college backs – some of England's loveliest gardens – after lunch, or meet to discuss science and philosophy in the pubs such as the Bun Shop or the Eagle. Philosophy and physics are no longer together. One pub is now a road, the other closed for redevelopment. Today at one end of the backs there is a gigantic ugly building complex ringed by roads and parking. Many of the walks are closed or charge admission because of the pressure of tourism. Everywhere, the roar of traffic permeates, and the intercollegiate rowing races that were one of the symbols of the dreamy old England no longer begin in a lush green setting under the song of larks ascending: the start is now under a concrete freeway bridge. The larks remain, but cannot be heard.

Recent plans by the Cambridgeshire county council have been based on the assumption that road traffic will increase markedly in the next decade. The plans are for improved road links, the construction of new parking garages (including one in a major public park) together with some improvements to rail and cycle transport. This car-first view of regional government conflicts with city policy, but reflects current national policy in England, which is based on the expectations that car numbers will rise sharply and that massive new expenditure on roads is needed. One cause of the rise in car numbers in England is the trend toward the use of company cars, as part of salary packages. This trend, which depends on commercial fashion and national fiscal policy, has filled commercial Cambridge with new cars, though academic Cambridge's parking lots remain populated by older vehicles – the university does not give out cars as compensation.

The city engineers are less enamored of cars. Relatively recently, laudable attempts have been made by the city council to provide designated cycleways, with some success, but often this has been done at the expense of pedestrians, by allowing bicycles on sidewalks. Bicycling into the city from outlying villages is very dangerous, as the roads are mostly designed only for cars.

Less squalid is the example of Zurich, where they *do* measure daily ozone levels. This too is a wealthy city, more so than Cambridge, and its residents are well able to afford several vehicles to each family. The society has chosen otherwise. It has perhaps the finest public transport system on earth, and the system is used. The key to the system is integration. Buses meet trains, trains disgorge by tram stops, ticketing is simplified or virtually abolished by passes and honesty. Above all, service is frequent and regular. Inhabitants of distant villages can rely on post-buses every ten minutes; trains are strictly on time. It is not cheap to the society – a Zurich bus driver in spring 1988 earned a salary of U.S. $47,200 per year (Union Bank of Switzerland figures). The recent investment in rail transport, for a region of about one million people, is of the order of several billion dollars.

Part of the attraction is the ease of the system. Quite simply, public transport is less hassle than a car for most people. There are also inducements. There are plans for harsh taxes on fossil fuel. Each automobile must pass a very strict annual test of its mechanical safety and of the effectiveness of its catalytic converter. Once licensed, a commuting automobile must find a parking place – not easy unless the owner has a privately owned space. Much parking has been removed, and city lots have been closed. Shops do not typically provide parking, nor do offices. Moreover, many Swiss return home for lunch; hunting twice a day for a parking spot is tedious. In contrast, the integrated public transport system, with very frequent services (every ten to twenty minutes in outlying villages, as often as every three minutes in the city), is much more attractive.

There are problems, of course. Zurich is still thick with vehicles. It has experienced an explosion of wealth and building over the past decade, and with that wealth comes the ability to buy more and bigger cars and more owners demanding facilities for their cars. Bicycles are common in the suburbs and outlying villages, but central Zurich is dangerous for cyclists, with few cycle tracks. Furthermore, many public transport schemes are disruptive or expensive. Nevertheless, the system has saved the city from much worse. Zurich has little of the degradation by wealth shown in Cambridge. Most Swiss public transport is electric, and most of that electricity does not come from fossil fuels. The Swiss Official Timetable, listing all train, ship, and rural bus services in the nation, is one of humanity's more impressive documents. It represents the integrated efforts of many different private companies and public corporations who jointly have created the cleanest and most pleasant transport system on Earth. A feature of transport policy in the Germanic nations is that they depend heavily on rail for moving freight. The United States and the United Kingdom have freeways dominated by large trucks and semi-trailers. In Germany, in contrast, some road freight is used, but much freight goes by rail in electrically powered trains, via the wide network of tracks reaching most factories. This is much less invasive of the countryside and less polluting.

Finally in this tale of four cities, comes Harare, the capital of Zimbabwe. Harare is a rapidly growing metropolis, with a busy core, spread-out low-density suburbs, and closely spaced high-density townships. During the dry winter months, air pollution is severe in the high-density and industrial areas. All vehicles and most of their fuel are imported, though vehicles are locally assembled and Zimbabwe was one of the first countries to reduce its fossil fuel imports by adding alcohol derived from crops to its gasoline. Transport is a major part of the nation's foreign expenditure.

The low-density suburbs depend on the car. The suburbs, which are inhabited by the rich elite of government officials, senior managers, and the remaining whites, are beautiful. They are regions of lawns set in forests, with individual lots ranging up to a hectare or more. The trees planted in the gardens are a sink of CO_2 that counteracts the emissions during commuting. Nevertheless, commuting times are long and expensive to a country that can ill-afford foreign currency. The high-density suburbs, in contrast, depend on bicycles, buses, and aged taxis – a typical pattern in poor countries. In the early mornings of still winter days the diesel and cooking smoke hangs over the townships like a dark, almost solid, catafalque.

Harare's town planning has been relatively sensible, and there are few real slums. But little attention has been paid to environment. The successful struggle to house an influx of poor people has come first. The obvious solutions are to reduce the need for commuting by placing housing closer to work places, to improve cycleways, and to introduce integrated light-rail, electric tram, and trolleybus services and perhaps an urban subway, using the high level of local mining expertise. Electric transport, apart from being less polluting, also has the advantage that it can economically be built in poor nations, unlike the gasoline engine which is best built in huge factories producing millions. Bicycles and shoes are even easier to construct.

These four cities illustrate the varying effects that town design has on energy use and also on pollution. The Swiss example shows that it is possible for a wealthy society to move around with only a small emission of greenhouse gases or atmospheric pollutants, as less than 2% of Switzerland's electricity, which drives its public transport, comes from fossil fuels. The investment needed in public transport is large, but it is offset by the savings in the capital cost in roads and automobiles as well as in imported gasoline.

To succeed, public transport needs to be frequent, reliable, comfortable, safe, and integrated. The Swiss system achieves all of these. It is also aided by the legislative pressure on the automobile, which has helped to tilt the balance of convenience to the user away from the car and toward the public system. In particular, restricting parking is effective in reducing traffic flows. In contrast, the Cambridge system, relying on cars in an old town, is markedly less pleasant or convenient. In general, pedestrian sidewalks cost less than bicycle paths, and bicycle paths cost less than roads. Cambridge has attempted to help its cyclists but at the cost of its pedestrians. By designating many of its sidewalks as bicycle paths, it has moved some bicycles off the roads, giving priority *first* to cars, *then* to bicycles, and *last* to pedestrians. Any town design that is aimed at reducing CO_2 emission and pollution would reverse this order of priority by encouraging those who walk or cycle to work by allocating them space at the expense of cars and by providing efficient public transport, also at the expense of cars (e.g., by putting bicycle lanes and streetcar tracks on streets, by selling publicly owned city-center car parks for redevelopment as housing or office space, and by imposing taxes on parking space).

The Swiss experience also gives lessons about town layout. Many Swiss villages are a complex mix of closely juxtaposed factories and residential areas. In contrast, for example, to Saskatoon, many Swiss towns do not segregate factories and dwellings nearly so rigorously. This proximity allows workers to return home for lunch, often by bicycle or on foot, and reduces commuting distances and traffic jams. It has also the effect of disciplining air or noise pollution by factories. People who live near factories, especially in the vigorous Swiss democracy, rapidly find out if pollution occurs and object to it. The result is that standards of pollution control and air quality are generally high and are enforced. Indeed, a draconian but effective way of eliminating pollution would be simply to ban all industrial chimneys higher than 15 meters (50 feet) and to insist that cities draw drinking water downstream of their own drainage. By making pollution visible in the polluter's backyard, high standards of emission control are attainable.

Energy conservation by sensible redesign of our towns may become one of the most effective ways of managing the atmosphere. This sort of conservation can be achieved over a few decades by careful legislation and without large-scale economic disruption. It may also improve the quality of our lives.

But there is a limit to conservation. Harare is very wasteful in its energy use, but most Zimbabweans use little energy. If the poor nations, especially China, adopt motor cars and abandon their bicycles, the effect will far outweigh all gains from conservation in the rich nations. For the poor, the options are not easy. There is much waste of energy in the poor nations; burning wood or coal in fires is very inefficient. But for most people, conservation is possible only if there is something to conserve. It is little use to add superior insulation to one's hut if there is no need for insulation, and calls for energy efficiency in automobiles sound obscene to those who are lucky if they possess shoes. The poor nations need more energy, not less, if they are to be allowed a measure of wealth. Conservation usually takes skill, and poor people do not usually have skill to spare. They need energy, lots of energy, and that energy must be moderately cheap.

6.3 The energy competition and the greenhouse

Conservation measures, if rigorously implemented, can in future reduce per capita energy use very substantially in the rich nations, but not in poor countries. The recent experience of the rich nations shows that total energy use in cars, houses, domestic appliances, lighting, and many industries can be reduced without great disruption, perhaps by as much as one-third. This is because in the richest nations (e.g., Switzerland, Sweden, Canada) total numbers of cars, houses, and domestic appliances are unlikely to increase greatly, so that improvements in efficiency will reduce total energy use. In the poorer and more polluting second tier of industrialized nations, such as the United Kingdom and Spain, car and appliance numbers may increase (e.g., most British houses do not yet have a dishwasher) but the increase may be counterbalanced by reduction in energy use by each house or appliance. In contrast, in the poor nations (China, India, Nigeria) the demand for appliances such as refrigerators and washing machines is enormous, and even the most rigorous conservation will not prevent a massive increase of energy use.

In the preceding discussion it was argued strongly that net emission of greenhouse gases from fossil fuels must *soon* be reduced, to a level equivalent to 2 billion tons of carbon as CO_2 per year, and then, soon thereafter, to nothing. If this target is accepted, it imposes a fierce competition among the various processes that emit greenhouse gases. For example, every ton of CFC emitted is equivalent to the emissions of 10,000 tons of CO_2 from fossil fuels; every ton of methane lost from a pipeline is equivalent to 60 tons of CO_2 from oil or coal that cannot be burnt; every ton of coal that is burnt is equivalent to roughly two tons of oil and gas that cannot be burnt; every forest that is destroyed is an oil field that cannot be used or a power station that must be closed down.

Seen in a different way, one can envision an ultimate limit – say 400 ppmv of CO_2, 2000 ppbv of CH_4 – beyond which the greenhouse effect will have consequences to the climate that are intolerable to human beings. We can choose to reach this limit either rapidly by burning coal and forest or slightly more slowly by burning more oil or gas (if CH_4 emissions are controlled) and slowing deforestation. We can reach the limit most rapidly of all by continuing to emit CFCs. Each ton of CFC added to the air means that we cannot, in future, burn thousands of tons of oil or gas.

The industrial paths to the limit, to the most different atmosphere that humanity can accept, are emissions of CFCs, coal burning, oil burning, gas burning. Linked to these industrial paths and speeding our approach to the limit is deforestation. These paths compete with each other. Each needs to be examined for its usefulness and benefits.

6.4 The chlorofluorocarbons and their effects on energy policy

The greenhouse effects of the chlorofluorocarbons have been given far less publicity than their effects on stratospheric ozone, and consequently there has been less political response to the competition between CFCs and fossil fuels than might have been expected. To recapitulate, each molecule of CFC that is released into the air is equivalent in greenhouse effect to roughly 10,000 molecules of CO_2 emitted by the combustion of fossil fuels, though the exact number depends on the specific CFC. CFCs have atmospheric lifetimes of around a century, so this effect will decay slowly, persisting for longer than most humans live. If CFC emission continues at present rates, the atmosphere will reach a state equivalent to a doubling of CO_2 by 2025–2030; if CFC emission is halted tomorrow, this doubled-CO_2 state will be reached some decades later. In other words, halting CFC emission will allow a slightly longer period for the reduction in fossil fuel burning and significantly ease the economic disruption that the reduction will involve.

The history of attempts to reduce CFC emission is interesting because it illustrates the complex problems that will have to be overcome in any more general attempts to control emissions caused by burning fossil fuel and by industrial processes. CFC production is a relatively minor part of the industrial economy, yet it has been very difficult to control or reduce output. The problems in controlling CFCs center on the usefulness of the chemicals, the profit that can be made from them, the strength of the interest groups that promote them, and the complexity of achieving an international political response.

CFCs, because of their safety to the human physiology, are widely used in applications that range from the frivolous through the useful to the essential.

Table 6.2. *Electrical power in selected countries, 1986*

	Installed capacity (megawatts)	Power Source (%)		
		Fossil fuel	Hydropower	Nuclear & geothermal
Australia	33,478	87.2	12.1	0.7
Bangladesh	1,339	91.2	8.8	
Brazil	44,749	9.3	90.6	0.1
Canada	98,400	18.5	66.3	15.2 (15.6)[a]
China	87,000	77.5	22.5	
France	92,100	11.9	17.7	70.4 (74.9)
East Germany	22,059	89.0	1.5	9.5
West Germany	82,660	66.5	4.1	29.4 (34.3)
India	54,689	71.0	26.5	2.5
Japan	173,329	61.8	12.9	25.1 (27.8)
South Korea	19,607	53.6	5.8	40.6 (50.2)
Malawi	160	2.5	97.5	
Poland	29,773	97.3	2.7	
South Africa	24,727	96.2	0.6	3.2
Sweden	33,100	5.1	44.0	50.9 (45.1)
Switzerland	15,210	1.8	59.4	38.8 (41.6)
United Kingdom	66,512	78.8	1.4	19.8 (21.7)
United States	719,444	72.1	11.4	16.0 (19.1)
USSR	321,671	76.4	13.5	10.1 (12.3)
Zimbabwe	1,539	47.3	52.7	

Note: Certain countries stand out as "dirty" in CO_2 emissions (and other pollutants): Australia, Bangladesh, China, former East Germany, India, Poland, South Africa, the United Kingdom, the United States, and the USSR are all over 70% fossil-fuel-dependent. "Clean" nations include Brazil and Canada (but their hydropower has other environmental effects), France, Malawi, Sweden, and Switzerland. 1000 megawatts (MW) = 1 gigawatt (GW).
[a]Percentages in parentheses are provisional 1989 figures for share of total power derived from nuclear energy.
Source: Encyclopedia Britannica Yearbook 1989; Nuclear Canada, May 1990.

Frivolous uses, which can easily be eliminated, include especially their use in spray cans. But sometimes spray-can CFC substitutes themselves are damaging to the atmosphere: for instance, in the United Kingdom some dairy cream is at present sold in cans with nitrous oxide as the propellant, a gas that is both a strong greenhouse absorber and a factor in stratospheric ozone chemistry. The propellant is explicitly labeled on the cans, giving the impression that because it is not a CFC it is environmentally friendly, or "green." Slightly less frivolous uses of CFCs include use as foaming agents, for instance in house insulation and disposable coffee cups. Here too, substitutes are available.

The more important uses of CFCs and halons are in refrigeration, air-conditioning, and fire extinguishers. Here substitution is more difficult. To some extent, changes in behavior patterns may help. One of the world's more incongruous sights is the Washington bureaucrat in July, sweltering in a suit or uniform that is made according to a pattern devised and suitable for nineteenth-century London as he crosses the brief space from one air-conditioned building to another. It would clearly be much more logical to outfit the inhabitants of the Pentagon in swimsuits, suitably striped to indicate rank, and to replace coffee-breaks with 10-minute dips in the pool. But sometimes humanity is not logical: we will continue to use air-conditioning, and the denizens of Washington will not forgo their epaulettes or three-piece suits, even in July.

One method of controlling CFC release is to prevent emission to the atmosphere by recycling CFCs from old products. This may be particularly useful in refrigeration and air-conditioning installations, where CFC emission could be strictly controlled by good engineering to make accidental release less likely and by financially attractive payments for recovered gas when the equipment is junked. For instance, anecdotal evidence from a garage specializing in automobile air conditioners implies that the recharge of air conditioners is not always wholly necessary, and that when it is done routinely, the gas is vented to air. Closed-system design, with units being replaced on mechanical failure, followed by recapture of CFCs from junked units sent back to the factory for recycling, would drastically reduce emissions. Substitution is also possible. All gases can, in principle, be used as refrigerants, and many (propane or ammonia for instance) have good room-temperature properties. Carbon dioxide or nitrogen can be used in place of halons as a fire extinguisher. Other products can replace CFCs in industrial solvents. To conclude, the emission of CFCs to the atmosphere can be almost wholly eliminated by a mixture of substitution and closely monitored recycling.

The CFC problem illustrates the broader problems of controlling atmospheric pollutants. For CFCs, substitutes or alternative usage are available but demand some thought and financial incentives; exactly the same is true for all other atmospheric emissions by humanity. The history of the attempts to control CFC emissions is therefore very important, because it illustrates the problems that are likely to be encountered in obtaining wider control of CO_2, CH_4, acid rain, and so on.

The following account of the history of attempts to control CFCs relies in part on the recent account by Stephen Schneider (referred to in the reading list at the end of the previous chapter) of the negotiations behind the 1987 Montreal protocol on emissions and on Derek Elsom's (1987) account. Attempts to control CFC output began in the early 1970s, shortly after their detection in the atmosphere by Lovelock and the realization by Molina and Rowlands that they would deplete stratospheric ozone. In 1978 the United States, which was the world's largest producer of CFCs, acted unilaterally to control CFCs by banning their nonessential use in spray cans. Canada, Denmark, Finland, Norway, and Sweden followed to varying degrees. The effect of these restrictions is illustrated in Figure 3.9. The European Community responded in 1980 by issuing a directive that placed a ceiling on production and controlled the use of CFCs in spray cans. However, there was substantial transatlantic disagreement on what exactly should be done. By 1985, the U.S. share of the market for CFC-11 and CFC-12 was much smaller than that of the Europeans, and there was considerable tension between the Europeans, who wished to maintain their new competitive edge, and the U.S. environmentalists. The United Kingdom and France, in particular, opposed the regulation of CFCs. However, in 1987 both Britain and France were absent from the inner circle of the rotating European community presidency; they were replaced by Belgium, Denmark, and West Germany, who all favored control. Finally, the Montreal protocol was drawn up, an agreement to cut CFC production by 50% by the turn of the century.

The potential effects of the 1987 Montreal protocol are illustrated in Figure 6.4. It soon became clear that the protocol was wholly inadequate and that there was a need for much stricter control. Indeed, after the agreement in Montreal, some countries unilaterally introduced more rigorous controls on production. Other nations, especially poorer countries, expressed concern even at the relatively modest Montreal targets. This history has been described at some length because it illustrates the great difficulty in reaching international agreements, even among nations that are scientifically well-informed, and over an issue that is of relatively minor importance to the industries of those nations. Despite the major environmental implications, it took a decade to gain an initial international agreement, and even then the agreement was inadequate.

For the United States, the CFC story illustrates the danger and sacrifice of unilateral action – the early and unselfish American controls were exploited by other nations less concerned about the effects of CFCs and more interested in gaining a competitive advantage. After the Montreal agreement there was concern that newly industrializing nations would exploit Western restraint by increasing their own CFC production. If gaining agreement about a relatively small matter such as CFCs is so difficult, how much greater from the U.S. point of view is the danger of unilateral action and the difficulty of negotiation over fossil fuels.

The position of those nations that were, reportedly, slow in accepting controls on CFC production is also interesting. For instance, it was British scientists who first measured atmospheric CFCs and who discovered the ozone hole over Antarctica. Britain is also a major user and producer of fossil fuels, with a strong coal industry and large oil and gas fields in the North Sea as well as ownership or part ownership of two of the largest multinational oil companies. Given the greenhouse competition between these fossil fuel interests and CFC production, it is interesting that the United Kingdom chose initially to protect a minor product of its important but smaller chemical industry, against the long-term interests of its fossil fuel producers and consumers. The reasons for this British stance are not obvious, but may include poor communication between scientists and civil servants, inadequate thought by the oil companies, and the generally low status of science in British intellectual society.

More recently, British policy on the environment has become much better

(a)

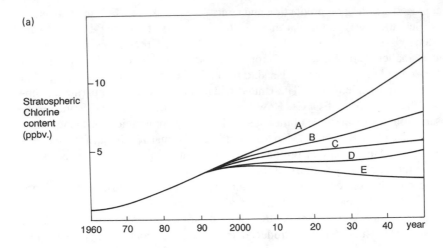

(b)

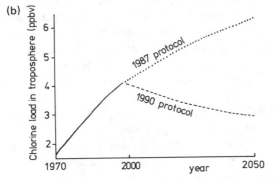

Figure 6.4. Projected atmospheric chlorine concentrations, which are closely linked to CFC emissions. (a) Chlorine concentrations in the stratosphere. Scenarios A through E are all optimistic projections. Scenario A assumes compliance by rich countries with the 1987 Montreal protocol on reducing CFC production and modest use of CFCs in the poorer countries. Scenario B assumes that all nations comply with the 1987 protocol, as does scenario C with somewhat more optimism. Scenario D assumes that all nations achieve a 20% reduction in CFC emission by 1990, a 50% reduction by 1994, and a 100% reduction by 2000. Scenario E assumes that this is achieved and that there is also control of methyl chloroform, CFC-22, and carbon tetrachloride. (b) Chlorine concentrations in the troposphere. The upper dotted curve assumes worldwide compliance with the 1987 Montreal protocol. The lower dotted curve assumes worldwide compliance with the 1990 London agreement. ([a] From Deutscher Bundestag 1988. [b] From media reports after the London agreement, quoting U.K. government sources. See also Table 6.3.)

Table 6.3. *Past and probable future concentrations of CFCs in the air*

	Concentrations (in pptv)				
	1960	1970	1980	1990	2010
CFC-11	11	60	173	275	286
CFC-12	33	121	297	468	544
CFC-113	0.2	2.3	15.3	51	72
CFC-114	0.2	1.4	3.8	7	10
CFC-115	0	0.2	2.1	5	6
HCFC-22	1	10	54	111	567
halon-1211	0	0.1	0.5	1.8	1
halon-1301	0	0.1	0.6	3.2	4.8
CCl_4	75	85	95	95	120
CH_3CCl_3	5	55	105	150	230
Total Cl_x (in ppbv)	1.01	1.55	2.52	3.5	4.5

Note: pptv = part per trillion (10^{12}). ppbv = part per billion (10^9). The estimates here assume a 95% phase-out of CFCs by the year 2000, with 50% by mass being added to the emission of HCFC-22 and related substitutes for CFCs, in concordance with the 1990 London agreement. Numbers here for Cl_x differ from those in Figure 6.4a because of differing assumptions, and refer to the stratosphere. Chlorine levels of 4.5 ppbv will cause significant ozone depletion in the stratosphere early next century, and stratospheric ozone levels will be particularly vulnerable to disruption if volcanic eruptions, such as the eruption of El Chichón in 1982, eject large quantities of aerosol into the stratosphere.
Source: Brausseur et al. (1990).

informed. The change in British policy on CFCs bore fruit in June 1990, with agreement in London to revise the Montreal protocol. This agreement considerably shortened the timescale of the elimination of CFC emissions, so that by the year 2000 most production of CFCs should be phased out and emissions (e.g. from old refrigerators and car air conditioners) declining.

This agreement was bought at a price: the cost includes the acceptance that substitutes should be produced that are less damaging than CFCs, but nevertheless not good for the atmosphere; it also includes significant technical financial aid from the richer nations to the poorer, especially India, to help in the conversion. The personal intervention of the British prime minister, who was excellently informed this time, played a major role in securing this agreement and in persuading the rich nations, particularly the United States, to aid the transition. It is noteworthy that although most Western political leaders now profess to support measures to protect the environment, few have taken the trouble to become scientifically well-informed, and most are content simply to support the often trivial issue of the day. In the long term, effective global action depends on well-informed leaders, as the British prime minister demonstrated over CFCs. The Montreal protocol and the leadership in its revision may serve as a model for future agreements on the more serious problems of CO_2 and CH_4 emissions; it is to be hoped that other leaders will become as well-informed as she was.

Whatever the reasons behind the behavior of individual nations, the history of the Montreal agreement demonstrates the need for the fullest possible information to be available to foreign policy makers in each nation. Environmental issues are likely to become a major cause of international disagreements and diplomats will in future spend much of their time negotiating over the complexly intertwined problems of trade and the environment. It is therefore important that national foreign offices such as the U.S. State Department be as well-informed as possible about the scientific issues of the environment and be staffed by scientifically literate personnel.

CFCs will take a century to disappear from the atmosphere even if emission is halted tomorrow. Carbon dioxide will be with us for much longer.

6.5 The coal industry

The next problem to discuss is coal. Coal was the foundation of modern industrial society. The roots of our wealth lie in coal. One of the world's more romantic and historical sights is a Rhodesia Railways 20th-class Garratt heavy steam locomotive thundering across the veld of Matabeleland hauling a train of coal, or a 15th-class engine racing past the elephants, drawing old colonial passenger coaches to the Victoria Falls. The Zimbabwe government, with a fine appreciation for history and eye for tourism, protects these magnificent engines and also the elephant and antelope they frighten. They are an excellent history lesson. The dirt is obvious, in the smoke from the locomotives, as is the impact of humanity on nature.

Burning coal produces heat energy, CO_2, and a wide array of other waste products including acid gases that contribute to acid rain, particulate matter in the atmosphere, radioactive radon in the atmosphere, and solid waste that needs to be disposed of on the surface. One million tons of oil produce about 42 petajoules of energy; one million tons of hard coal produce about 28 petajoules; and one million tons of soft coal produce about 14 petajoules. Per unit of energy, CO_2 emissions are in the rough ratio 0.75 for coal, 0.62 for oil, and 0.43 for natural gas. These numbers vary according to the quality of the coal, but in general coal produces much more CO_2 per unit of energy than oil or gas.

China produced 23% of the world's coal in 1986, marginally more than North America (also 23%), and the USSR produced 16%. Western Europe produced about 8% and other centrally planned economies, mostly in Eastern Europe, produced about 15%. For North America, Western Europe, and to some extent the USSR, coal is an optional fuel, but China and Eastern Europe have little present alternative. Any reduction in Chinese coal use must depend upon either much greater efficiency in Chinese use of energy in the face of rapidly growing demand, a switch to hydroelectric and nuclear power, major new oil discoveries, or economic collapse, as there is no foreign currency to pay for energy imports.

Similarly, in Eastern Europe there is no money available to import cleaner energy, unless Western help is given, as it will be within Germany.

Coal reserves are large, so large that they cannot properly be quantified, and widely distributed. At present rates of consumption, most major coal-producing nations have reserves for several centuries or more of burning, enough probably to make the planet a very different place from today. Many recent papers by resource economists envision the next century as a time when coal will be the primary source of energy for humanity.

Some forms of coal pollution can be reduced, but at the expense of markedly increased consumption per unit of energy output. In other words, reducing problems such as acid rain usually means increasing CO_2 output.

Control of coal emissions began seriously with the British Clean Air Act of 1956. This significantly reduced urban coal smoke in Britain, but its effects have now been largely counteracted by the less visible pollution from motor vehicles. The problem of coal pollution in Europe is now pan-European: a black acidic snowfall in Scotland in 1984 came from a mix of European pollution and British power station ash. In the United States, Clean Air Acts of 1963 and 1970 initiated controls on pollution, but by the late 1980s the continent-wide scale of the problem had become apparent and was reflected in Canadian concerns over acid rain caused by U.S. use of coal for generating electricity. Ironically, one of the chief successes of early clean-air legislation was to force industries and utilities to use high smokestacks in order to inject waste into the winds and so to disperse pollutants widely, on the assumption that the atmosphere has an "assimilative capacity," or ability to accept new ingredients without itself changing. Local intense pollution was transformed by this policy into regional haze over wide areas, but total output of pollutants was not reduced. As already mentioned, a height limit of 15 meters (50 feet) for smokestacks would have dramatic effects on local willingness to accept pollution.

More recently, there have been attempts to reduce the problems of coal use. One measure has been to use scrubbers to remove various pollutants from coal smokestacks, at the general cost of higher CO_2 emissions. Another has been to burn coal with a low sulfur content. This reduces SO_2 emissions and hence acid rain. Coal in the U.S. northern Appalachian and Middle Western coalfields tends to have more sulfur than coal in central Appalachia and the Rocky Mountains. Pyrite in coal, which contains 30–70% of the sulfur, can be removed through crushing and washing. However, low-sulfur coal may rise in price as demand increases, and more intensive cleaning is more difficult and expensive.

Another option is fluidized-bed combustion. Finely divided coal is burned in a granular bed of limestone or dolomite, together with ash from previously burned coal. Sulfur dioxide reacts with the bed to form calcium sulfate, which can be removed. This method both reduces the SO_2 output of the waste gas and improves efficiency of combustion, but is difficult to apply to large units.

Flue-gas desulfurization systems remove SO_2 from the flue, typically by wet scrubbing, spraying a slurry of limestone or lime into the flue gas. This produces solid waste, often a wet sludge, that must be disposed of, thus removing the waste from one part of the environment, the atmosphere, and placing it some-where else. Flue-gas desulfurization is very expensive, but is the preferred way of retrofitting old coal-fired power plants. In some cases, flue-gas denitrification is also carried out.

There have been some extraordinary schemes for removing CO_2 from power station flue gases and pumping it into deep ocean water, with unknown environ-mental consequences. However, CO_2 is the main waste product of any process that obtains energy by oxidizing carbon in coal. Typical well-scrubbed flue gas consists by weight of about 25% CO_2, 61% nitrogen, 6% oxygen, and the rest water. Removing the CO_2 can be attempted by a process known as the alkanolamine-monoethanolamine process (MEA), which strips CO_2 out of the gas. The removed CO_2 can then be transported as high-pressure liquid by pipe-line to old oil fields and gas fields and injected at depth. However, the cost of this procedure is very high, doubling the cost of electricity generated even as-suming suitable fields are available for the dumped CO_2. Injection is difficult because the acid gas attacks pipes.

In reality, emissions from coal-fired power stations remain very dirty. In the mid-1980s in Europe, the United Kingdom, which is heavily dependent on coal for its electricity, annually emitted about 3.7 million tons of sulfur dioxide. In contrast, pre-unification West Germany was emitting about 2.7 million tons de-spite having a richer, larger economy. Czechoslovakia annually emits over 3 million tons, though much smaller and less rich, and former East Germany about 4 million tons. Travelers on the main railway line north from London to Edin-burgh can see one after another the giant coal-fired power stations and the envi-ronmental damage caused by the coal mines that feed them and the gases they produce. Under European Community pressure, the United Kingdom decided, as the largest emitter in the Community prior to German unification, to desulfur-ize flue gases on several power stations, but later this decision was reconsidered under pressure from potential investors after a decision to sell, or privatize, the public utilities. Had the full desulfurization program been implemented, it would have cost several billion dollars. However, without this expenditure, it will in the near-to-medium term be easier to scrap the coal-fired stations completely. A final option, often proposed by political leaders, is to accept acid rain but to counteract its effects by adding lime to lakes and streams. This procedure does improve conditions in some lakes badly affected by acid rain, but it can damage forests and appears only to have a temporary effect. Furthermore, liming lakes does not help dying forests or reduce CO_2 emissions.

Other waste products from coal are less discussed. Desulfurization produces a nasty sludge in large quantity. Radon is a common by-product of coal-fired power stations, which are typically more radioactive in normal operation than nuclear power stations of equal output. Coal burning also produces large quantities of ash that have to be disposed of in landfill or in radon-emitting building blocks or by other methods. In the United States, 570 acres per year are used for disposing of flue-gas desulfurization waste, and 1440 acres a year for ash dispersal. The United Kingdom dumps some ash at sea, with severe effects on marine life. Coal mines are also major sources of waste material, and old coal-mining districts are often a dangerous terrain of mine dumps liable to landslip. One slip in Wales killed 116 children in school and 28 adults. Coal mining is a lethal business worldwide, and single mine disasters can claim several hundred to thousands of lives. Injuries are common in coal mining, as is disease following long exposure to coal dust. Finally, coal mining and coal burning, especially in China, give rise to CH_4 release to the air from gassy mines and inefficient burning.

To sum up, coal is an extremely unattractive fuel. Its use emits CO_2 in large quantities, more so than oil or natural gas; it is also the cause of acid rain on a large scale; it produces huge quantities of undesirable solid by-products; it kills and injures many miners. Given that the capacity of the human community to accept the burning of fossil fuel is limited, it would be better to abandon coal entirely, as soon as possible.

6.6 Oil, natural gas, and hydrogen

Oil is the main source of energy for Western Europe, North America, Japan, and the poor countries. The USSR depends roughly equally on natural gas and oil, with significant coal use also. Worldwide, oil is the chief source of humanity's energy and of much of the atmospheric pollution, though as a source of CO_2 it may be equaled by coal. Natural gas is less widely used than oil and only has a minor role in transport, but it is regionally important, especially in the Soviet Union. Per unit of energy, the use of natural gas produces less CO_2 than oil or coal, but the significant CH_4 losses from gas transmission make it a major greenhouse source.

Hydrocarbons have been known since biblical times, but their large-scale use as an energy source dates back only about 130 years, to early wells drilled first in Ontario and then in Pennsylvania. Half of all known global oil reserves originated during a 20–30-million-year time span in the Cretaceous. Much of this is in the region around the Persian Gulf. The western Siberian gas fields, which overlie Cretaceous sediments, similarly dominate the world's natural gas reserves.

The world's hydrocarbon reserves remain large. At present rates of reduction, proven reserves will supply the world with oil for more than 30 years. Much oil has not yet been found, so it is probable that there are, at current rates of usage, reserves for at least 50 and more likely 100 years to come. In the United States, however, reserves are much smaller and will only last one to two decades. Worldwide, at present production rates, natural gas reserves that are already proven will last nearly 60 years. If likely discoveries are included, there is enough natural gas for one to two centuries of use at present rates. These estimates do not include the vast reserves of heavy oil and methane in permafrost and oceanic hydrates. Excluding political interruption of supply, the world as a whole has an adequate supply of hydrocarbons for at least another generation; that is, if the atmosphere allows. However, in the United States and Western Europe, reserves will be largely depleted by the year 2000, and these nations and Japan will have to either import or find substitutes for hydrocarbons – a serious peril to their security.

The pollutants produced by oil use have already been described in detail. To recapitulate, they include CO_2, which is distributed worldwide, CO, NO_x, tropospheric O_3, and various locally distributed pollutants, especially in smog. No action can be easily taken to reduce CO_2 emissions except to reduce vehicle use and to make vehicles run more efficiently. But a variety of devices, including catalytic converters and lean-burn engines, can be used to reduce other noxious emissions. Three-way catalytic converters, which have been in use in the United States and Japan for some years, can remove 97% of nitrogen oxides and 90% of hydrocarbon emissions. However, they are easily damaged by lead in gasoline and many may not function effectively. As mentioned, some countries (e.g., Switzerland) now monitor the effectiveness of their action, refusing to license vehicles that fail stipulated tests. Lean-burn engines also reduce emission, with the additional advantage that they tend to be more economical of fuel, in contrast to catalytic converters which tend to increase fuel consumption. Compared to traditional engines, lean-burn engines may use up to 15% less fuel; in contrast catalytic converters on traditional engines reduce combustion by-products but emit more CO_2.

Natural gas (mostly CH_4) typically produces less CO_2 per unit of energy output than either coal or oil, but at the risk of large CH_4 losses in transmission. Gas supplies include "sweet" gas, typically poor in sulfur (e.g., in the North Sea) and "sour" sulfur-rich gas (e.g., in some wells in western Canada). Transmitting and burning this gas produces variable amounts of CO_2 (depending on the composition of the gas), CH_4 (depending on transmission losses), SO_2 (depending on how sweet or sour the gas is), and minor components such as radon if the gas is burnt immediately. The CO_2 output per unit of energy is about two-thirds that of oil, or three-fifths that of coal, so in CO_2-greenhouse terms natural gas is a preferable fuel.

However, the CH_4-greenhouse effects of transmission losses are very substantial. In the United States and the United Kingdom gas losses may average 2–5%. In the Soviet Union they are probably very much higher, possibly as high as 10%. Each 1–2% loss of gas (CH_4) very roughly has the same immediate greenhouse effect as burning an amount of oil or coal that is equivalent in weight to the total amount of gas that is successfully transmitted. This means that, in terms of the immediate greenhouse effect, natural gas may be much more strongly warming the planet, per unit of energy delivered, than coal or oil. Over about eight years, the 1–2% CH_4 loss is converted to CO_2, which has a much smaller effect, so the long-term impact of natural gas use is less than that of generating energy from oil or coal. To summarize: in the short term, natural gas use has a strong greenhouse impact; in the long term, it has a smaller impact than alternative fossil fuels.

Methane release from natural gas transmission has other, wider effects on tropospheric chemistry. Because of the complex interplay between CH_4, CO, O_3, and OH, abundant CH_4 leakage from natural gas transmission may slow the destruction of CO or promote increased levels of tropospheric ozone, both effects damaging to vegetation. There is also a distant but real risk of a marked reduction of OH because CH_4 is too abundant, which may have widely unpleasant consequences.

In the next few decades, natural gas may be an important interim means of generating energy during long-term programs for reducing greenhouse emissions. High-pressure gas supplies to power stations can be built so that losses are negligible. Combined-cycle gas power stations have thermal efficiencies of about 45%, compared to 38% for coal-fired stations. In the short-to-medium term, replacement of coal by natural gas as a main supplier of electricity would do much to reduce CO_2 emissions and to eliminate acid rain.

Oil and natural gas are the main currencies of energy in the world, especially in areas – North America, the United Kingdom, and the poor nations – where electricity is little used for public transport. Hydrocarbons can power individual vehicles; they can be carried to remote locations and used there; they can be refined to provide a high enough energy density to be useful as aircraft fuel; they can be stored reasonably safely; and they are enormously flexible, capable of providing heat, air conditioning, light, and especially transport. In short, hydrocarbons are wonder fuels, capable of being used wherever, whenever, and however they are needed.

There is only one obvious comparable alternative to hydrocarbons: hydrogen. Hydrogen can easily be generated, electrically, from water. It is a high-density currency of energy that rivals hydrocarbons in many ways, and surpasses them in some. Like hydrocarbons, hydrogen can be stored and transmitted relatively easily. It competes directly with hydrocarbons in most uses: it can fuel ships, trains, remote engines, and factory plants. It can be turned back into electricity with ease. Many vehicles, especially taxis, now run on propane gas: engines can be devised to run vehicles on hydrogen. Similarly, aircraft can operate efficiently on hydrogen because of its high energy density.

The chief problem with hydrogen use is safety. Humanity has had a century to become used to handling gasoline and natural gas, both dangerous and explosive fuels. The extensive use of hydrogen, a highly flammable gas, would require yet stricter and different standards of safety. Most probably, new types of fuel tanks would be needed – for example, cellular tanks that are rapidly self-sealing and possibly cryogenic. However, hydrogen spills do have the useful property of rising rapidly into the air, unlike gasoline.

The chief advantage of hydrogen is that it is clean. Properly burnt, it produces only water, from which it was originally made, although nitrogen oxides would also be likely in exhaust gases. As a currency of energy in a society that is attempting to reduce pollution, it is therefore ideal: energy goes into water to produce hydrogen; hydrogen carries that energy until it is liberated in a hydrogen-powered engine; and the return product is water, closing the cycle. Of course, this is ideal. In reality, the vehicle being powered may crash, with a loud bang, and the hydrogen-fueled jet engine will produce nitrogen oxides. Nevertheless, it is an attractive alternative to oil and gas.

Perhaps the most suitable first use of hydrogen technology is in aircraft and ships. Air travel in particular is using rapidly increasing amounts of fuel as more and more people choose to travel long distances for business and pleasure. Airlines are familiar with technological innovation and high degrees of safety. Using hydrogen as fuel is a challenge and will need much careful design of fuel tanks and delivery systems, but it is certainly possible and would significantly reduce the problems of the atmosphere.

This discussion of the merits of hydrogen as a currency of fuel has implicitly assumed that the hydrogen can itself be manufactured in a way that does not pollute the air. If not, the pollution is merely transferred, for instance, to the site of the power station that makes the electricity used to produce the hydrogen. Sources of electricity that do not produce CO_2 include solar power, tidal power, hydroelectric power, geothermal power, and nuclear power. They are discussed later. Biomass can also be used to generate electricity, but the environmental demands are high. Alternatively, it may be advantageous to generate electricity by a little-polluting fossil fuel, such as natural gas supplied by pipelines tightly controlled against leakage, and to use that electricity to produce hydrogen for transport in a system that in total is cleaner than existing oil-fueled vehicles, or to make the hydrogen directly from natural gas. Finally, there are other means of producing hydrogen, not via electricity, that may become economical. Hydrogen, therefore, is a very attractive currency of energy that merits much more attention.

Another possible currency of fuel is alcohol, especially methanol. A model study of the use of methanol in Los Angeles showed that the results would include reduced low-level ozone formation and generally better air quality. The use of methanol in stationary combustion would also reduce NO_x emissions. However, formaldehyde emissions into the air might become a problem. Some nations, for example Brazil and Zimbabwe, already use alcohol-gasoline mixes, which also improve air quality slightly, but the environmental cost in land devoted to growing crops for conversion to methanol is high. In general, hydrogen offers much better prospects than alcohol.

Electric cars, especially for urban commuting, offer enormous potential improvements in the urban environment. Their range and speed is limited by the inadequacy of modern batteries, but even with existing technology, electric cars could replace the gasoline car for most urban uses, especially if plug-in recharging parking meters are installed at office parking lots and on urban streets. Unfortunately, such facilities are at present very rare, and urban planning typically, though unintentionally, discriminates against electric vehicles, by assuming typical speeds and fuel needs of gasoline engines in road designs.

To conclude, oil and natural gas are major sources of CO_2 and other pollutants. Unless gas transmission is very strictly controlled against leakage, oil is probably a better fuel in immediate greenhouse terms than gas. However, if gas leakage can be reduced to minimal levels, then natural gas becomes the most attractive of the fossil fuels. For instance, electric power could be generated directly at the gas field and then transmitted by wire, allowing the closure of large numbers of oil- and coal-fired power stations. Unfortunately, long-distance transmission of power (e.g., from Siberian gas fields to Europe) is also wasteful of power, so the greenhouse saving is not obvious: more natural gas might have to be used, and CO_2 and CH_4 emitted, to offset the waste of electricity in transmission than would be saved by reducing transmission losses in gas pipelines. One possible use of the giant gas fields of western Siberia would be to generate electricity and hydrogen on-site, above the gas pools, and to reinject waste CO_2 extracted from flue gases into the deep strata below the gas fields. There are major problems in this, but it is not impossible and could supply the Soviet Union and Europe with large amounts of energy. Tactics such as these, together with substituting natural gas for oil, or both for coal, may substantially reduce greenhouse emissions. However, the fundamental problem remains. These are all fossil fuels; in most uses they add unwanted gases to the air. Inevitably, the use of fossil fuel will have to be reduced or terminated, probably within the next few decades.

Three decades from now we may be forced by depletion and atmospheric constraints to a society in the rich nations that uses no fossil fuels at all, with hydrocarbon use permitted only on a small scale in the poorest nations, where substitution is financially or technically impossible. The century of the internal combustion engine may be closing; in the next century we may commute less, using public transport or electric vehicles when we do. We may even restore some of our asphalt-clad wastelands to nature. For the United States and Europe such a shift would greatly improve national security. If they do not shift, and instead retain their dependence on petroleum, they will become increasingly insecure as their domestic oil supplies run out and they depend ever more heavily on Middle Eastern oil.

The future of oil and natural gas use depends on the intelligence of the oil industry. Consider the reaction of an imaginary oil executive: Sir Cornelius van der Plonk, head of the giant Royal British-Pecten oil corporation. Perhaps, when he realizes the implications of the greenhouse, he will order his powerful and well-funded advertising department to put every penny into campaigning against the greenhouse competition: against deforestation, against CFCs, and against coal and acid rain. Public-interest advertising is something all giant multinationals like doing to improve their image. Simultaneously, he will order his more cunning executives and research scientists to move discreetly into other energy strategies, particularly electric vehicles, solar energy, and the use of hydrogen (after all, the firm likes to be called an energy company, not an oil company). Finally, he will use all the political influence at his disposal to persuade the leaders of the major industrial countries to subsidize the shift into new technology, to protect domestic oil, and to limit imports of cheap oil from OPEC nations. Oil companies are interested in profit; with care, the shift away from oil can be made profitable and attractive to the multinational energy companies.

Such a transition will take decades, but it may be time to move away from oil as the prime currency of energy. In a well-managed Earth, oil can still be used, but in balance against recapture of CO_2 by forest. The prime use of oil may shift

Table 6.4. World sources of electricity, by region, 1987

Region	Fossil fuels		Hydroelectric		Nuclear		Geothermal		Total
	GWh	%	GWh	%	GWh	%	GWh	%	GWh
Africa	202,875	81	43,494	17	3,930	2	359	0.1	250,658
North and Central America	2,190,118	66	598,957	18	532,531	16	19,889	0.6	3,341,495
South America	89,442	23	293,983	75	7,436	2	0	0	390,861
Asia	1,403,797	69	356,248	18	268,126	13	6,400	0.3	2,034,571
Europe	1,427,274	54	490,834	19	695,606	27	6,863	0.2	2,620,577
USSR	1,258,115	76	219,825	13	186,984	11	0	0	1,664,924
Oceania	127,877	78	34,244	21	0	0	1,950	1	164,071
World	6,699,498	64	2,037,585	19	1,694,613	16	35,461	0.3	10,467,157

Note: 1 GWh = 1 gigawatt-hour. 10^6 GWh = 3.6 exajoules = 3.6×10^{15} joules at 100% efficiency. 1 gigawatt (GW) = 1000 megawatts (MW) = 1 million kilowatts (kW) = 1 billion watts (W). A typical domestic electric kettle uses 1–2 kW of power. One GWh would power between half a million and a million such kettles, for an hour, or 2 million 50W light bulbs during a 10-hour night.
Source: World Resources Institute (1990).

Table 6.5. *Comparative carbon dioxide release from production of thermal and electrical energy by various fuels*

Fuel	CO$_2$ release (in kg of carbon per 10^6 BTU of energy)	Ratio of CO$_2$ release to CO$_2$ release by methane
For thermal energy production		
Natural gas (methane)	14.2	1.00
Oil	21.0	1.48
Coal	26.0	1.83
Shale oil (minimum value)	30.0	2.11
For electrical power		
1978 mix of fuels (coal, oil, nuclear, etc.)	63.0	4.44
Coal in conventional coal-fired stations	72.0	5.07
Coal in fluidized-bed stations	72.0	5.07

Note: The units of CO$_2$ release have one kilogram of carbon per million British thermal units of energy. For those not used to the ways of the U.S. energy industry, 1 BTU = 1.055 kJ (kilojoules). See also Table 3.3.
Source: MacDonald (1982).

Table 6.6. *United Kingdom emissions of SO$_2$, NO$_x$, and CO by source, 1986 (in thousand tons)*

	SO$_2$	NO$_x$	CO
Domestic	200	57	483
Commercial, public service	130	45	11
Power stations	2600	783	48
Refineries	170	38	4
Agriculture (fuel use)	10	3	1
Other industry	570	179	75
Rail transport	0	37	14
Road transport	50	784	4748

Note: Most SO$_2$ – 2.6 million tons – is emitted by power stations, while road transport emits much NO$_x$ – 0.8 million tons – and CO – 4.7 million tons.
Source: Harrison (1990).

from energy to its role as the basis of the chemical industry. In the long run, it may be that the industrial nations will reserve the use of domestic oil, bought at protected prices, to sustain the supply of plastics (most of which should be recycled), solvents, and so on, but otherwise distribute all their energy via elec tricity (for static applications, trains, and local commuting) or hydrogen (for long-distance aircraft and remote machines).

Eventually – say by 2030 – industrial societies may have no option but to follow this path or some variation on it. The nature of the transition depends on present behavior. If there is rapid action to halt deforestation and CFC and coal production, the transition can be relatively gradual and painless or even profitable to the large energy companies. There is little doubt that Royal British-Pecten and all its real sisters, given political encouragement and cooperation from Detroit, can transform themselves and our society within a few decades. The benefits would be large, not only for the security of the planet as a whole, but also for the economic interests and security of the industrial nations. The price of oil would fall in the medium term; as the large industrial users switch away from oil, it might be possible, under agreement, for the densely populated poor nations briefly to exploit cheap oil if usage were balanced against reforestation. Ultimately, the climate determines oil usage, and it is in the nature of humanity to attempt to set the climate to its own liking.

6.7 Hydroelectric and tidal power

In contrast to fossil fuels, which contain stored solar energy that may have been collected millions of years ago, hydroelectric power is a nearly direct way of tapping solar energy, renewable from season to season as long as the Sun shines, water evaporates, and rain falls. Tidal power is even more immediately renewable, tapping the Earth-Moon system's potential energy from tide to tide and month to month, and will last for aeons. Both forms of power are, superficially, very attractive, because they do not have obvious drawbacks. However, there are many hidden problems.

Hydroelectric power provides a substantial amount of the world's electricity, ranging from 9.5% in the United States to two-thirds in Canada and virtually 100% in Zambia, Paraguay, and Norway. The technology is relatively simple, long-lasting, and reliable. Resources are large, and only a small proportion of the total potential hydroelectric power output has yet been captured. One estimate is that if all hydropower that could reasonably be exploited were distributed evenly, each person in the world would be supplied with as much electricity as the average Briton now uses.

The largest potential hydroelectric power sources are in Zaire, Canada, the United States, Brazil, China, India, and the Soviet Union. In Zaire, enormous quantities of power, about 120 gigawatts (GW), could be obtained from the Congo River (see Table 6.2 to put this in context: West Germany's installed power today is roughly 80 GW). Similarly, in Brazil about 100 GW can be supplied from Amazonia. Other rainforest nations, such as Gabon, Cameroon, and Colombia, also have huge hydroelectric resources. The boreal forest nations, Canada and the Soviet Union, have similarly large hydropower reserves. Canada can potentially harness about 140 GW, equivalent to the output of about 120 nuclear reactors and very roughly enough to supply about one-fifth of all U.S. electricity. If all readily usable Canadian and U.S. hydropower resources were used, perhaps nearly half of the U.S. electricity supply could be hydroelectric. In the Soviet Union, resources are even greater and could supply all the electricity currently used.

There are, however, very serious problems in using hydroelectric power. The least discussed is probably the danger of dam failure, which does occasionally occur though mostly in poor nations, so it is not reported. Typically, large dam failures, such as a failure in India some years ago, kill thousands of people. Warfare is also a problem: during the Zimbabwe civil war there were persistent fears that the Kariba Dam might be blown up. The flood that would have resulted would have severely tested the lower Cahora Bassa Dam in Mozambique and would have killed a large number of people downstream, perhaps up to a million. Another example is the detonation of a nuclear device over the Aswan High Dam, which could kill many tens of millions, destroying a nation.

More serious to the biosphere is the impact of the dams themselves, especially those that are sited either in high-rainfall areas or in wild areas that are attractive to dam builders because there is little local political opposition to the flooding. For instance, the Kariba Dam flooded about 5100 sq. km, including some of the wildest and most "natural" land on Earth, a precious fastness of rhino, elephant, and complex and previously undamaged ecosystems. The dam is a tourist attraction, but had this region not been flooded, it might now have become a greater economic asset as a game park. The dam on the Volta in Ghana, built at enormous cost with borrowed funds, flooded over 8000 sq. km, nearly the area of Cyprus, displacing 78,000 people from 700 towns and villages, the plan being to supply electricity cheaply to aluminum smelters. The two dams, put together, supply as much electricity as a large nuclear complex.

Canada has recently built a number of gigantic hydroelectric power schemes, especially in northern Quebec. In 1985 the installed capacity was nearly 60 GW, producing about two-thirds of Canada's total electric power. A substantial amount of this power is exported south to the United States, especially to the northeastern states, where Canadian hydroelectricity has helped to make up for the cancellation of nuclear power plants. The environmental price paid in Canada for the U.S. choice not to have nuclear power has, however, been extremely high. The dams, especially in northern Quebec, have altered the hydrology of a very large area, comparable to the area of a smaller European country such as Switzerland. Much forest (roughly, 10,000 sq. km) has been flooded. In most cases the trees were not cleared before flooding took place. This large store of carbon means that the area is probably, although there are no measurements, a large-scale emitter of methane, and will be for many decades or centuries as the carbon in the flooded forest and peats is converted anaerobically to CH_4 in the shallow lakes. In consequence, although this is a renewable energy project, it is probably also a major greenhouse source.

Perhaps less damaging but also disturbing are some of the rainforest projects in Brazil. The Tucurui Dam has displaced many thousands of people and flooded or disturbed three Amerindian reserves. The reservoir is roughly 20 meters deep when full, 16 meters deep when drawn down. It was selectively cleared of forest, partly by defoliants. Most probably, the majority of the decomposition of the tree trunks that remain will be aerobic, but some methane is likely to be emitted. Forest loss is considerable – the surface area of the Tucurui Lake (Fig. 7.5) is about 2200 sq. km when full, formerly rainforest, and deforestation has occurred around the lake. Eventually, the Tocantins River project, of which the dam is part, will involve 8 large and 20 small dams, forming a chain of lakes 1900 km long. The Tucurui Dam (see Fig. 7.5) produces 4 GW of power. Another dam, the Balbina Dam near Manaus, will flood 2000 sq. km, in which the forest is to be cleared by herbicide. By the year 2000 about 22 GW of power is expected to be produced from Amazonia, about equivalent to the output from 10–20 large nuclear complexes, at the cost of flooding many thousands of square kilometers of forest.

Hydroelectric dams in rainforest cause other problems in addition to the direct effect of destroying forest. Decomposition of vegetation in the large lake created by Brokopando Dam in Suriname produced major pollution problems, and workers had to wear gas masks for two years to protect them from hydrogen sulfide. During the decomposition the water became acidic, causing corrosion of the turbines. Later, weeds grew in the reservoir and had to be controlled by herbicide spraying. Other problems of large shallow reservoirs created by flooding rainforests include malaria, from the mosquitoes that breed around lake edges or on muds exposed during times of drawdown, and bilharzia (schistosomiasis), a serious parasitic disease that affects tens of millions of people in the tropics. Finally, and perhaps most devastating, the access and new development created by the dam-building projects opens up huge areas of rainforest to exploitation, including cutting, burning, and conversion to cattle ranches.

In Asia, where human population levels are generally higher, hydroelectric power can cause major disruption and dam failure can be devastating. The Pa Mong project in Vietnam is reputed to have involved the resettlement of half a million people. In Sri Lanka, the Victoria Dam, built by the United Kingdom, is reported to have displaced 45,000 people.

China, perhaps, best illustrates the complex trade-offs involved in deciding to use hydroelectric power. At present, about three-quarters of China's power is from fossil fuel, mostly coal; the rest is hydroelectric. Chinese hydroelectric potential is vast, over 200 GW, in contrast to the total present electricity supply of around 90 GW. Even a partial use of the hydroelectric resources could allow the coal-fired power stations to be shut down. Full utilization would allow China to shift virtually completely away from coal as a fuel, using hydroelectric power for heat, light, public transport, and industrial machinery. However, the cost would be high, not just in finance but also in environmental terms. Important ecosystems would be flooded, very scenic areas would be destroyed, and there would be a high risk of earthquake-induced failures that could kill many millions of people. Perhaps the best solution is not to install huge single dams, but cascades of smaller dams, weirs, and tunnels. Unfortunately, soil erosion from the deforested headwater terrains is so intense that in many rivers such schemes would have short lifetimes before being choked by sediment.

India, Nepal, Burma, and Thailand also have very large hydroelectric potential, especially in the Himalayas. Indonesia, too, has great potential. Unfortunately, in each case, there will be environmental damage.

The Victoria Falls in Zimbabwe well illustrate the conflicts. If a low weir were to be built along the lip of the falls, and the water diverted through tunnels into a power station, the effects on the biosphere would be minimal. The only damage would be to a few acres of rainforest growing opposite the waterfall and kept wet by the spray – these could easily be sustained by artificial mist. In contrast, the Kariba Dam further down the Zambezi River stores large quantities of water, providing security of supply in the dry season, but has destroyed a huge area of wildlife habitat. Humanity values the waterfall as one of the wonders of the world. It is beautiful in the extreme, and so far undamaged by the type of despoliation that afflicts Niagara. In consequence, Kariba was flooded but the Victoria Falls survive, a net loss to the biosphere but a preservation of beauty for humanity. In general, hydroelectric power offers these trade-offs. The best sites are also the most spectacular and beautiful to humanity. The poorer sites, where inundation is over a wider area, destroy large chunks of the biosphere, as at Kariba, in Amazonia, and especially in Canada.

Finally in this section comes a brief mention of tidal power. Certain regions, such as the Severn estuary in the United Kingdom and the Bay of Fundy in Canada, have massive tidal fluxes of water. In the Severn and the Bay of Fundy, tides range up to 40 feet or more (12–15 meters). Barrages across the Severn and part of the Bay of Fundy would supply large amounts of power to nearby industrial regions – southern Britain and New England. Unfortunately, the tidal ecosystems in these areas are rich and complex and would be almost wholly altered. Human beings would have to adjust too: tides at Boston would rise significantly if Fundy were exploited. It will be very difficult to assess the net effect of such changes, whether it is worthwhile producing power this way. Tidal power has one advantage over hydroelectric power in a changing world: it would not be affected by climate change unless sea level rises drastically. In contrast, many hydroelectric schemes, especially in the western United States, could be threatened by severe changes in rainfall distribution.

6.8 Solar and biomass power, winds, and waves

The Sun is the biosphere's source of energy, and enough sunlight falls on each square meter to power several light bulbs if the conversion to electricity were efficient. Even in Britain, which is damp and northern, the solar energy falling annually on the external surfaces of an average house far exceeds the energy expended inside it. Solar energy is therefore a very attractive source of power. Fossil fuels are simply stored solar energy. In tapping sunlight directly, however, there are several problems, including night, clouds, and the low density of the energy compared to traditional human sources.

Solar energy can be tapped in various ways and for many purposes. These include solar heating and air conditioning, solar generation of electricity, photolytic production of hydrogen, and biological gathering of solar energy in bio-

Table 6.7. *Hydroelectric power: selected countries*

	Total installed electric power, 1986, all sources (GW)[a]	Installed hydroelectric power, 1985 (GW)	Potential hydroelectric power (GW)
Africa			
Angola	0.6	0.4	23
Cameroon[b]	0.6	0.5	23
Congo[b]	0.1	0.1	11
Egypt	5.8	2.7	3.2
Gabon[b]	0.2	0.1	18
Ghana	1.2	0.9	2.0
Mozambique	1.8	1.5	15
Zaire[b]	2.2	2.1	120
Zimbabwe	1.5	0.6	3.8
N. and Central America			
Canada	98.4	57.5	141.4
Costa Rica[b]	0.9	0.6	9.1
Mexico	24.1	6.6	19.7
United States	719.4	88.0	183.2
South America			
Argentina	16.2	6.0	21.2
Brazil[b]	44.7	36.9	106.5
Colombia[b]	5.8	3.8	94.3
Peru[b]	3.6	2.0	60
Venezuela[b]	12.5	4.4	32.9
Asia			
Burma[b]	0.9	.2	30
China	87.0	26.5	219.6
India	54.7	15.1	100
Indonesia[b]	8.5	.5	81
Japan	173.3	34.3	81.9
Thailand[b]	7.6	1.8	20.1
Europe			
Austria	15.7	10.2	6.4
France	92.1	21.8	7.7
Norway	23.7	23.0	19.6
Sweden	33.1	15.7	11.3
Switzerland	15.2	11.4	n.a.
USSR	321.7	61.3	437.3
Australia	33.5	6.6	6.1
New Zealand[b]	7.4	4.3	5.5
Papua New Guinea[b]	0.5	0.1	29

[a] gigawatt (GW) = 1000 megawatts (MW) = 1 million kilowatts (kW) = 1 billion watts (W).

[b] Rainforest nation.

Sources: World Resources Institute (1988), *Encyclopedia Britannica Yearbook 1988*.

mass. Wind energy and wave energy are also solar. So is hydroelectric power, but it has been considered separately.

Sunlight warms water, which can then be used for domestic heating. The heat can be stored overnight, for example, in pools of water. A simple solar heating system typically has a heat collector on the roof of a house, heating water. The water can be stored in a tank at around 40°C or less and extracted either directly or via circulated air to heat the house. The system can also act as a supplementary source together with another power source such as electricity. It can also be used to warm the evaporator of a heat pump, and so to drive an air-conditioning system. In hot regions, such as the U.S. Southwest, solar air conditioning could much reduce the load on hydroelectric stations during hot weather, effectively conserving water in the reservoirs.

Electricity can be generated directly from sunlight in various ways. These include direct heating of thermic fluids (which is expensive), photovoltaic cells, and solar ponds. In recent years, substantial advances have been made in improving the efficiency of photovoltaic cells. In concentrator modules, using lenses or mirrors to focus sunlight onto the cells, in units that track the sun across the sky, efficiencies have reached 38%. Nevertheless, despite the hope of rapid progress, photovoltaic electricity is still very expensive, costing 35 cents per kilowatt-hour or more. Photovoltaic cells will probably become much less expensive in future, and large arrays in desert areas could provide enormous amounts of power with little or no penalty to the biosphere. The Sahara, for instance, could power Europe by day, or the U.S. Southwest could power North America's daytime activity. Unfortunately, it is very difficult to store large amounts of energy for nighttime use. Until it becomes possible to do so, perhaps with superconductors, or it is possible to exchange energy across time zones (e.g., from the Sahara to the Gobi Desert over the Bering Strait to the U.S. Southwest), there are major limitations on photovoltaic power.

Thermoelectric generation, using solar heating, is a promising technology. An 80-MW power plant in the Mojave Desert of southern California generates the bulk of the world's solar electricity. The plant has mirrors tracking the Sun and focusing light onto pipes filled with oil. The oil is heated above 400°C and then used to produce steam for electric generators. The cost of the electricity is about 8 cents per kilowatt-hour. In hot, sunny places, such plants could contribute greatly to electricity supplies at times of peak summertime loads.

Solar ponds have been developed in Israel and to some extent circumvent the problem of night and dull weather. In them, a salt gradient is imposed on a black-bottomed pond about one meter deep. This creates a density gradient that stops the pond from convecting. Hot water collects at the bottom and retains its heat at night and over cool cloudy days, insulated from the air by the overlying water. In one pool of 1200 sq. m a temperature of 103°C was attained. Such ponds can be used to generate electricity. One pond in Israel, with an area of 250,000 sq. m, is fitted with a 5-MW turbine connected to the national grid. A pond of one square kilometer can be used to produce, reliably, about 6.4 MW of electric power; a complex 12.5 km by 12.5 km would produce 1 GW, annually saving over 40,000 tons of oil. In some regions, solar ponds could be installed very cheaply on old salt lake beds, especially in California, Australia, parts of the Soviet Union such as the Aral Sea or the eastern gulf of the Caspian, and northern Africa (e.g., Tunisia, supplying Italy) and Botswana. Seawater can be used, so there is no water-supply problem in most areas. The cost is much lower than other forms of solar energy and the system is an extremely attractive option for countries with suitable geography. For instance, much of California's summer electricity supply could be provided this way. Ponds could also be used to preheat water for thermoelectric power. If salt lake beds are used, there is no loss to the biosphere.

Photolytic production of hydrogen can be carried out by the use of organic photosynthesizers. Bacteria or plant chloroplasts can achieve this, and in principle it could be a major source of hydrogen for transport: sunlight is free, water is common, and bacteria breed quickly.

Biomass is the most obvious way of collecting solar energy. In principle, growing plants in order to use them as fuel is atmospherically neutral, as the CO_2 liberated by burning equals the CO_2 taken up in growing the plants. In practice, the burning produces harmful by-products and, more important, the area used to grow the plants is removed from both the natural biosphere and the agricultural land inventory, so the net cost is considerable.

Two nations, Brazil and Zimbabwe, have used biomass to produce alcohol that has been used as a vehicle fuel. In both cases the schemes were successful, but it has been difficult for them to compete with the low oil prices of the late 1980s and with the rewards of using the land for other purposes, such as food. In general, biomass stores less than about 5% of solar energy, as a maximum. Usually the real figure is much less, 0.1–1.0% being typical. Given the world demand for food, it is probably not efficient to use land to produce energy or fuel in this way. However, there may be some exceptions. For instance, in many deforested regions the natural vegetation needs shade and cannot grow in the open when young. One way to reforest is to plant fast-growing exotic species to shelter the slower-growing local vegetation. After a few decades, the fast-growing species can then be cropped carefully and used as biomass, when the local trees are established.

Wind and wave energy can be used in some nations, such as the United Kingdom and the nations with coastlines on the Pacific. Wind power is relatively cheap to run but very erratic. The installations demand capital and are generally considered eyesores. Nevertheless, the potential is large and it can make a useful contribution to displacing fossil fuels in some countries. Wave power is somewhat more reliable, but still underresearched. Several methods of collecting it have been developed. One example is the Cockerell raft; another is known as the Salter duck. Both bob up and down on the waves, extracting kinetic energy. The potential for this type of energy gathering is very large. Unfortunately, wave power has high capital cost and low running cost. In a throwaway society, accounting standards demand rapid write-off of capital cost, over 20–25 years, although the installations may last for 40 years or more. As with nuclear power, this means wave installations are not favored: we prefer our creations to be disposable.

6.9 Geothermal energy

Geothermal energy resources are considerable, but not renewable in human terms. The extraction of geothermal power is a mining process, and it may take thousands or tens of thousands of years for the extracted region to heat up again. Furthermore, there is often considerable pollution around geothermal stations, either from gas or liquid discharge and from radioactive radon. Nevertheless, in some areas, geothermal power is an excellent alternative to fossil fuels. This is particularly true in regions with recent volcanic activity, such as California, New Zealand, Iceland, Italy, Nicaragua, and El Salvador (which generates about 40% of its electricity this way).

6.10 Nuclear power

Nuclear power is the most controversial alternative to fossil fuels. It is extremely attractive in that it is atmospherically clean provided there are no accidents. It bears a heavy burden of association with nuclear weapons and with the history of Chernobyl and Three Mile Island; it produces nuclear waste; it is expensive. Technology has produced the problems of global change, and in nuclear power, technology provides an answer, at least to the energy supply of humanity. Many environmentalists, perhaps the majority, reject nuclear energy for exactly this reason: it is technological, and it comes out of the same Pandora's box as the causes of our global problems.

Put most bluntly, the argument is whether it is better to use fossil fuels and accept the climatic consequences, or to use renewable but damaging sources such as hydroelectric power, accepting the flooding of forests and wild land, or to use nuclear power and accept the consequences for human beings. For example, the South African reactor complex near Cape Town produces about as much power as Lake Kariba. The argument centers on which is preferable, the destruction of wild land or the risk to the humans around the reactor.

6.10.1   Historical background and reactor design

Nuclear power was made possible by the work of Ernest Rutherford around the turn of the century, followed by the discovery of the neutron by his associate, James Chadwick, in 1932. Neutrons can be used to bombard atomic nuclei and make them split, releasing energy. Fermi showed that low-energy (slow) neutrons are in general more effective than high-energy (fast) neutrons in causing nuclear reactions, and high-energy neutrons can be most effectively slowed down by scattering collisions with light elements, such as hydrogen or carbon. This slowing-down material is called a *moderator,* and one of the first moderators to

Table 6.8. *Nuclear reactors*

Country	In operation (1987) Number of units	Megawatts	Under construction Number of units	Megawatts	% of national electricity supply
Argentina	2	935	1	692	13.4
Belgium	7	5,477			66.0
Brazil	1	626	1	1,245	0.5
Bulgaria	5	2,585	2	1,906	28.6
Canada	18	12,142	4	3,524	15.1
China			2	1,188	
Cuba			2	816	
Czechoslovakia	8	3,207	8	5,120	25.9
Finland	4	2,310			36.6
France	53	49,828	10	13,410	69.8
East Germany	5	1,694	6	3,432	9.7
West Germany	21	18,947	4	4,047	31.3
Hungary	4	1,645			39.2
India	6	1,154	8	1,760	2.6
Iran			2	2,392	
Italy	2	1,120	3	1,999	0.1
Japan	36	26,888	12	10,692	29.1
South Korea	7	5,380	2	1,800	53.3
Mexico			2	1,308	
Netherlands	2	507			5.2
Pakistan	1	125			1.0
Poland			2	880	
Romania			3	1,980	
South Africa	2	1,842			4.5
Spain	9	6,529	1	990	31.2
Sweden	12	9,646			45.3
Switzerland	5	2,932			38.3
USSR	56	33,616	28	25,098	11.2
United Kingdom	38	10,294	4	2,520	17.5
United States	106	92,982	13	14,844	17.7
Yugoslavia	1	632			5.6
World	417	297,927	120	101,643	

Note: During 1987 22 new reactors were connected to national grids, a rate of one reactor every 17 days.
Source: Baum (1988), from International Atomic Energy Agency annual report, 1987, GC (XXXII):835.

Table 6.9. *Inventory of longer-lived radionuclides that will be present in British high-level waste after 1000 years*

Radionuclide	Half-life (years)	Activity (Bq)
Fission products		
^{99}Tc	2.14×10^5	4.4×10^{15}
^{93}Zr	1.5×10^6	1.1×10^{15}
^{93m}Nb	13.6	1.1×10^{15}
^{126}Sn	1×10^5	2.0×10^{14}
^{79}Se	6.5×10^4	1.4×10^{14}
^{135}Cs	3×10^6	9.7×10^{13}
^{107}Pd	6.5×10^6	4.2×10^{13}
^{151}Sm	90	2.4×10^{13}
^{129}I	1.6×10^7	1.1×10^{13}
Actinides		
^{234}U	2.45×10^5	1.1×10^{13}
^{236}U	2.34×10^7	5.9×10^4
^{238}U	4.47×10^9	5.8×10^4

Note: Intermediate-level wastes contain other, short-lived radionuclides. These include (with half-lives): ^{60}Co (5.3 yr), ^{63}Ni (100 yr), ^{137}Cs (30 yr), ^{154}Eu (8.6 yr), ^{239}Pu (2.4×10^4 yr), ^{240}Pu (6.5×10^3 yr), ^{241}Pu (15 yr), ^{241}Am (433 yr), as major constituents. Waste reprocessing removes Pu, etc., for further use. 1 Becquerel is one disintegration per second.
Source: Chapman and McKinley (1987), from data of Hill and Lawson (1980), NRPB-R108, National Radiological Protection Board, London.

be used was heavy water, deuterium oxide, which is a trace component of all natural water and can be concentrated.

In a nuclear reactor, a neutron hits an atomic nucleus in the fuel and causes it to split, or fission, producing fission products and more neutrons that go on to hit other atomic nuclei in the fuel. For example,

$$^{235}U + {}^1n \rightarrow {}^{147}La + {}^{87}Br + 2\ {}^1n$$

is a typical reaction, in which a neutron (^1n) hits an atom of uranium 235 (^{235}U), producing lanthanum 147 (^{147}La) and bromine 87 (^{87}Br) and also two more neutrons that can be used to hit other uranium fuel atoms. The fission products are radioactive themselves and decay, mainly by emitting beta particles (electrons). For instance, after the reaction above, ^{87}bromine decays to the gas krypton, and then mostly to ^{87}rubidium and finally to stable strontium. This rubidium-to-strontium decay is the same as the decay that geologists use to date old rocks, such as those in the Canadian Shield, because nearly all rocks, especially granites and volcanic rocks, have naturally radioactive rubidium in them.

Table 6.10. *Waste arising from a projected 240 GW-per-year nuclear power program*

Type of Waste	Volume (cubic meters)
Reprocessing waste	
Vitrified high-level waste	1,120
Concentrates, resins, etc.	10,780
High-alpha technological wastes	13,900
Low-alpha technological wastes	27,800
Operational waste	
Ion exchange resins	35,100
Other (filters, solids, casings, etc.)	9,140
Decommissioning waste	
Intermediate-level waste (steel and concrete)	21,140
Low-level waste (steel, concrete, etc.)	62,300
Secondary wastes (resins, concentrates, etc.)	12,660

Note: For comparison with high-level wastes, volume of Olympic-sized swimming pool (50m × 25m × 2–3m) = 3125 m³. For comparison with low-level wastes, volume of building containing that pool, 80m × 40m × 15m = 48,000 m³.
Source: Chapman and McKinley (1987), from data of Swiss Project Gewaehr 1985, NAGRA NGB-85-09, NAGRA, Baden, Switzerland.

To build a nuclear reactor three components are needed: fuel, moderator, and coolant. The fuel must fission to supply neutrons, and the most common fuel is ^{235}U, which is a natural isotope of uranium. Most uranium, however, is ^{238}U – natural uranium is 99.285% ^{238}U and 0.715% ^{235}U. Uranium 238 cannot be used directly as a fuel, but in reactors it captures neutrons and produces ^{239}U, which then decays to plutonium 239, which can be used as a fuel. The natural isotope ^{232}Th of another element, thorium, behaves similarly to make ^{233}U, which can also be used as a fuel. In fast reactors high-energy neutrons (the word *fast* refers to the energy of the neutrons) are used to breed plutonium from ^{238}U. A fast breeder reactor uses fast neutrons to breed plutonium slowly.

The moderator is typically a light element or a compound that contains a light element. Three materials are used: water, heavy water, or carbon (graphite). Most reactors have solid fuel elements arranged as a lattice within the moderator, which can be either solid or liquid. Ordinary, or light, water is the most common moderator, but must be kept under high pressure (150 atmospheres or more) to prevent it from boiling, in the same way that an automobile radiator is pressurized to prevent boiling. The pressures and temperatures used are much lower than in steam in coal-fired power stations. Heavy water (deuterium oxide) is similar to light water, but because of the properties of deuterium, natural, unenriched uranium can be used as fuel. This means that complex fuel enrichment facilities are not needed. However, heavy-water-moderated reactors have to be much bigger than reactors that use ordinary water. This means that they are too big to be used for military propulsion (for instance in nuclear submarines), but such reactors have the advantage that if any unexpected power increase occurs, it takes longer for the huge mass of moderator to heat up. Canadian CANDU (*Can*adian *d*euterium-*u*ranium) reactors use heavy-water moderators that allow the normal Canadian molluscan decision-making process (a long committee meeting, broken by a good lunch) in the event of a problem. In contrast, some other reactors need quicker responses by duty operators. However, the production of heavy water is very expensive.

A third common moderator is graphite (carbon). Graphite reactors operate at fairly high temperatures. Prolonged neutron irradiation of graphite causes the buildup of stored energy in the graphite, and this energy, called Wigner energy, must be released carefully. If it is released too rapidly, the graphite heats and burns in air or even CO_2, as at Windscale (Sellafield) in Britain in 1957 when an early plutonium-producing reactor was disastrously destroyed. Modern British graphite-moderated reactors are designed to anneal the stored energy continuously, so they cannot behave disastrously in the same way. Another problem with graphite occurred in one of the Soviet Union's RBMK reactors, which operate with graphite at temperatures in excess of 700°C. In these conditions, graphite ignites spontaneously on contact with air (just as anthracite burns). If there is an inadequate containment structure (as in many Soviet and Western reactors), the result is Chernobyl.

Coolants have to carry the heat away from the reactor and must not themselves become radioactive when passing through the core of the reactor. Gaseous coolants include carbon dioxide and helium. Carbon dioxide is a good coolant, but does produce minor amounts of carbon 14 (^{14}C), which can be vented to air. This annoys atmospheric chemists attempting to study CH_4 and CO_2, because the ^{14}C from reactors partly counterbalances the reduction in atmospheric natural ^{14}C caused by fossil fuel use and makes calculations imprecise. Helium is ideal as a coolant, but difficult to use in engineering terms and expensive. It is unreactive and not a radioactive hazard, but it is used in only a very few reactors.

Water or heavy water cool most reactors. In most reactors, the coolant water is pressurized. In boiling-water reactors, however, the water is allowed to boil in the core, which operates at somewhat lower pressure. The water must be kept very pure to prevent radioactive contamination in impurities. Liquid heavy water can be used as a coolant, but it is expensive and, like ordinary water, must be kept under high pressure. Fast reactors use no moderator, and the best coolant is liquid sodium, which does not need to be pressurized – a valuable safety feature, given the chemical reactivity of sodium.

Control systems in reactors are used to bring about small changes in reactivity needed to start up the reactor, to change power levels, and to shut down the reactor if need be in an emergency. The most common control system consists of rods of neutron-absorbing material such as boron, held above the reactor.

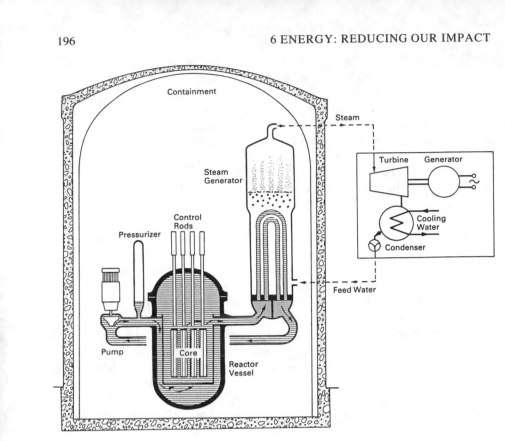

Figure 6.5. Pressurized-water reactor. (From Rippon 1984.)

These rods can be driven into the core, or dropped in if there is a power failure. In the Soviet RBMK design, control rods are power-driven into the core in an emergency. They are not dropped rapidly under gravity. They take 20 seconds to enter.

6.10.2 Reactor types

There are many different reactor types, with different mixes of fuel, moderator, and coolant. Most are light-water reactors, especially in the United States, France, Germany, and Japan. A few, especially in Britain, are gas-cooled. Canada alone uses heavy-water reactors that allow natural uranium to be the fuel.

Pressurized-water reactors (PWRs) are small, with cores that can be only two to three meters across. This makes them suitable for nuclear submarines. Liquid ordinary light water is the moderator and coolant, which must be kept under pressure. The fuel is slightly enriched in ^{235}U (to 3%), which means that major enrichment facilities have to be set up, at considerable cost. A typical modern PWR has an output of about 1.3 GW.

Boiling-light-water reactors are similar, but the water is at lower pressure and allowed to boil. The typical electrical output is around 1.2 GW. Boiling-water

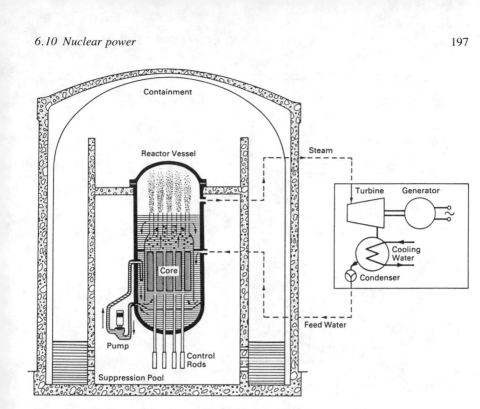

Figure 6.6. Boiling-water reactor. (From Rippon 1984.)

reactors are designed with good negative void coefficients: this means that if reactor power increases (e.g., in an accident) and more boiling occurs, the reactivity of the reactor drops. This is a self-stabilizing effect, an important safety factor.

Gas-cooled reactors are characteristic of the British nuclear program, begun in 1956. Early reactors used uranium fuel clad in a magnesium alloy called Magnox. Later, in advanced gas-cooled reactors (AGRs), oxide fuels clad in steel were used. Initially, the AGRs proved to be construction disasters (exacerbated by inept planning) but later AGRs have much better records. Intrinsically, this is an excellent reactor design. Each AGR complex produces about 1.25 GW of power, with a thermal efficiency of about 42%, high for power stations. AGRs have a graphite moderator and carbon dioxide coolant gas at about 40 times atmospheric pressure. In the Magnox reactors, a loss-of-coolant accident would lead to a slow increase in core temperature. The AGRs operate at higher temperatures and need a back-up cooling system in an emergency, although the rate of temperature increase is relatively slow compared to some other designs. The presence of a large mass of hot graphite is a cause for concern. High-temperature gas-cooled reactors can also be thermal (not fast) breeder reactors, cooled by helium, which is an inert gas. Only prototype reactors have been built, but they have given promising results and have good intrinsic safety against coolant loss.

Graphite-moderated, water-cooled reactors of the RBMK type are a Soviet

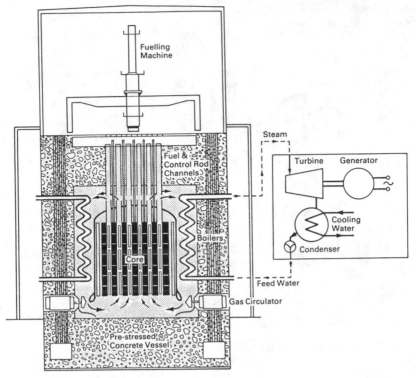

Figure 6.7. Advanced gas-cooled reactor. (From Rippon 1984.)

design that is now being terminated. They have several interesting features. These include inadequate control rods, no secondary shutdown system, a graphite moderator at 700°C or more which, as mentioned above, will burn spontaneously in air, a positive void coefficient (the opposite of boiling-water reactors) so that at lower power the reactor can very rapidly "runaway," and a large but inadequate containment structure. Finally, it is now known that in the RBMK design the operators were able to override safety systems and to withdraw more control rods than permitted and therefore operate the reactor beyond the danger point.

Canadian CANDU reactors have a development history that is very different from all other reactors. Canadian nuclear research began nearly a century ago, with Rutherford's work in McGill University, but the present nuclear program began in the early years of the Second World War, when much British expertise was transferred to Canada. At the end of the Second World War, Canada and Britain were cut off from the U.S. military research. Britain decided to build its own nuclear weapons, while Canada became the only nation to maintain a major nuclear research program that was not directly linked to military needs. Consequently, it has produced unusual reactors based on a long civilian tradition.

At the end of the Second World War, Canada alone had the ability to produce large quantities of heavy water. It used this ability to develop the heavy water

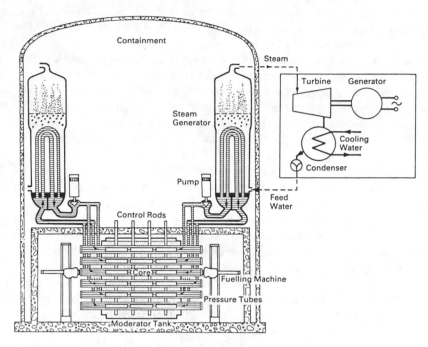

Figure 6.8. CANDU heavy-water reactor. (From Rippon 1984.)

reactor, which uses natural, unenriched uranium fuel. Eighteen of these CANDU reactors have been installed in Canada and have had extremely good operating records. Recently, smaller modular reactor systems have also been designed, for a variety of users.

The reactors use heavy water both as moderator and coolant, though the two are physically separated. The moderator is contained in a cylindrical steel vessel 26.8 mm thick (a little over an inch), called a calandria. This vessel is 6 m (about 20 ft) long and 7.1 m (about 25 ft) in diameter. Passing horizontally through the calandria are 380 tubes, each containing another inner tube carrying fuel, with helium between the two tubes. This arrangement allows the moderating heavy water in the calandria to be maintained at relatively low temperature (65°C) and pressure; the coolant is under 88 bars (about 87 atmospheres) pressure, 250°–293°C (in contrast, a PWR operates at 150 bars, 295°–330°C, rather more severe conditions). The pressure tubes are both a safety feature and a problem in the design. In contrast to many other reactors, in which depressurization would affect the whole system, in the CANDU the failure of a pressure tube is a local event that does not lead to the need to shut down the reactor. However, this is also a problem: instead of one big disaster, many small failures occur and in the early CANDUs the pressure tubes had shorter-than-expected lifetimes. This was very costly. In more recent reactors, the pressure tubes are expected to perform better: time will tell. The void coefficient is slightly positive (see the discussion of Chernobyl), but because of a phenomenon exhibited by ^{238}U known as Doppler broadening, the net effect as temperature rises is a negative feedback. The Pickering CANDU reactor complex has eight reactors, each producing 515 MW of power, a total output of 4.1 GW.

In the CANDU reactor, an accidental rise in temperature (e.g., from a failure in the coolant system) would cause the moderating heavy water to warm. If the fuel has been in the reactor for a long time and contains significant ^{239}Pu, increase in moderator temperature could increase the rate of fission of the ^{239}Pu. However, the moderator has a large mass and heats slowly, no matter how rapid the power increase is. Thus the time available to cure any problem should be long.

The British steam-generating water reactor (SGHWR) is related to the CANDU reactors, but uses light water as the coolant, which is much less expensive than heavy water. In principle this is an efficient reactor design. A prototype reactor, built at Winfrith in Britain, has a good record, though there were problems in designing a larger commercial model, as well as safety concerns.

Apart from their initial wartime beginnings, and profits from the sale of plutonium for weapons, heavy-water-moderated reactors have not had the support of major military investment (e.g., for submarine power plants or for weapon-making). That the CANDU reactors have survived, given the comparatively modest level of Canadian support, implies that their advantages of natural fuel, operational efficiency, and general safety are considerable. Recently, however, the government of Ontario, where many of the reactors are installed, has declared a moratorium on future construction.

6.10.3 Nuclear safety and accidents

The main radiation hazard to most people is background radiation, especially from domestic radon in houses. Nuclear reactors in normal operation, coal-fired power stations, metal-smelting factories, mines and quarries, and the medical uses of radiation, including X rays, all add to the natural background.

Radiation dose is expressed in millisieverts, a unit that takes into account the nature of the radiation. For most people, the natural background radiation is around 2 millisieverts each year. Of this, about three-quarters is from terrestrial sources, including the natural elements that make up our bodies, and domestic sources such as radon from buildings and ground. Cosmic rays add to this, creating radioactive carbon 14 in the air and striking our bodies. Natural exposure to radiation varies greatly; people who live in granitic regions or in sealed houses are much more exposed to radon; people living at higher altitude (e.g., in Denver) are more exposed to cosmic rays. In Switzerland, the annual average exposure is about 2.4 millisieverts, but natural levels vary to above 5 millisieverts a year.

Artificial radiation exposure is mostly medical, mainly from X rays. For many people, this exposure is about 0.4 millisieverts a year, or about one-fifth of the natural level. Added to this is nuclear fallout from bomb testing (0.02 millisieverts) and exposure from nuclear power (0.001 millisieverts). Other major sources include coal-fired power stations and heavy-metal-smelting factories. Many in dividuals, especially miners and medical technicians, can have much higher exposure.

The main releases of radioactivity caused by nuclear power come from mining, enrichment (minor), reactor operation or failure, and waste disposal. Mining operations can involve major radon gas risks, which can be countered by masking, ventilation, or automated mining. The appreciation of radon risk in uranium mines and processing has helped to improve safety standards generally, because all mining (e.g., of coal and gold) and the occupation of most buildings and homes involve radon risk. Fortunately, most radon has a very short life (a few days) and the risk can be reduced effectively in mines and in buildings and homes. In overall risk of accident, uranium mining is considerably safer than coal mining, by a factor of about 100. Conversion plants change the output of the mill at the mine (yellowcake – U_3O_8) to a compound, uranium hexafluoride, which is used in enrichment plants. The fluoride, once enriched in ^{235}U, is then converted back to uranium dioxide (UO_2) for fuel, the waste being depleted in ^{235}U. After irradiation, fuel can be reprocessed to recover uranium, plutonium, and fission products. The uranium may be somewhat depleted but can be reused; the plutonium needs long-term storage for use eventually in fast reactors, or disposal; the fission products need disposal. Until recently, especially in military programs, safety regulations in nuclear fuel plants were much too lax, and some workers were exposed to high levels of radiation. In some plants these lax standards were in effect until comparatively recently (up to about 1980 or even later; one hopes not now). These exposures have been associated with cancer and leukemia in children of workers in the Sellafield plant in England, although the case is not proven, and the link has not been seen in U.S. plants. More recently, annual dose rates have been brought down to a few millisieverts, about double the natural background levels, but it is not clear what the long-term effects of such dosage or similar exposure rates in many radon-rich houses will be. Another possible cause of childhood leukemia is exposure to toxic chemicals and to low-frequency electric fields, or acquiring viruses when moving to new communities, for instance near power stations. The degree of risk is still not understood, but there appears to be a small but significant risk of contracting a variety of cancers from 50–60-hertz power and from exposure to power distribution lines. Chemical exposure may also have been significant in causing disease in workers and their families.

Most public danger comes from nuclear accidents. The first important accident occurred in 1952, in Canada. Fortunately, and with some courage, it could be cleaned up to the extent that the reactor was restored to service (one of the cleaners was Jimmy Carter). The three major accidents have been at Windscale (now renamed Sellafield) in England, at Three Mile Island in the United States, and at Chernobyl in the Soviet Union. Some poorly reported military accidents have also taken place as well as what was reputedly a major waste accident in the Soviet Union.

The Sellafield-Windscale accident took place in 1957. The reactor was one of Britain's two original air-cooled plutonium-producing reactors, a very early design. The reactor had a graphite core, in which stored or Wigner energy built up

Table 6.11. *Relative risks of nuclear power, coal, hydropower, and conservation*

Source	Hazard	Deaths per GW per yr
Nuclear	Routine discharge	0.05
	Accidents (Chernobyl)	1.0
	Mining	0.1
Coal	Cancer, etc., in general public	10
	Mining accidents	2
	Mining disease	8
Hydro	Dam failure	10
Conservation	Radon buildup in houses	100

Note: These figures are very dependent on circumstance: a major coal-mining disaster, Chinese coal-induced illness, another major reactor accident, or a large dam failure could all increase the relevant figure greatly. For instance, one of Britain's worst disasters was the Aberfan landslip, in which waste from a coal mine tip buried a school filled with children. Serious dam failures or accidents have occurred in India and the Alps. Conversely, radon buildup in houses can be controlled and the danger from this source can be greatly reduced. Oil and natural gas accident figures are not given here, but are also high; for instance, several hundred people recently died in a major Soviet gas explosion, and many people are killed in oil-rig or exploration disasters. Geothermal projects around volcanoes have experienced a variety of accidents, some with serious fatality. A comparative study of safety at the Sizewell B PWR station and the Canvey petrochemical industrial complex, both in the United Kingdom, showed disturbingly high risks for the petrochemical complex, including a significant probability of a major accident capable of killing tens of thousands of people. For Sizewell B, the risks were four orders of magnitude less. For the victims, whatever the cause of the accident, statistics such as these have little meaning: energy should be as safe as possible.
Source: Bennet and Thomson (1989).

during irradiation. At temperatures below 100°C the energy accumulates; at higher temperatures it is released by self-annealing in the graphite. What happened was that the energy was suddenly released and this caused the graphite temperature to rise to such a high level that the graphite burned and the reactor's core was destroyed. The result was a major accident and a fire that released large quantities of radionuclides into the air. These radionuclides included, notably, iodine-131, which is notorious for passing from air to grass to cows to milk to children's thyroids. The release from Sellafield-Windscale contaminated the nearby pastures of Cumbria, and was significant as far away as Holland.

Table 6.12. *Common types of nuclear reactor*

Name	Fuel	Moderator	Coolant	Country of origin	Comment
Pressurized-water reactor (PWR)	Uranium dioxide (slightly enriched)	Water	Water (non-boiling)	United States (also France, Japan, West Germany, etc.)	Most common reactor type, for power and military use in submarines. Three Mile Island reactor
Boiling-water reactor (BWR)	Uranium dioxide (slightly enriched)	Water	Water (boiling)	United States	
Gas-cooled reactor (GCR)	Uranium (natural)	Graphite	Carbon dioxide	United Kingdom	
Advanced gas-cooled reactor (AGR)	Uranium dioxide (slightly enriched)	Graphite	Carbon dioxide	United Kingdom	
Pressurized heavy-water reactor (CANDU)	Uranium dioxide (natural)	Heavy water	Heavy water (nonboiling)	Canada	
Steam-generating heavy-water reactor (SGHWR)	Uranium dioxide (slightly enriched)	Heavy water	Water (boiling)	United Kingdom	Terminated
Graphite-moderated boiling-water reactor (RBMK)	Uranium dioxide (slightly enriched)	Graphite	Water (boiling)	USSR	Chernobyl reactor
Fast breeder reactor	Uranium dioxide and plutonium dioxide	none	Liquid sodium	United Kingdom, France, USSR	
High-temperature gas-cooled reactor	Thorium-enriched uranium carbide	Graphite in fuel-moderator mixture	Helium	United Kingdom, United States West Germany	Thermal breeder

Graphite fires are very difficult to put out, and the public was not initially made fully aware of the seriousness of the Sellafield-Windscale incident while the fire burned. Nor was the danger of the contamination by fallout made fully public, although a large quantity of milk was destroyed. Indeed, some documentation did not emerge for 30 years. Apart from the immediate effects on the victims, who were not advised to take adequate precautions, the longer-term consequence of this secrecy was that there was no full debate about the methods for dealing with a graphite fire and especially for handling the problems of fallout. Had the debate taken place, and had its conclusions permeated the safety thinking of authorities responsible for radiological protection, the response to Chernobyl might have been much more effective. As it was, the important British experience was less help than it should have been to those dealing with Chernobyl. The moral is that there should be full disclosure of information as soon as possible when an accident occurs, even at the risk of some public panic.

The Three Mile Island accident took place in 1979 in one of a complex of two pressurized-water reactors near Harrisburg, Pennsylvania. The accident originated with a failure of the feed-water system. The reactor shut down, but fission-product decay heat was not transferred out of the reactor because the feed-water pump had cut out. A progressive series of failures followed – everything that could go wrong did – and some time later the water in the core boiled and part of the core was uncovered for several hours. The high temperatures that resulted seriously damaged the fuel elements, resulting in the release of volatile radioactive fission products into the sump on the floor of the reactor containment. From there the contaminated water was pumped to an auxiliary building and the radioactive volatile products, especially xenon, iodine, and krypton, escaped into the air. The iodine release, most important biologically, was about 16 Curies (a Curie is a unit of radioactivity, defined as 3.17×10^{10} disintegrations per second).

In the reactor, the high fuel temperatures caused a reaction between steam and zirconium, producing hydrogen, which escaped into the containment building. It was thought at the time that there was a risk of a major hydrogen explosion in the reactor, but this was unlikely as insufficient oxygen was present in the pressure vessel. Hydrogen did burn, 10 hours after the beginning of the accident, causing a pressure rise in the containment building. About 16 hours after the start of the accident the primary coolant pumps were restarted. The water flow through the reactor cooled the core and brought the accident under control, but the reactor had been destroyed.

Three Mile Island was caused by incompetence, malfunction, and inadequate design, combinations that will inevitably occur again, to a degree that will vary according to the efficiency of regulating bodies. The long-term effects of the accident are still being debated, but most probably there will be one to two extra cancer deaths in the region caused by the release of radioactivity. In contrast to the reactor at Sellafield-Windscale, which had no containment, the containment structure held. The financial cost of the lost reactor and the clean-up was huge.

Furthermore, the accident virtually terminated U.S. construction of civilian nuclear power reactors, although the U.S. military submarine program continued.

The Chernobyl accident, which took place in 1986 in the Ukraine, was more deadly. The Chernobyl No. 4 reactor was an RBMK graphite-moderated, water-cooled design. As mentioned in section 6.10.2, this design has several features of concern, including a positive void coefficient at low power, allowing rapid runaway, inadequate control rods, no secondary shutdown system, a hot graphite moderator that can ignite on exposure to air, and a poor containment structure. The containment structure at Chernobyl was substantial, including a 3-meter (10 ft) thick biological shield as a cap, and outer walls 1.5 meters (5 ft) thick. Although, overall, the containment was below Western standards, it compared with older reactors in the West.

The accident was initiated at 01.23.10 hr on 26 April 1986, by operators who were attempting to test the emergency core-cooling systems. Because of their actions, the problem of the positive void coefficient was maximized and bulk boiling took place almost simultaneously throughout the core. The operators decided to trip the reactor by driving control rods into the core at 01.23.40 hr, but by 01.23.43 hr the reactor power rose exponentially in a fraction of a second. The power rise shattered the fuel elements and five seconds later there was a steam explosion, followed by a hydrogen explosion. The reactor containment was shattered and the reactor burned for 10 days: a worst-case accident.

Thirty-one people died as a direct result including some who courageously gave their lives deliberately to reduce the blaze and to save a wider death toll. Some 135,000 nearby residents were evacuated, and several thousand will probably die early. Further afield, the dosage was of the order of a few millisieverts. On average, a few millisieverts is a small dose, but averages are of little meaning when fallout is concerned. Some regions had very little fallout at all; in other areas, where rainfall occurred as the radioactive air mass passed over, there was intense contamination. In general, the Polish and Scandinavian governments responded quickly but there was inadequate detailed meteorological monitoring and not enough long-term monitoring of the specific zones of high contamination. Some governments in Western Europe appeared more concerned about soothing their populations than protecting them, repeating the mistakes made after the accident at Sellafield-Windscale.

The excess cancer mortality will be difficult to distinguish from other cancer mortality (this part of Eastern Europe is appallingly badly polluted from other industrial operations) but may reach several thousand additional cancer deaths. Numbers of deformed or handicapped children may be born in the area, or pregnancies may miscarry. The area around Chernobyl is now evacuated. In death or cancer toll, comparable polluting events include the Spanish rape-oil disaster of 1981–5, which probably had rather greater mortality (400 dead, 20,000 affected), and the 1984 chemical accident at Bhopal, India, which was more immediately devastating.

The two major nuclear accidents that have occurred in civilian reactors have been described in detail because there are close parallels between these catastrophes and the similar but worse problems that have occurred in other forms of energy generation. The Chernobyl reactor was of a design that, one hopes, would not have been licensed in Western countries. Similarly, other energy industries in the poorer or centrally planned nations have been designed with little thought for safety, and large-scale disasters involving major loss of life have occurred. Death tolls include 5000–15,000 in a hydroelectric dam failure in 1979 in India. Coal-mining accidents in China are common, many with a few fatalities each, a few with hundreds or thousands killed. The Tangshan earthquake in 1976 killed tens of thousands of coal miners. A methane gas blowout in the Soviet Union recently killed several hundred people. In comparison with these, the nuclear record, even including the figures for Chernobyl, is excellent. This, of course, would sound hollow to a Ukrainian farmer. In the rich nations, the civilian nuclear record of human injuries and deaths is nearly impeccable when compared with the deaths and injuries in coal-mining accidents and the poor record of the oil industry.

Chernobyl, which means "Wormwood" (Revelations: 8:10–11), had other consequences. It removed a large area of contaminated farmland from the human economy. This area is comparable to the area of a large hydroelectric dam and reservoir or to the area spoiled by coal mining over some years. The Chernobyl area is not, however, removed from the biosphere. The plants and animals that grow in it may include mutant forms in numbers that are much larger than in areas where the radiation is at natural background levels. Nevertheless, the abandoned Chernobyl area represents a net gain to the natural biosphere, unlike the areas flooded for hydroelectricity, covered by coal waste, or polluted by oil spills. The area is now a nature reserve, one of the largest in the European part of the Soviet Union. A second important consequence was that the Chernobyl disaster accelerated the changes taking place in the Soviet government.

The experience of nuclear accidents is better than the record with other forms of power but still not acceptable. The most obvious improvement is to design reactors so that they are passively safe – they can be walked away from. Passive reactors are the opposite of Chernobyl; they are designed so that the system becomes safer, not more dangerous, as it is abused. An example is the Canadian Slowpoke design, a small reactor that reduces its reactivity as it heats, or the Swedish PIUS design. Another improvement would be to design containment structures for large reactors that are far better than the inadequate containment at Chernobyl. The Three Mile Island containment *did* hold. Ideally, containment buildings should be designed to withstand the maximum internal and external risks. Internal risks include hydrogen explosions and steam explosions. External risks include earthquakes and aircraft impact. Most reactors today should withstand aircraft impact. One Canadian study investigated the cost of putting reactors underground. This would be expensive, but it would make containment much more secure. Andreí Sakharov also suggested this. If all large reactors were deeply buried, the atmosphere would be much, much safer.

6.10.4 Nuclear waste

It is commonly said that the problem of nuclear-waste disposal is unsolved. Geologically speaking, this is incorrect, as the geochemistry of the problem is relatively well understood. Indeed, the study of nuclear-waste disposal has indicated how poorly designed is most chemical and industrial toxic-waste disposal. Politically, however, the statement is correct, because no communnity is willing to accept nuclear waste, though all communities are willing to accept electricity made from fossil fuels that discharge their waste into the most fragile of all disposal repositories, the air.

Nuclear waste consists of fission products, such as rubidium, strontium, cesium, zirconium, niobium, and samarium isotopes, and also actinides such as uranium, plutonium, and thorium (Table 6.9). The actinides are potentially useful as fuel for breeder reactors, and therefore for most of them the best option is safe long-term storage, not disposal. A few actinides can be used, such as in domestic smoke detectors, which contain small radioactive sources, and in medicine. The fission products need permanent disposal.

There have been many studies of the disposal of civilian, nonmilitary, waste. These include major research projects in Switzerland, Sweden, and Canada. The simplest designs emplace the waste products in tunnels in rock. More complex designs involve the emplacement of waste in canisters in boreholes drilled from the tunnel floors. Because the volume of nuclear waste is small compared to the volume of rock handled by most mines, such a repository would have a surface impact comparable to that of a small mine.

Research into waste disposal has concentrated on the fates of the various radioactive isotopes as the canisters age and chemical species enter the environment. Interestingly, some of the elements involved – for example, rubidium-strontium, uranium-lead, samarium-neodymium – are those that have been studied for many years by geochronologists. Most of the geological dating of old rocks, upon which geologists depend for their investigation of the geological histories of the Precambrian shields, is based on the understanding of how these elements behave in geological conditions. Other elements, such as zirconium, are also used by geologists as immobile trace elements in geochemical studies.

Two studies, the NAGRA study in Switzerland and the KBS-3 study in Sweden, modeled the long-term history of nuclear waste under a variety of release scenarios and types of containment failure. In the central scenario in the Swedish study, maximum release, coming in about 10 million years, was around one-thousandth to one-ten-thousandth of a millisievert per year of exposure on the surface. The Swiss NAGRA study gave similar but lower results. In Canada, a safety assessment code known as SYVAC produced similar doses, about one-ten-thousandth of a millisievert in about 10 million years being the most likely outcome of waste disposal in crystalline rocks. To put this into context, the natural background level, as stated above, is about two millisieverts, but varying greatly.

The Swiss, Canadian, and Swedish studies have concentrated on disposal in hard crystalline rocks. Other approaches, especially in the United States, include disposal in salt, clays, basalt, or volcanic ash. Each of these types of repository has its advantages, but typically the advantages are balanced by equal problems, such as disruption of salt domes and ingress of fluid or porosity in volcanic rocks. Some military disposal programs have been dangerously negligent. Furthermore, safe long-term storage in crystalline rocks is compatible with later retrieval if the actinides are needed for breeding and further burning. In general, proper storage in crystalline rocks at depth is expensive, but more attractive than some other disposal strategies.

Natural analogues support these conclusions. The best-known example is the Oklo uranium deposit in Gabon, Africa. Over 2 billion years ago, a number of natural nuclear reactors reached criticality here, using 1000–2000 tons of uranium fuel and producing about four tons of plutonium. The reactors operated at about 400°–600°C and 800–1000 atmospheres of pressure. Study of the waste products shows that the uranium isotope ratios, which measure the degree of burn-up, are constant locally but characteristically different between reactor hearths, indicating that little or no uranium has been transported from hearth to hearth. Typically, transport distances have been limited to a few meters. Another natural analogue is the uranium deposits of the Athabasca basin in Canada. These are extremely rich in uranium but typically have very little surface sign, although they are at very shallow level. Consequently, they are difficult to find; it is not possible simply to fly over the area with a radiation counter and find uranium. The uranium has stayed in place, unless shifted as boulders by ice sheets. This natural evidence therefore agrees with the Swiss and Swedish conclusions. More generally, geological evidence shows that although uranium and its products are mobile, in the correct setting they can be contained safely.

One possible problem with nuclear-waste disposal is geological accident, such as the eruption of a volcano or a large earthquake rupturing the repository and allowing fluids to enter. For this reason, active tectonic zones are not suitable locations for waste disposal; the western United States, Japan, and large areas of China are unsuitable. Workers on the subject have stressed the advantages of building regional storage centers for high-level civilian nuclear waste, under strict control. Thus for Europe, one or two well-designed storage sites in geologically safe areas would suffice. In North America, a single site in a Precambrian shield terrain could handle the continent's waste. These storage facilities would not need to be large; the physical size of the waste is small – tiny when compared, for instance, to the volume of high-level chemical waste.

Perhaps the strongest lesson from the intense research into nuclear-waste disposal is that current disposal techniques for nonnuclear, chemical wastes need much improvement. In general, the standards demanded of chemical disposal are much lower, though the waste may be more dangerous. Most cities have landfills from which toxic elements such as mercury or cadmium rapidly enter aquifers. Similarly, clean-up practice around petrochemical installations often involves deep-level injection of chemicals into aquifers, with little appreciation of the final fate of the water in the aquifers. These are local problems and outside the scope of a book about global change, but the implications of the detailed work on nuclear-waste disposal is that industrial-waste disposal in general needs to be greatly improved.

6.10.5 Economics of nuclear power

Nuclear power has been attacked strongly on the grounds that it cannot compete with fossil fuel power in an open market. It has also been attacked on the grounds that it cannot compete with the financial benefits of conservation of energy. Both attacks are justified, but in limited circumstances. The comparison with fossil fuels is based on the assumptions of high interest rates and that the cost of fossil fuel is purely the extractive cost, with no environmental charge. The comparison with conservation is correct in the rich nations in the early stages of conservation, when conservation is inexpensive. In the poor nations, for most people energy use is very low; even though this energy is wastefully used, there is little scope for conservation unless education and technical knowledge improve, as well as wealth.

The comparison with fossil fuel costs was well illustrated by a controversy in the United Kingdom in 1989 that led to the abandonment of plans to build further nuclear reactors in Britain after the completion of the Sizewell B plant. The following costs refer in part to the unusual British experience from a mix of gas-cooled reactors, and are probably above Canadian but below U.S. experience. They do, however, appear to reflect the real costs of nuclear power, which have been long hidden.

During a public inquiry into the construction of pressurized-water reactors, the base figure for nuclear electricity of 2.24 pence per kilowatt-hour unit was given. (One British penny was then worth roughly 1.6 U.S. cents. One kilowatt-hour is equivalent to boiling a North American electric kettle for roughly an hour, or using a 100-watt light bulb for 10 hours.) This cost was based on the assumption of a 5% real return on capital to the public utility. The British government resolved to privatize the utility, and the return expected was raised to 10%. Because of the high capital cost of a nuclear reactor, this raised the base cost to 3.8 pence. In contrast, on the same basis, coal-fired power cost 3.4 pence. Inflation set in during the inquiry and the figures increased to 4.2 pence per unit for nuclear power against 3.7 pence for coal-fired power.

Next, the decision came to depreciate the nuclear installation over 20 years, not 40 as previously planned (40 years is about the lifetime of a modern PWR). This was done to meet British accounting standards; the effect is equivalent to being forced to pay off a house mortgage in 20 years, not 40. At the same time, the full costs of all research were loaded onto the nuclear station. These decisions, particularly the change to a 20-year depreciation, raised the cost of nuclear electricity to 5.4 pence per unit. Then, decommissioning and waste disposal were costed in at realistic levels, adding another 0.2 pence per unit and raising the cost to 5.6 pence. Finally, it was assumed that the reactor would not perform to specifications and that there would be major delays and cost overruns, raising the total cost to 6.25 pence per kilowatt-hour. In contrast, the cost of 3.7 pence per unit for coal-fired power assumes little research, no environmental cost, and no cost overruns. Shortly after these figures were published, these assumptions were confirmed by a government decision to relax plans for flue-gas desulfurization.

It is worth repeating that short-lifetime depreciation, which raises nuclear costs so much, also makes it uneconomical to gather energy from other nonfossil sources, such as from waves. For both nuclear and wave energy, initial costs are high but running costs are low; for fossil fuels the opposite is true, and the standard short-lifetime method of accounting therefore favors fossil fuels. The British decision to abandon its nuclear program was not based on environmental issues. To the contrary, it was a triumph of the throwaway society against long-term investment.

A comparison between the figures is interesting. Nuclear power costs 6.25 pence per unit, or, if costed over 40 years, about 5.5 pence. In a long-term, standardized program where research costs are spread over more reactors, the cost is about 5 pence per unit. Decommissioning, surprisingly, is not a major cost when distributed over 20 or 40 years. The main cost, that of capital, depends on accounting procedure for write-off time. It is the assumption of short lifetimes and high interest rates that makes nuclear power, solar power, wave power, and various other nongreenhouse sources of energy uneconomical. This is especially true in countries such as the United Kingdom, which in recent times have had much higher interest rates than Japan or Germany, discouraging capital-intensive long-term investment. At interest rates below about 7%, very roughly, nuclear power is financially advantageous; above this, coal is cheaper assuming environmental costs are ignored.

Coal-fired power costs 3.7 pence a unit. With full flue-gas control, it would cost about 4.5 pence a unit. If CO_2 were also scrubbed and reinjected into the ground (e.g., in North Sea hydrocarbon fields), the cost would be around 9 pence a unit or more. This is probably close to the real cost of coal-fired electricity. Oil-fired electricity may be slightly less expensive. In the short run a cheaper alternative to reinjecting CO_2 might be reforestation on a scale that counterbalances CO_2 emissions. This would be a much cheaper way of making fossil fuel use acceptable: on this basis the cost of coal-fired power would be around 5 pence per unit, more or less depending on how much was spent on research and on reforestation. Over time, costs would rise as replanting moved from the more easily reforested regions to more difficult areas.

One celebrated analysis, by Keepin and Kats (1988b), examines the nuclear option for reducing CO_2 emission. In one scenario they consider a sixfold expansion of world nuclear-power capacity between 1986 and 2025, which would mean an increase of about 1350 GW from 270 GW in the mid-1980s to 1600 GW. (To put this in context, total installed world electrical power capacity at present is about 2600 GW, of which hydropower accounts for about 600 GW and fossil fuels and nuclear most of the rest.) The sixfold expansion could be achieved by adding one gigawatt of nuclear power (a large reactor complex typically produces one to two gigawatts) every 7.5 days throughout the period, in contrast to the 1980–6 rate of one gigawatt every 14 days.

The Keepin-Kats assumption of a linear building rate for new reactors conflicts with how technological skill and economic wealth normally increase – they behave strongly exponentially. In addition, the building capacity of society in 2020 may be much greater than today. So the one-gigawatt-every-7.5-days assumption is an overstatement of the impact of the immediate change needed. Nevertheless it correctly implies that the rate of expansion of nuclear capacity would have to increase sharply if nuclear power is to increase to 1600 GW by 2025. Their more realistic exponential model assumes around 100 GW a year of new installations until after the year 2000, followed by a rise in installations to around 400 a year in 2025, when around 100 reactors each year will be coming to the end of their lives.

The cost of this, in present-day U.S. currency, would be around $150 billion a year (assuming $1.5 billion per GW, which is probably a low estimate). The Keepin-Kats analysis thus illustrates the high cost of switching to nuclear power, and indeed, the high cost of maintaining the present level of electrification worldwide, even with fossil fuels, given the costs of mining coal and of replacing existing fossil-fuel power stations as they age. Electricity is expensive, which is why most Chinese and Indian people have very little.

The $150-billion-a-year price tag for the nuclear option, however, should be adjusted for the very large saved cost of not replacing fossil-fuel power stations and the lower cost of nuclear fuel over the next 40 years. The net *extra* cost is difficult to assess, given the various accounting and depreciation methods. Furthermore, it is difficult to predict the cost effects of changes in reactor design and of future improvements in construction techniques. Consequently, the net extra cost of switching from a fossil fuel economy to building a nuclear capacity of 1600 GW may be low, or may be as high as, say, $100 billion a year. For comparison, total worldwide military expenditure is around $700 billion a year. Britain spends roughly $5 billion a year on candies and chocolates.

In Keepin and Kats's analysis, the effect of building a 1600-GW nuclear capacity would be to reduce CO_2 output by a cumulative amount of 2.335 gigatons of carbon in CO_2 a year. However, they point out that an even greater reduction in CO_2 output, 2.918 gigatons a year of carbon in CO_2, could be obtained by an equal investment in efficiency; not in nuclear power. They therefore conclude that it is cheaper and better to invest in reducing energy use than in nuclear power stations. The Keepin-Kats calculation is probably optimistic in its appraisal of the cheapness of efficiency: the deeper efficiency cuts into electricity use, the more expensive it becomes. Even so, it is a strong demonstration of the value of reducing CO_2 emissions by investing in efficiency first, before nuclear power or any other option.

Two nations have already moved down the nuclear path, France and Belgium. In early 1974, at the time of the oil crises, France embarked on a major nuclear program based on PWRs. In 1974 and 1975, 13 GW were ordered, in 1976 and 1977 12 GW more and from 1978 to 1981 5 GW more each year. This rate of conversion to nuclear power, per capita, is considerably faster than the more relaxed timescale of the Keepin-Kats 2025 scenario discussed above. The cost, however, has been enormous. By the end of 1987 the debt accumulated by the French electric utility was 221 billion francs, roughly $35 billion. At 10% interest, say, this means a real cost of $3.5 billion a year. This cost to the French nation can be contrasted with the annual French military budget of around $20 billion. The debt is heavily subsidized. One estimate – though probably wildly exaggerated – of the real cost of French nuclear electricity is as high as 46 U.S. cents per kilowatt-hour (contrast this with the British costs above).

Despite this high cost, the impact of nuclear power when seen through the air is striking. In east-central England, the air is gray-brown with the high, dispersed effluent from coal smokestacks. For mile after mile in central England the landscape, as seen from a London–Edinburgh train, is dominated by huge coal-fired power stations emitting smoke into the dull air. In Scotland, the train passes close to a clean bright building in green sheep fields: an AGR nuclear station. In Europe, similar contrasts can be seen. To an observer on a Swiss mountain, the air to the northeast, where eastern Germany and Czechoslovakia lie, is dark and heavily polluted. To the west, the air is clean, at least until the French have a nuclear accident. French fossil fuel use in vehicles is still large, but in recent years there has been major investment in electrically powered public transport and in very rapid trains. The enormous financial debt from reactor construction remains, but France today is a much cleaner and more wealthy country than its neighbor, the United Kingdom, which has chosen to rely heavily on coal for its electricity. The difference between British and French air is nearly a third of a million tons of coal burned daily. France has significantly and successfully reduced its CO_2 output, to the benefit of all the world. The United Kingdom has not, nor have most other industrial nations.

6.10.6 Fuel reserves

The world's uranium reserves are large and will sustain a major nuclear program for many years. Just as the global energy needs could be met solely from coal for centuries, or from oil and natural gas for many decades, there is no sign of a shortage of uranium. Indeed, the opposite is more likely to be true, because discovery is only beginning.

For most mineral resources, the history of discovery and exploitation has followed a consistent pattern. Initially, small rich deposits are found. Then, with improving techniques and the exploration of more remote areas, giant deposits are discovered. Finally, at a late stage, many minor deposits are exploited. Costs tend to follow the pattern too. Initially, they are high but moderated by the profitability of the rich early deposits. Then they fall in real terms as the giant deposits are exploited. Finally, prices rise as the deposits are exhausted and supply is reduced, enabling the exploitation of the small, expensive, residual deposits. Copper, for instance, initially came from high-grade deposits; then the large porphyry deposits of North America and Chile were found, bringing down prices, and these giant deposits are still being exploited. Coal, similarly, is now obtained from giant fields. Hydrocarbon extraction is in a more mature phase, at the stage where it is unlikely that new giant deposits will be found, but the giant deposits of Arabia and Siberia still contain many years of reserves.

In contrast to oil, uranium reserves appear to be at an early stage of discovery. Some of the largest and highest-grade deposits, such as those at Olympic Dam, Australia, or in the Athabasca basin in Canada, have only recently been found. To date, most exploration has been based on relatively simple techniques. Many large deposits, comparable to the rich Cigar Lake deposit, must lie undiscovered underground in the Athabasca basin. In many countries no exploration has been carried out at all. Consequently, if uranium discovery follows the normal pattern of other resources, it is likely that future costs of extraction will fall as new giant deposits are found. If so, the resources are probably adequate to sustain the world's nuclear reactors for some time, probably many decades to centuries, without recourse to breeder technology. If breeder reactors are used, the time period is multiplied tenfold.

One rich mine of nuclear fuel is a giant deposit, so far little explored but extraordinarily cheap to exploit, whose use would significantly reduce the proliferation of nuclear weapons. This is the military stockpile of nuclear devices, especially in the United States and the USSR. These devices contain uranium and plutonium, and mechanically age as they are kept. The threat of nuclear holocaust would be much reduced if a large proportion of the world's nuclear weapons were cast into reactors, to burn.

6.10.7 Proliferation

The most worrying of all aspects of nuclear power is probably not accident but deliberate construction of nuclear weapons using plutonium diverted from civilian reactors. To do this, sophisticated fuel expertise is needed. As long as an irradiated fuel remains unreprocessed, the plutonium in it is inaccessible. An alternative route to a nuclear weapon is to obtain uranium 235 from natural uranium by enrichment, so even without nuclear power, nuclear weapons can be made by any country with access to uranium and with enrichment expertise and capacity. One process in particular, laser enrichment, may make it easy for nuclear weapons to spread to countries without nuclear power, unless that process is carefully regulated. This, it has been pointed out, is the poor nation's route to a bomb. An even cheaper route is to use petrochemical expertise to build chemical weapons, often called ''the poor man's nuclear bomb.''

To date, the risks of proliferation have been reduced, with partial success, by tight controls on reactor fuels, by international agreement. With a few exceptions, these constraints have been accepted by virtually all countries, and are preconditions to the export of nuclear power reactors. However, it is obvious that any nation with enrichment skills does not need to divert civilian nuclear fuels in order to make a bomb. Therefore the best way to stop the proliferation of nuclear weapons is to control not only fuel reprocessing but also enrichment facilities. In this context, the CANDU reactors are very attractive, because they operate with natural uranium fuel. Canada is the only nation that has possessed a sophisticated nuclear industry for half a century and yet has wholly renounced nuclear weapons. However, proliferation can only be securely guarded against if all spent fuel is returned, under rigorous control, to a few well-monitored repositories.

6.10.8 The future of nuclear power

This discussion of nuclear power has been lengthy because of the potential importance of nuclear energy in controlling the atmospheric greenhouse. Nuclear power is out of Pandora's box: it offers enormous benefits; there are enormous risks.

For the rich nations, an obvious way to reduce CO_2 emissions is to replace all fossil-fuel power stations with nuclear reactors, and to make electric public transport more attractive than gasoline-driven private transport. This is the path on which France is embarked, and to a lesser extent Sweden, Taiwan, Japan, and Switzerland (which has large hydroelectric resources). The analysis by Keepin and Kats of the cost of nuclear power came to the contrary conclusion that this path is expensive, and it would be better to spend on conservation, not on nuclear electricity. To some extent, their conclusion must be correct at least in the early stages of conservation when savings are easy to achieve. However, as efficiency cuts deeper and if the electric demand from public transport rises, electricity must come from somewhere. For a society that seeks to emit no net CO_2 to the air, at present the most obvious large sources of power are hydroelectricity and nuclear power. Until solar power is available, we will have to choose one or the other.

The conclusion from this analysis is that conservation coupled with a switch to nuclear power is the best immediate option for the rich nations. Under this option, as coal-fired power stations age and are replaced over the next 20 years, they would be replaced by a new generation of nuclear reactors. The safety record of nuclear power, compared to that of coal and oil, is superior, and the waste disposal problem, again compared to coal and oil, is smaller. The immediate cost of nuclear electricity is probably slightly higher than the immediate cost of power from coal or oil, but the full environmental cost of nuclear power is much less than that of fossil fuel power. The experience of France and Belgium, and to a lesser extent of Sweden, Switzerland, Japan, and Taiwan, is that it is financially possible to run a prosperous state with a major energy contribution from nuclear electricity.

This inference, in favor of nuclear power in the rich nations, is very controversial. Sweden, for instance, which now generates nearly half its electricity from reactors, has decided to abandon nuclear power, for environmental reasons, in the light of Chernobyl. However, from the point of view of the poor nations, the dependence of the wealthy countries on fossil fuel means that the air of the rest of the world becomes the waste dump of the prosperous. Poor nations will suffer or die in a changing climate, rich nations can adapt. The not-in-my-backyard debate about nuclear power epitomizes all pollution debates: whether to accept local or regional risk (e.g., a reactor leak or accident) or to disperse the risk and waste on a worldwide scale (e.g., CO_2 emissions).

For the poor nations the nuclear option is more complex. Most poor nations are inhabited by a small elite that uses power wastefully and a large populace that uses virtually no energy at all. Electricity, whether from fossil fuels or nuclear fission, is too expensive for them. In a later chapter it will be argued that rural electrification in poor nations is essential if vegetation is to be saved and human population growth is to be restrained. If this energy is to come from fossil fuels, then any savings in CO_2 emission in the rich nations will be more than offset by the increase in CO_2 emissions from poor countries. Nuclear power and hydroelectric power are the only immediate answers. Reactors like the CANDU reactor, which do not use enriched uranium, are suitable if the spent fuel is strictly controlled. For example, one CANDU complex could supply all the near-future extra electricity needs of Zimbabwe, Zambia, Botswana, and Mozambique, for a cost equal to one year's expenditure of Canada's annual overseas aid budget. To give another example, in China, 20 reactor complexes of the same size as the large CANDU complex at Pickering, Ontario, would produce the equivalent of China's present electric power output. A program of this sort, linked with careful hydroelectric expansion and electric public transport, could transform China from being one of the dirtiest nations on Earth to becoming one of the cleaner.

Nuclear power today is mostly produced by reactors that owe their basic designs to the early enthusiasms of the 1950s. Engineering as a profession learns from its failures, and 1990s reactor designs will be significantly different from 1950s designs. There are close parallels with bridge building. In the 1820s and 1830s bold engineers began to build with iron and steel. Larger and longer bridges were built, until in 1879 the great Tay bridge in Scotland collapsed together with the train upon it, with heavy loss of life. (The disaster, incidentally, also produced what is generally agreed to be the worst poem in the English language, by William McGonagall.) The response from the engineers was not to give up building bridges, though few now read poetry, but to be very conservative in their appreciation of safety margins. The result was the Forth bridge, also in Scotland, opened in 1890. This bridge is so overengineered that it has needed virtually no maintenance to this day, except painting. Since then, with few exceptions, bridges have not fallen down, except in rare cases of radical design leaps or poor earthquake resistance. After Chernobyl, the design failures and risks of the early reactors have been exposed. Overengineering, especially in containment structures, is needed, but the record of engineering history is encouraging and implies that a reactor failure on a similar scale need not happen in the next generation of nuclear power plants.

This conclusion, so strongly in favor of nuclear power, must be qualified by the lessons of the bad past. The first lesson is organizational: neither the Soviet nor the American approach to nuclear power has worked. Nuclear power organizations should be large enough to be skilled, they should be long-lived enough to benefit from a learning curve of knowledge, and they should not be driven by short-term objectives, whether profit or the demands of a plan. Only long-lived organizations build up expertise and become wiser after error: the essentials of a learning curve.

Before the Sellafield-Windscale accident, it is said that Sir John Cockroft, then Britain's leading nuclear physicist and one of Rutherford's students, ordered that filters be placed in the discharge stack above the reactor. This order was implemented, but met with derision at the time. Had the filters not been in place, the accident would have been much more serious. The incident illustrates the importance of intelligent leadership. Unfortunately, over the past decades the nuclear industry has not attracted the best and the brightest. Poorly led, it is weak in skill and insight. There are exceptions, but the industry is in need of a powerful infusion of thoughtful leadership and young skill that is aware of the mistakes of past years and prepared to correct them.

The corporations or state organizations running reactors should be as intelligently led as possible, by boards comprising a spectrum of political, scientific, and, especially, ethical and critical expertise. There should be powerful, independent watchdog inspectorates, able and willing to intervene in the interests of safety. Reactors should be passively safe, clearly and obviously secure against attacks of error and terror (i.e., built underground). Reactors should be incapable, without large-scale and easily detectable effort, of producing material for weapons. Once a good design has been achieved, reactors should be standardized to contain costs. Finally, the nuclear industry should be prepared to dispose of waste in an ethical (and expensive) way.

6.11 Clean energy: the move toward a managed greenhouse

The atmosphere can be managed only if there is a deliberate worldwide move toward clean energy, to a planetary economy with no net global emission of CO_2 or CH_4. These two gases should only be released on a planned basis, to manage temperature. How, in practical terms, can this be accomplished?

Eliminating net greenhouse emissions will be a long, complex, and expensive business. Expensive is not really the correct word. Rather, any effective solution will mean, in the long term, a redirection of the evolution of human society toward a culture in which the rich do not spend an hour each day commuting, and the poor do not scratch the soil for dung to burn. Nevertheless, to return from that optimistic vision, there is a series of immediate actions we can take, and plans we can make for the medium to longer terms, in order to reduce the impact of our energy use.

The following discussion is confined to sources and distribution of energy. Throughout the discussion, it should be remembered that the first way of reducing greenhouse impact is conservation, by improving energy-using machines and devices, and in the longer term by rearranging our life-styles. Given that a seri-

Figure 6.9. The effects of changing fuel mix. The diagram illustrates the total cumulative increase in atmospheric carbon, in billion tons (gigatons) of global emission after a starting date of 1990. Note that the vertical scale is logarithmic. The model assumes that the rate of growth in fuel use tapers off to zero by 2040 (i.e., a strong conservation program). Switching from the present fuel mix (coal, oil, nuclear, hydro) to natural gas only slightly retards the increase in atmospheric CO_2. Switching to coal or to synthetic fuel from coal (as oil is depleted in the West) slightly shortens the doubling time. See Fig. 7.9 for the local atmospheric effects of such a switch. At present (1990), many nations are following the first path, abandoning nuclear plans in favor of natural gas and conservation. Though this path does slightly slow the change in the atmosphere, it is nevertheless a path of rapid progression to doubling of atmospheric CO_2. (From MacDonald 1982.)

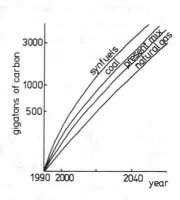

Figure 6.10. Greenhouse emissions by country, showing the 12 largest emitters. Emissions are expressed as percentages of the world's total anthropogenic emissions, as calculated by conversion to equivalent heating values. Note that this type of calculation depends on assumptions about the significance of varying atmospheric lifetimes, and will give different results according to the relative weighting accorded to each gas. For an estimate of emissions, see Table 6.14, which also gives conversion factors for instantaneous heating effects of emissions of methane (not, as here, allowing for its atmospheric behavior). See also James (1990). (Data and conversions from World Resources Institute 1990.)

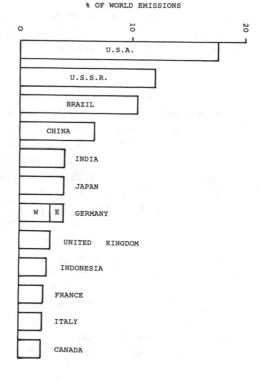

ous effort is made to use energy more efficiently, what can we do about reducing
the impact of the sources?

The most immediate ways of reducing the global greenhouse are through reducing or at least slowing the growth of the concentrations of the trace gases.
For CFCs, the 1990 London agreement probably represents the best attack on
the problem that we can hope for in the present political system, so there is little
scope now for further progress. We must await the further rise and eventual fall
in concentrations and hope there is no ozone catastrophe or greenhouse warming
effect above expectation.

Methane offers much more scope for immediate action, and because the lifetime of methane in the air is so short (7–10 years), the results of effective action
will be measurable very quickly, within the term of most governments (i.e., in
time for the next election). Around one-quarter to one-third of atmospheric methane (perhaps 130–160 million tons a year) comes from fossil sources, according
to the evidence from carbon-14 measurements in New Zealand, the United States,
and Canada. These fossil sources include emissions from warming permafrost,
but it is very likely that a substantial part of the discharge to the air comes from
leaky natural gas distribution pipelines. One estimate (Table 6.13) is that in 1987
pipeline leakage of methane was 53 million tons of methane, worldwide. This is
probably a conservative estimate, as it is based on a relatively low figure for
leakage in the Soviet Union. Removing 53 million tons from the annual methane
sources would, after a decade, reduce global methane levels by about 10%, or
150–200 ppbv. A large part of the leakage almost certainly comes from the
chronically leaky Soviet gas industry. Bringing Soviet pipelines up to Western
standards would not take long. It would not be cheap to do this, but not horrendously expensive either, and the work could be rapidly done.

In the near to medium term, the need to distribute gas via pipes could be much
reduced, especially by burning it at the wellhead to generate electricity. There
are trade-offs to consider here. Transmitting electricity is wasteful, and in some
circumstances it may be more efficient in greenhouse terms to pipe gas in tight,
high-pressure pipelines to power plants near the consumer. Low-pressure pipes,
such as those that run under city streets to domestic users, are very leaky indeed.
There is a strong argument for closing all the older urban low-pressure pipeline
systems and switching to electric cooking and heating. Again, however, there
are trade-offs: it may be more efficient to waste methane from pipes than to
generate more CO_2 at an electricity plant. Of course, there will be other losses
too – gas cooking is a gourmet chef's amenity that will be sadly missed, and the
residents of Millington Road, in Cambridge, England, advertised in the scientific
journal *Nature* as the center of the universe, may miss their romantic but weak
gas street lights. Perhaps even in the greenhouse world there will be place for a
little frivolity.

Worldwide, a vigorous effort to reduce methane losses from gas systems will
probably be the quickest and cheapest way of beginning to reduce the atmospheric greenhouse. The countries most involved will be the Soviet Union, the
United States, Canada, Mexico, Algeria, Saudi Arabia, the Netherlands, the

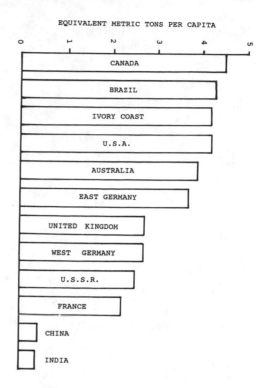

Figure 6.11. Per capita greenhouse emissions for selected countries. Emissions are expressed as equivalences to tons of carbon in carbon dioxide. (See note in Fig. 6.10 caption on the validity of such calculations. Data from World Resources Institute 1990.)

United Kingdom, France, Iran, Venezuela, and Australia. In some of these countries, such as the Soviet Union, Canada, the Netherlands, and the United Kingdom, it may prove possible to burn the gas at the wellhead, separate the CO_2 from the flue gas, and reinject it into the reservoir, to produce greenhouse-free power. Financially, encouragement by relatively small tax incentives should make such projects commercial. Indeed, the Soviet Union could become a major source of Europe's electric power.

Controlling the impact of the energy industry on carbon dioxide will be a much more difficult business. To be successful, control measures will need carefully thought-out policies and international cooperation on a scale not hitherto known, as well as political willingness to take complex decisions in which there is no clear good and bad, only shades of better and worse. Industry lobbies, representing, ultimately, jobs and hopes of present prosperity, and environmental pressure groups, representing visions of the future or of an idealized past, will compete for the direction of society, each group seeking to further its own interest.

The discussion must turn on politics. The first prerequisite is political will. By the end of 1990, 22 countries had adopted or were considering programs to slow CO_2 and CH_2 emissions. Environmental politics are at present rapidly changing, as both scientists and the public gain a better understanding of the reality of our predicament. Since human society is a very heterogeneous mélange, it is necessary to consider the world region by region.

Table 6.13. *Estimated leakage of*
CH$_4$ in selected countries

	Leakage (million tons)
Algeria	3.7
Canada	7.8
Mexico	4.5
United States	15.0
Venezuela	1.0
Iran	1.1
Saudi Arabia	4.8
France	1.5
Netherlands	2.2
United Kingdom	1.8
Soviet Union[a]	3.7
Australia	0.8
New Zealand	0.4
World	53

[a] The Soviet figure is almost certainly much understated.
Source: World Resources Institute (1990).

Western Europe, Japan, and rich Asia are capable of becoming the first regions to generate energy without emitting greenhouse gases. These nations are technically and administratively capable of vigorous conservation, and are also able to build or operate nuclear power plants; in many nuclear power is already a major source of energy. Opinion in Europe is balanced between interests strongly in favor of nuclear power and public opinion strongly against. Those against are, for the most part, environmentally aware, together with some supporters of fossil fuel interests. The outcome of the debate is uncertain: it may lead either to very rigorous conservation and very expensive energy or to nuclear power.

It is possible that in France, Japan, South Korea, and Taiwan, realization of the environmental cost of fossil fuel, coupled with national self-interest, may prompt a shift to the generation of virtually all electricity by nuclear power within the next two decades. France already has an electricity industry that is heavily nuclear, and the atmospheric improvement this has caused has already been mentioned. However, the French nuclear industry is based on PWRs, designs from the first, pre-Chernobyl generation of the nuclear industry. These reactors will surely become a source of increasing concern. Eventually, they may have to be replaced or in some way made much more safe than they are now. In reunified Germany the environmental debate is vigorous and well-informed. Most probably, there will be strong moves to conserve energy, to improve the rail and public transport systems (which are already excellent), and to close the filthy coal in-

Table 6.14. *Estimated anthropogenic emissions of CO_2 and CH_4, 1987*

	CO_2 per capita (tons of C in CO_2)	CH_4 per capita (tons of CH_4)
Africa	0.9	0.03
Ivory Coast	9.1	0.03
Gabon	3.0	?
South Africa	2.3	0.08
North and Central America	3.8	0.15
Canada	4.3	0.40
U.S.A.	5.0	0.17
South America	5.8	0.07
Brazil	9.1	0.08
Colombia	4.6	0.04
Asia	0.8	0.04
Bahrain	9.4	0.15
Burma	4.0	0.08
China	0.6	0.03
India	0.4	0.04
Laos	22.4	0.09
Malaysia	3.1	0.03
Thailand	2.1	0.10
Europe	2.4	0.05
France	1.7	0.08
East Germany	5.4	0.04
West Germany	3.0	0.04
Luxembourg	6.1	0.13
Netherlands	2.5	0.19
Switzerland	1.7	0.04
United Kingdom	1.5	0.08
USSR	3.7	0.07
Australia	4.0	0.27
New Zealand	1.8	0.43
World	1.7	0.05

Note: These estimates are, in many cases, based on very poor data sources. Even in well-documented countries, information is lacking. For instance, Canadian government estimates of CH_4 emission differ significantly from those on which the estimate here is based. Nevertheless, the table is a rough guide to global greenhouse emissions. To contrast the instantaneous heating effects of CO_2, quoted in tons of carbon in CO_2, with CH_4, quoted in tons of methane, multiply methane number by roughly 16. The table demonstrates the greenhouse importance of methane output in some nations, such as Canada, the United States, Saudi Arabia, the Netherlands, Australia, and New Zealand. Burma has recently been renamed Myanmar.

Source: World Resources Institute (1990).

dustry of former East Germany as soon as possible, to the great benefit of all Europe. At present, public opinion in Germany is strongly against nuclear power, and the demise of the East German reactors in particular will much improve the security of Europe. Paradoxically, however, the greenhouse debate in Germany, well informed as it is, may lead to a new nuclear power program based on alternative reactor technologies, such as gas, pebble-bed, or heavy-water reactors.

In the United Kingdom, home of the most valuable knowledge in Europe about alternative reactors, present plans are for the eventual decline of nuclear power and increasing dependence on fossil fuels, including natural gas. The nuclear industry, as a result of the initial difficulties with AGRs and the subsequent controversial choice of PWRs, is now demoralized and losing skill. There is little real evidence of a forceful national program to reduce greenhouse emissions and clean the air of trace pollutants. A U.K. Department of Energy forecast shows a rise in CO_2 emissions from 601 million tons in 1985 to 1037 million tons in 2020. Current (1990) plans are for a major investment in roads and a massive increase in road traffic, but little expenditure on energy efficiency.

If Europe follows the French pattern, not the British, and there is a shift to nongreenhouse sources of power, the change must be coupled with environmental sensitivity, strong energy conservation policies, and safe reactor design. The world cannot risk another Chernobyl; Europe, especially, is too small and too densely populated. If, however, Europe and Japan do convert to nuclear power and succeed in making it safe, the world as a whole will be granted some relief in the growth rate of the greenhouse increment.

Outside of Western Europe and Japan the problem is more complex. In the Soviet Union and its neighbors, coal is a basis of the economy. Most of these nations are impoverished and heavily indebted. Poland and Czechoslovakia are examples – poor, and massive coal producers. It is in the interests of Western Europe to help, if only to save Germany from acid rain. The best help may be in advice about conservation and in financing the design and construction of safer reactors and better-run gas fields, and in buying their power. Marshall Plan aid transformed Western Europe 40 years ago. This aid may now be passed eastward, and one of the best uses for such aid would be power-plant construction and large-scale electrification of transport. One possible mechanism of this transfer would be to build Eastern European reactors on specially designated land diplomatically under the control of the European Community. Extraterritoriality has many attractions, not least that it means the reactor fuel could be closely watched, to avoid its diversion to bombs. The Soviet gas fields might become important sources of hydrogen and electricity, especially if CO_2 could be reinjected on site into the strata.

China is a nation that lives under a pall of coal dust. The trees have gone for fuel or buildings: Paul Theroux calls a tall tree one of the rarest sights in China. China has neither the wealth nor the expertise to install and safely run nuclear power on a large scale. It does, however, have the ability to make the Earth's atmosphere very unpleasant. To assist China, Japan is the obvious source of finance, France, Canada, and Britain the obvious sources of expertise. Perhaps it is Japan's turn to offer Marshall Plan-type aid in the electrification of all China, urban and rural. This will be a massive project, but the cost is comparable to several times the present annual Japanese overseas aid budget, and would be spread over one to two decades. As in Eastern Europe, the reactors could be under extraterritorial control.

In North America the problem is different. The United States now has a deep aversion to nuclear power, because of its very poor track record of expensive, poorly designed, and badly run civilian and military reactors, and politically powerful coal interests. In part because of the cancellation of nuclear plants in the United States, Canada has become a major exporter of power, much of it hydroelectric, from flooded forests. In particular, the James Bay project, in Quebec, has, arguably, damaged the biosphere on a global scale. However, it is in Canada's interests to alleviate acid rain from U.S. coal-fired power stations. A possible Canadian strategy would supplement the export of hydroelectric power by building CANDU reactors, carefully sited underground but relatively close to U.S. industrial regions, and persuading the United States to close its own coal-fired power stations.

Of the poor nations, India and Pakistan are well versed in nuclear technology and close to the large hydroelectric potential of the Himalayas. Given that global change may mean major changes in India's climate, it is in India's interests sharply to reduce its use of fossil fuel. It is planning a large nuclear program, 8 GW by the year 2000. As in China, rural electrification is essential if the Himalayan forest is to be regrown. The Indian government is a democracy: it is likely to take action as environmental pressures grow, especially if given aid.

Much African energy comes not from fossil fuel but from the remains of Africa's woodland. There is virtually no wealth; most nations are in debt and without skill. Political systems are frequently unstable, and conflict or corruption are common. Here, more than anywhere else on Earth, is a need for outright aid. Aid can provide capital for electic installations with low running costs. The first priority should be rural electrification, providing aid as cheap power to poor people. Solar power, where possible, is the obvious first choice as a power source. Beyond that, or where solar power is too expensive in running cost, nuclear power will be needed. Could Africa handle nuclear reactors? Many Western authorities feel that it could not. They make the point that Africa lacks the essential technical skill and especially the political stability to run a nuclear reactor.

There is, however, an encouraging example to be followed. Thirty years ago the Central African Power Corporation was set up to generate and distribute electricity from Kariba Dam to what was then Northern and Southern Rhodesia. Wars, political alienation, and closed borders have all occurred over the 30 years, but the generation of electricity from the dam, which is effectively extraterritorial, has quietly continued. Effective extraterritoriality of the reactors could follow a similar pattern. The Commonwealth of Nations and the Francophonie (the French-speaking community) could enable this by financing and monitoring the power plants and siting them on diplomatic territory. A network of half a dozen such corporations in sub-Saharan Africa, based on nuclear reactors, could be linked to an extensive rural distribution system. Such a network would do much to restore the trees and climate of Africa. South Africa already has nuclear power in the Cape but is otherwise dependent on coal, both for electricity and for gasoline. The country is one of the world's larger producers of coal and in the eastern Transvaal has some of the world's most acidic rain. The vegetation shows the damage. There is thus a strong argument that the local and global environment would benefit if South Africa abandoned coal, in favor of conservation, natural gas, and nuclear power based on local uranium fuel.

In Latin America, Brazil and Argentina may be persuaded by financial inducement and national pride to adopt nuclear energy and perhaps to shelve some damaging hydroelectric plans for Amazonia. As in all other nations, nuclear power reactors should be under strict controls, especially given the attitudes of Brazil and Argentina toward nuclear weapons. The other nations of Latin America, for the most part, are both seismically active and oil producers. There is little chance that they will abandon fossil fuels, though they may be persuaded to reforest the Andes and to install hydroelectric power stations.

Such changes are possible, but will be expensive. The best source of finance might be a mixture of aid – for example, from Japan to China and India and from the West to Africa – and taxes on the use of fossil fuel. Most of the aid would be spent within those rich nations that build the reactors, so that the net fiscal impact on the donor economies would be small, especially if there were a parallel decrease in military spending.

The response of the market and of oil-producing nations would be to drop oil prices and attempt to attract industry to nations offering cheap fossil fuel energy. It is for this reason that if nuclear power is to be adopted as a ''clean'' source of energy, the poorer nations must be electrified by the industrial economies. Without help from the richer nations, fossil fuel consumption in the poor world would rise, eventually to equal or surpass the consumption of the rich nations today. The imposition of selected countervailing import taxes by the rich nations on manufactured goods sold by those nations using cheap fossil fuel energy (or emitting CFCs) would also help to restrain the use of cheap fossil fuel (or CFCs) by poorer nations. Ultimately, though, if nongreenhouse energy is not supplied to the poor, the efforts of the richer nations will be in vain, just as the courageous unilateral U.S. attempt to control CFC production failed.

6.12 Conclusion

A safe global energy regime is one that, on balance, has no net emission of greenhouse gases, especially CO_2 and CH_4. Such a regime could be accomplished by a mixture of policies, including conservation, first, in association with a switch from fossil fuels to hydroelectric power wherever the environmental effects are not too damaging, solar, wind, and wave power where appropriate, and baseload nuclear power. Eventually, solar power may become the dominant source of energy. Widespread electrification from nonfossil sources, linked to a rapid expansion of electric public transport, could very substantially reduce CO_2

Figure 6.12. Prospect of Cambridge, 1990.

and CH_4 emission. The development of hydrogen as a currency of energy, generated either from nuclear electricity or from photolysis, could replace fossil fuels for transport systems, such as aircraft, where electrification is impractical. Some fossil fuel use could be tolerated, especially in poorer nations, but only when set against CO_2 recapture by reforestation.

The cost of such a transition need not be large. Under present policies, the world's capital stock of energy-using equipment is replaced every 20–30 years. Power stations come to the end of their lives, aircraft age, cars rust. Although it can be argued that in high-interest-rate economies nuclear power is intrinsically more expensive than fossil fuel power if no allowance is made for environmental cost, that differential cost is not great when compared to military spending or even to expenditure on such trivial items as candies and chocolates. Overseas aid expenditures that have been traditionally used to support damaging projects in poor nations could more wisely be put into encouraging reforestation and supplying poor people with CO_2-free power sources. Transport systems, like power stations, age and need to be replaced. The extra cost of converting to systems that are not fueled by fossil sources is therefore not large, if the conversion is scheduled to correspond to the natural aging and replacement of the existing capital stock.

Conservation is extremely cost-effective, and the most efficient transition to a greenhouse-neutral economy would include vigorous conservation measures. Most probably, this would mean no net growth in energy use, or only slow growth as increasing energy use in poor nations counterbalances decreasing use in rich nations.

To conclude: with intelligent use of conservation and of nuclear, solar, and hydroelectric power, and with a switch to electric public transport and to hydrogen as a currency of energy, it is quite possible to imagine a global economy in 2020 that has no net production of CO_2 or CH_4. Furthermore, the transition need not be disruptive and need not involve gigantic extra cost or herculean efforts.

Reading list

Bennet, D. J., and J. R. Thomson (1989). *The elements of nuclear power*. 3d ed. Harlow, U.K.: Longman. New York: Wiley.

Brown, G. C., and E. Skipsey (1986). *Energy resources*. Milton Keynes, U.K.: Open University Press.

Chapman, N. A., and I. G. McKinley (1987). *The geological disposal of nuclear waste*. Chichester: Wiley.

Elsom, D. (1987). *Air pollution*. Oxford: Blackwell.

Gould, P. (1990). *Fire in the rain: the democratic consequences of Chernobyl*. Baltimore: Johns Hopkins University Press.

MacDonald, G. J., ed. (1982). *The long-term impacts of increasing atmospheric carbon dioxide levels*. Cambridge, MA: Ballinger.

McLaren, D. J., and B. J. Skinner, eds. (1987). *Resources and world development*. Chichester: Wiley.

Okken, P. A., R. J. Swart, and S. Zwerver, eds. (1989). *Climate and energy: the feasibility of controlling CO₂ emissions*. Dordrecht: Kluwer Academic.

Porter, G., and W. Hawthorne, eds. (1980). Solar energy. *Philosophical transactions of the Royal Society, London*, A295: 345–511.

U.S. Department of Energy (1989). *Clean coal technology demonstration program*. DOE/EIS – 0146D. Washington, DC: Department of Energy.

7
Managing the Earth's vegetation

The birds of the air have fled and the animals are gone.

Jeremiah 9:10

The cow is no longer an animal: it is a machine that has an input (grain feeds) and an output (milk). . . . by robbing the cow of the last remnants of her autonomy as a cow, man has, at the same time, robbed himself of his human identity and thus become like a piece of livestock himself.

Václav Havel, *Letters to Olga,* No. 118

The actions of human beings and the policies of human governments now dictate the vegetation of the planet. With the exception of the residual and retreating wild areas, the biological productivity of the planet is controlled by the needs, aversions, or accidental neglect of human beings. Governments spend extraordinary sums of money organizing the biosphere. Together, the European Community and the United States spend about $80 billion or more annually on vegetation policy. This expenditure not only affects the ecosystems of Europe and the United States, but has major impact elsewhere. Other nations, such as Canada, Australia, the Soviet Union, and Brazil, also have large-scale vegetation-management policies.

In virtually all cases, the management of vegetation by governments is carried out to secure food supply or to aid farmers and loggers. In almost all programs, there is virtually no thought given to the climatic or biological consequences of ecosystem management or disruption or to the long-term consequences for the security of humanity and the biosphere.

Agricultural policy in most nations is an expensive business of subsidy (whether direct or indirect), quota, and legislation. The aims are variously to grow more, to grow less, to maintain rural population, to reduce rural population, to compete with other subsidizing governments or blocs, or simply to reelect local politicians. The logic of subsidy has its surrealism. Canadians attempt to grow cucumbers in Newfoundland for export to the United States; the United States fulminates against certain Arab nations and shortly thereafter declares them to be "favored" nations for wheat exports; ships full of identical Canadian and European cheddar cheese pass at night in the North Atlantic to supply eager consumers on the opposite side; pigs are mysteriously created, ex nihilo, on the green fields of Ulster in the United Kingdom, and then cross into the Irish Republic collecting a subsidy, only to vanish and reappear the next morning in Ulster (or is it the other way around?); a ship in a German port off-loads food from one end, and via a conveyor reloads the same food at the other end to gain an export subsidy; African farmers give up planting, bankrupt, because they cannot sell their crops in the face of free grain dispersed as aid, while Canadian provincial premiers demand that more grain be sent to Africa as aid to reduce Canadian stocks and raise prices in Canada.

With the exception of New Zealand, agricultural trade in all the rich nations is now centrally planned and is not structured according to market forces, although the fiction of a food market still exists. The farmers of the rich nations now derive a very substantial part of their income directly or indirectly from government (about two-fifths in Europe and about one-third in the United States). Simultaneously, overproduction in the rich nations has helped to impoverish farmers in poor nations, depriving them of markets, depressing prices, and making it increasingly difficult to improve farming methods, so that more and more land is damaged in low-capital, low-skill farming that is steadily encroaching on the few remaining wild areas. Where selective markets are created by rich nations as aid to the poor, these distort local agriculture in the poor nations and intensify the pressure on the remnant wilderness.

There has been much recent discussion, especially in Europe, about decoupling agricultural subsidies from regulation through price support, so that price can play its classical role in a re-created agricultural market economy and subsidy can be used instead to reshape the rural environment. In a decoupled system, governments would recognize a healthy environment as a goal of agricultural policy that is equal to or greater in importance than the political needs for stable food supply and protection of rural incomes.

The discussion above has been general, but unsupported by evidence. In the following sections, some of the major points are argued in more detail.

7.1 The impact of agricultural policy on the biosphere

Over the past few centuries, as populations have risen and technical ability has increased, there has been a massive transfer of productive land out of the natural, wild biosphere and into the sphere of human activity. Table 7.1 lists the land areas converted into regular cropping since mid-Victorian times. Over the last century and a half an area that is comparable to twice the total area of Europe excluding the Soviet Union has been converted to cropland. This land has not

Table 7.1. *Land areas converted to and from regular cropping (in million hectares)*

World region	1860–1919		1920–78	
	To crops	From crops	To crops	From crops
Africa	15.9	—	90.5	—
North America	163.7	2.5	27.9	29.4
Central America and Caribbean	4.5	—	18.8	0.4
South America	35.4	—	65.0	—
Middle East	8.0	—	31.1	—
South Asia	49.9	—	66.7	—
Southeast Asia	18.2	—	39.0	—
East Asia	15.6	0.2	14.5	8.4
Europe (excluding Soviet Europe)	26.6	6.0	13.8	12.7
USSR	88.0	—	62.9	—
Australia/New Zealand	15.1	—	40.0	—
Total	440.9	8.7	470.2	50.9
Net area converted to crops	432.2		419.3	

Note: For comparison, the total area of Europe outside the Soviet Union is 473 million hectares.
Source: Richards (1986).

been taken from the deserts or ice wastes, but from some of the more productive parts of the biosphere.

The change in land use is, if anything, being intensified, with more and more land being withdrawn from the wild biosphere and placed under direct human management. In 1985–7, out of a total world land area of 130 million sq. km, nearly 15 million sq. km were cropland, 32 million sq. km permanent pasture, 40 million sq. km forest and woodland, and 43 million sq. km other land. Of all this, about 51 million sq. km were wilderness, much of it desert.

The impact of these changes is widespread, both in its effect on climate and in its more direct effects on humanity. Climatically, it has already been argued that the removal of rainforest and its replacement by pasture and cropland will markedly change the transfer of moisture and latent heat in the equatorial regions. This may radically alter the planet's climate, either directly or indirectly but more catastrophically by changes in the Earth's radiation budget as cloud cover and cloud structure change. More regionally, there is much evidence to support the contention that local climate, for instance in the Sahel or the Kalahari, has been altered and made drier and more erratic by deforestation and destruction of bush. Locally, intense soil erosion and soil decline tends to follow the removal of natural cover, so that many regions of historic cropland have now become scrub or semidesert.

As the natural biosphere has been reduced and replaced by agriculture, the

Table 7.2. *Some changes in land use in selected countries mid-1960s–mid-1980s*

Country	Population density (per sq. km)	% Change in cropland	% Change in forestland
Africa	21	+14	−9
Algeria	10	+11	+37
Benin[a]	41	+27	−23
Ivory Coast[a]	36	+51	−59
Madagascar[a]	19	+35	−20
Nigeria[a]	120	+6	−30
South Africa	29	+3	+7
Togo[a]	62	+27	−39
Zaire[a]	15	+15	−3
Zimbabwe	26	+27	0
N. and Central America	20	+8	−6
Canada	3	+12	+1
Costa Rica[a]	57	+29	−46
Cuba	94	+75	+44
Mexico[a]	45	+4	−19
Nicaragua[a]	32	+6	−35
United States	27	+6	−10
South America	17	+35	−7
Brazil[a]	17	+46	−6
Colombia[a]	30	+13	−23
Ecuador[a]	38	−1	−21
Venezuela[a]	22	+8	−13
Asia	112	+4	−5
China	119	−3	+25
India	274	+4	+11
Indonesia[a]	99	+19	−2
Malaysia[a]	52	+18	−17
Sri Lanka[a]	266	+17	−28
Thailand[a]	107	+55	−41
Europe	105	−5	+7
Albania	121	+37	−12
France	101	−8	+18
W. Germany	247	−3	+1
Italy	196	−20	+5
United Kingdom	233	+29	+23
USSR	13	+1	+8
Australia	2	+27	−23

Note: Percentage change is 1983–5 over 1964–6 area.
[a] Country with significant rainforest.
Source: World Resources Institute (1988), from UN data.

human food supply has become more and more precariously balanced on fewer and fewer genetic resources. Wild primates depend on many species of plants and animals. Humanity has chosen to depend on only about 90 crops, of which about 30 supply most of the world's food. Our animal protein comes mainly from five species: cows, sheep, goats, pigs, and chickens. Moreover, having concentrated on a food supply of limited diversity, by selective breeding we have reduced sharply the genetic variation within each of the most important food species and in some cases we have eradicated the wild stock.

For example, arabica coffee, which appears to be an essential ingredient in human intellectual and political discussions, evolved in southwestern Ethiopia. One plant shipped to Amsterdam in 1706, together with others taken from Yemen to Réunion by the French, appear to constitute the genetic base of most modern arabica coffee cultivation. Arabica coffee is susceptible to *Hemileia,* or leaf rust, and to berry disease. As southern Ethiopia is devegetated and turned into cropland, the wild genetic reserve of arabica coffee becomes scarcer, and there are fewer resources available to breed resistant strains. If arabica is lost, an alternative is robusta coffee, from equatorial Africa and Madagascar. These areas are, however, undergoing rapid deforestation, and robusta's genetic diversity is also rapidly being lost. It may be difficult, therefore, to counter new diseases that attack robusta coffee too.

Tropical moist forests are inhabited by many primates, and the foods that have coevolved with these primates are today a major resource for humanity – we too are primates. Many crops that we commonly use have come either from the moist forest itself or from regions peripheral but related to moist forest, including swamps, riverbanks, and small open spaces. These crops include: avocado, bananas, many beans, including common haricot, navy, French or snap beans, and scarlet runner beans, brazil nuts, breadfruit, cacao, cardamons, cashew nuts, cassava, citrus, cloves, coconuts, cola, cotton, cucumbers, guavas, jute, kapok, mangos, oil palms, papayas, passion fruit, pineapples, rice, robusta coffee, rubber, sago palms, sapodilla (chewing gum tree), sugarcane, sweet potatoes, tomatoes, and vanilla. Destruction of the moist forest and intense cultivation of associated regions, including swamps, endangers the genetic base of these crops, one of which – rice – is the most important single support of humanity. Protein sources too—chickens, for instance – come from these forests.

The cultivation of grasslands also endangers the genetic reserves of important food crops. Primitive wheats include einkorn wheat, now grown in Yugoslavian mountains and Turkey but originally from open oak forests and steppes in Iran, Iraq, and Turkey. Most of these regions are now suffering rapid population growth. Fortunately the wheat is still common along secondary environments such as roadsides. It is resistant to wheat rust and cold. Emmer wheat, the next step toward modern wheat, occurs in Armenia and Transcaucasia, and is grown in Ethiopia, Iran, and the Balkans. Drought in Ethiopia and the introduction of new farming techniques are reducing its diversity. Modern bread wheats include genetic material from *Triticum tauschii,* a wild plant that grows today on the edges of wheat fields and within them as a weed in Iran and Armenia. This plant has

(a)

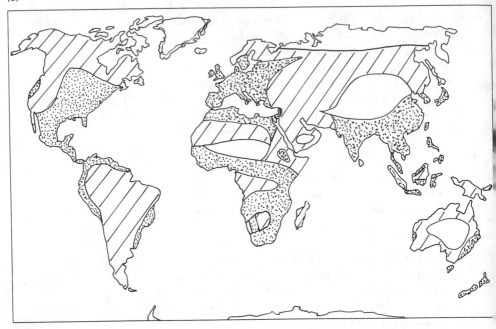

(b)

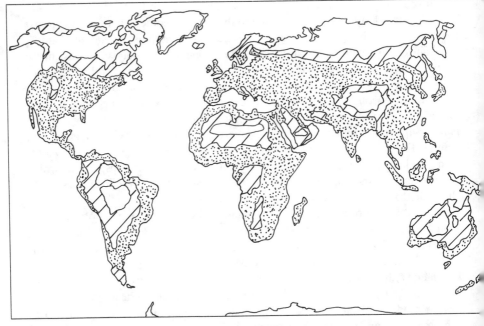

 Uncontrolled environments : Uninhabited

Near dominance : Urban / Intensive agriculture /
High degree of influence on
vegetation and soil

Partial control : Areas of Hunter - gatherers /
Pastoralists / Forest exploitation /
Shifting cultivation in forest

contributed the ability of wheat to grow in continental climates, away from the mild wet winters and hot dry summers of the Mediterranean. Any loss of diversity in these wild grasses that gave birth to modern wheat could have serious consequences for humanity if new diseases evolve or climate changes.

Maize has a parallel story. It comes from Central America and is related to teosinte and tripsacum. Modern maize, called corn in the United States, is genetically very limited, and most of the U.S. Corn Belt production is derived from crosses between only two of the 200 New World races of maize. Over 70% of the hybrid-corn seed production in the United States is based on a few inbred lines. This makes corn production very vulnerable to disease and pest, and the crops need heavy chemical protection. One serious disease could rapidly eradicate all commercial maize grown in the United States.

These few examples illustrate the importance of maintaining genuinely wild pools of genetic variation, so that if disease or pest attack our food crops there will be a choice of resistant strains to fall back upon. An alternative strategy is to maintain only the primitive cultivars of each crop. For instance, early maize, wheat, and potato varieties are collected and used in plant-breeding institutes. However, even this genetic material is becoming scarce and it is difficult and expensive to maintain viable seedbanks. The only secure way to sustain diversity is to sustain the wild habitats in which the crops originally grew. Because the ancestors of our crop plants coexisted and coevolved with animals, a successful habitat must include those wild animal populations too, grazers and predators.

An additional advantage of maintaining truly wild habitat is that it will contain many crop plants and animals that have not yet been utilized for food. The food plants that came from the tropical moist forest are only a tiny proportion of the tens of thousands of species growing in the forest, many of which will be suitable for human use. As the forest is destroyed and centers of diversity are reduced, not only is the security of our present crops being endangered, but we are also denying ourselves the opportunity of discovering others. The pecan tree was first cultivated in the 1840s, macadamia nuts in the 1860s, and the modern kiwi fruit only recently, but one cannot imagine good ice cream without them, especially if maple sugar becomes extinct from acid rain. More seriously, in a changing world, where new plant and animal diseases constantly appear and climate may shift radically, it is extremely important to maintain as much of our natural genetic heritage as possible. Furthermore, there is no space here to discuss the extraordinary importance of diverse forest plants in medicine. Humanity would be most unwise to allow the continuing extinction of natural diversity, but this is exactly what agricultural policy is doing. Region by region, the web of government controls and subsidies is dictating the planet's vegetation, almost invariably without thought either to the climatic effects or to the implications for the genetic stability of our food sources.

Figure 7.1. (a) Estimate of the occupation of the Earth's surface in 1800. (b) Estimate of the present occupation of the Earth's surface. (From Simmons 1987.)

7.2 Europe and the Common Agricultural Policy

The Common Agricultural Policy, colloquially known as CAP, has been devised by the European Community to ensure stability of food supply to European citizens and also to guarantee secure incomes to European farmers, many of them on tiny holdings of 10–20 hectares (25–50 acres). The CAP is extremely expensive and accounts for about two-thirds of the European Community's budget, amounting to $20–40 billion a year in the past few years, more if national and hidden subsidies are added. A newspaper report in mid-1990 quoted gross 1989 support at £31.5 billion (roughly U.S. $55 billion). This enormous CAP expenditure is rivaled by those programs in the United States and Canada with which it directly competes, and together the subsidy programs not only dictate the vegetation of Europe and North America but influence the biosphere throughout the world.

An extreme example illustrates this contention. Brussels, it can be argued, aided the destruction of rainforest in Thailand in order to increase (slightly) acid rain over West Germany. Of course, this was accident, not design, and the causes of forest destruction and acid rain are much more complex, but the chain of cause and effect is fascinating. One pillar of the CAP is its support for grain production, which is heavily protected by high prices and the intervention buying that has produced the various European food mountains. Rather than use Europe's excess grain, apples, and cauliflowers to feed the pigs of Holland and Germany, it was thought more conducive to a stable market to obtain the feed from other sources. In particular, the Europeans discovered that cassava could be imported from the Far East. This came largely from Thailand, and a trade of up to $1.5 billion a year soon built up. At one end of the chain, Thais were rapidly demolishing rainforest to grow cassava in order to feed the voracious European pigs. One source has described the urgency of this cassava boom as so frenzied that people actually drowned on the docks in Thailand in the scramble to load the freighters bound for Europe. Ships full of the cassava unloaded at Rotterdam. These imports drove French barley off the European feed market and into the CAP stores. The proper end of the chain, however, is the end of the 14 million Dutch pigs: the cassava residues were emitted into the Dutch manure mountain, or, more correctly, their national manure bank, described as a national treasure by one politician. Each year, the Netherlands emits about 100 million tons of manure (nearly 7 tons per human inhabitant), of which about 15 million tons come from pigs. Simply storing this ordure costs 2 billion guilders a year (one guilder is about 50 U.S. cents). As much manure as possible is carted around and spread upon the Dutch soil. The manure emits ammonia and methane. When it rains, a substantial amount runs off and into the North Sea. Here it sustains marine algae that produce dimethyl sulfide (DMS), of which the North Sea has now become one of the most productive regions on Earth. The DMS turns into SO_2 dissolved in rainclouds, and blows across Holland and Germany to increase the acidity of the local rainfall, an effect locally outweighted by ammonia from the manure. The contribution to German sulfur pollution is small, but surely not what the CAP intended. Thus the chain goes from chopping down Thai forest, to growing cassava, to feeding Dutch pigs, to manuring soil, to DMS, to acid rain, all paid for by the CAP at enormous cost.

In all, about 36 million acres in the poor countries, mostly tropical, are planted with crops that are grown to feed Western Europe's animals. The cassava crisis ended when the French grew so upset about the reduced market for their grain that they forced Brussels to reduce imports. This caused great disruption in Thailand, where cassava was by now established as a major source of revenue, and in compensation Brussels sent a check for roughly $50 million to Bangkok as slight recompense for a reduced market. The story illustrates not only the extent to which the CAP dictates vegetation in the poor nations, but also the insecurity and risk to the poor nations of depending on so ill-organized and undependable a customer. The Thai forest, of course, is gone now, and is not regrown quickly. Cassava was not the sole or major cause of its destruction, but it helped the process.

In Botswana and Zimbabwe, the Common Agricultural Policy has had similarly powerful effects. By encouraging the export of beef to Europe but imposing strict controls on foot-and-mouth disease, the Europeans have forced major changes in wildlife management, because wildlife is a reservoir of the disease. In the Kalahari, a fence built with European aid has cut off the Okavango Swamp from animal migration routes. Okavango is one of the jewels of Africa and is potentially one of the greatest economic assets on the continent because of its tourist appeal. Biologically, it is crucially important to the region's ecology. Its welfare is now heavily influenced by actions taken in Brussels to encourage the export of beef to Europe, although Europe periodically dumps excess beef even on South Africa, and in early 1991 had excess beef stocks of about two-thirds of a million tons.

The Common Agricultural Policy and the national farming policies that preceded it have had major environmental consequences within Europe. For example, in my own birthplace, Lower Saxony, 14 plant species became extinct in the intensive crop production area during the period 1870–1950. During the 1950–78 period, 131 species became extinct. Similar effects have been documented in the Netherlands. In one part of Germany hedgerows, which provide a valuable shelter for diversity, had a length of 133.4 meters per hectare in 1877, 93.75 meters per hectare in 1954, and 29.1 meters per hectare in 1979. In the United Kingdom 125,000 miles of hedgerows have been removed under a deliberate government policy of eradication. Fortunately, this policy has now been stopped, but British media frequently compare the landscape of East Anglia and central England to the Canadian prairies, a comparison that is an insult to any resident of Saskatoon.

One of the major changes over the past 40 years in Europe is the degree to which marginal land has been brought into crop production, despite the surplus of food. In Britain, for instance, land under rough grazing was 18,121,000 acres in 1961. By the 1980s it was 12,547,000 acres. This rough land was converted to tillage, under wheat, barley, sugar beet, and other crops, while previously tilled land was placed under asphalt or covered with housing, roads, shops, and industry, which in most societies can only exist, apparently, on the best and most productive soil. The pattern is not unique to Britain; it has happened throughout Europe and also in North America, for example in the San Francisco Bay area and around Toronto, with consequent major losses to the natural environment. The conversion of European land from natural or seminatural states to tillage and then to urban use has been expensively financed by the Common Agricultural Policy and aided by local and regional planners.

As an illustration of the lack of direction in the Common Agricultural Policy, Richard Cottrell, a Member of the European Parliament, has commented that

on one floor of the Commission headquarters in Brussels, zealots who want to clean up drinking water are zealously drawing up directives to control the use of nitrogen fertilizer, whilst downstairs a separate department is frantically writing out the cheques which stimulate farmers to use even more of the stuff. Somewhere in the middle you will find another group of distracted and disenchanted visionaries desperately wrestling with the entangled knots into which the CAP has tied itself, more or less in perfect isolation from their embattled colleagues elsewhere in the same building. (Cottrell 1987).

This is political language, written by a politician, but it well describes the degree of order and settled long-term planning of the world's largest and most expensive environmental management system. A final illustration confirms the judgment. Not long ago, Europe decided to dump excess butter on New Zealand, which is one of the world's larger exporters of butter. The New Zealanders, inventive as usual, resolved to turn it into lipstick.

The Byzantine complexity of the Common Agricultural Policy hides the fact that it is potentially an extraordinarily powerful tool for managing the biosphere, not only in Europe but also through its pervasive influence in the rest of the world. There is now a growing realization in Europe that the policy must be reformed and that the reformation should be based on decoupling price support from land management. To some extent, this decoupling is already beginning. For example, British farmers are now paid not only a compensatory sum for setting aside land as woodland, but also an annual management fee to maintain that woodland for the benefit of the environment. In a fully decoupled policy, the large subsidies would go only to biospheric maintenance, and prices would be freely set on an open market. The English landscape, in particular, would be much improved if the present agribusiness estates were replaced by small-scale individual holdings, environmentally managed. In this way, the European landscape could be reshaped without bankrupting the politically powerful farming lobbies.

A reshaped Common Agricultural Policy could become a major force in global vegetation management. In Europe, large areas of marginal land could be withdrawn from production and restored to the natural biosphere, with very little loss of security of food supply. If the large sums spent on price support were spent instead on restoring the environment, this switch could occur relatively rapidly, without damaging the income security of landowners and land workers. Prices

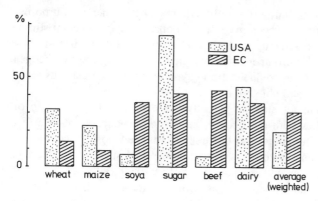

Figure 7.2. U.S. versus European Community government support for various farm products, in percentage producer subsidy during 1982–4, including monetary and non-monetary support. (From Franklin 1988.)

could revert to a free-market basis if stockpiles were only held as guarantees against weather, not as a financial device. For the poor nations, the Common Agricultural Policy could be structured to encourage increased profitability and technical sophistication of farmers in the tropics, by making it profitable to supply Europe with tropical products from carefully selected and environmentally benign sources. Increased technical sophistication of farming, especially in Africa, could lead to the withdrawal of much marginal land from cultivation and its return to nature.

7.3 North America

North American vegetation management policy is comparable in cost to the European CAP (estimates of the full cost in Canada and the United States range up to $60 billion a year) and is only mildly concerned with the environment. However, some aspects of agricultural policy are more helpful to the natural biosphere than the CAP, and overall the U.S. policies appear to have been less devastating in their impact.

The American system of supporting farmers depends heavily on direct or indirect payments, but the market in products is much more free, so U.S. food prices are lower than European prices in terms of the income of consumers. A major part of the agricultural support is through set-aside schemes. Farmers are paid to put land out of production. In 1983, 78 million acres were set aside by government programs, or 26% of the total crop area. In 1985, the Food Security Act made provision for powerful set-aside policies and for the creation of a conservation reserve in areas subject to soil erosion and in wetland areas. In 1987, 17 million acres entered this reserve.

Canadian support for agriculture is also massively expensive, about U.S. $8 billion a year, but is less related to specific land measures. Traditionally, much of the subsidy has gone to transport; more recently an array of ad hoc schemes has poured money onto the farms for political, not environmental, purposes, though much of the money has been given in response to environmental stimulus, especially drought. In general, the response to drought has *not* been to pay farmers to improve their environment, for instance by planting trees or hedges to reduce soil erosion. Instead, compensation payments have been made as gifts, with little or no environmental obligation.

The status of Canadian forestry has already been discussed. Ownership of much of the land still lies with the Crown, and wood producers are in effect heavily subsidized by being allowed to fell trees at very low cost. In contrast to the highly managed but small forests of Switzerland and West Germany, there is virtually no detailed environmental management of regrowth, although some re-planting programs have had limited success. Much of the forest is used for the paper industry, not for construction timber. Effective recycling of paper in the major rich nations, especially the United States and the European Community, would substantially reduce the demand for wood pulp, allowing Canadian forestry to concentrate on growing higher-quality timber for houses and buildings.

7.4 Soviet Union

Soviet agriculture is at present extremely ill-managed and undergoing large-scale reorganization or is in chaos. Historically, large collective farms have been unable to produce enough food adequately to feed the Soviet people, and major imports of subsidized wheat have made up the balance, the subsidy being paid by the United States, Europe, Canada, and Australia. One estimate is that the Western subsidy to the centrally planned economies in the former Soviet bloc, via agricultural exports, was around $23 billion a year. An injection of capital of this sort into the Soviet Union, if it were made into the farms and not by way of bread loaves, could reform the Soviet environment. Very much less land, properly used, would produce very much more food; the land taken out of production could be set aside for environmental purposes. Such a capital injection is unlikely, but a thorough reform of the Soviet agricultural system should take the restoration of the natural environment as one of its chief goals, equal to the need for security of supply.

Soviet forestry has already been discussed. As in Canada, the record is atrocious. The boreal forest, which is one of the most important components of the biosphere, has been severely damaged. Intelligent forest policy should be devoted to the cutting of less timber of higher value. Careful management of forest sharply reduces wastage, and the Soviet forest, like the Canadian, should be used to produce a sustainable yield of high-quality timber and far less newsprint and paper. A shift in forest policy of this sort will require cooperation between consumer nations and producer nations, to increase greatly the recycling of paper and old wood in order to reduce the demand for pulp, while at the same time ensuring that good forest management is properly financed so that high-quality timber can be produced with little waste. In time, the boreal forest may be able to supply most of the lumber market now supplied by the tropical rainforest.

There are relatively few nature reserves in the Soviet Union compared to, say, Tanzania. The Soviet Union has large tracts of moderately undamaged and even virgin natural vegetation, despite the extent of the mismanagement. As political barriers ease, these regions will become increasingly valuable as tourist attractions, accessible to visitors from Western Europe. Unfortunately, the European USSR has virtually no official nature reserves, and until more are designated and protected, the tourist potential will be limited. For example, in the vast tract of the European steppes, only about 200 sq. km of reserves exist, including 48 sq. km of the Kursk and Belgorod regions in the Black Earth land, 111 sq. km of fescue grass on the mouth of the Dnieper River (now inhabited by llamas, emus, and wild asses), and 16 sq. km in the Ukraine. Added to this now is the large accidental biosphere reserve at Chernobyl. There is an urgent need for the designation of much larger reserves and natural parks in the Soviet Union, including the restoration to nature of some land now in collective farms. The potential financial reward from tourism is enormous if North American experience is a guide. If properly farmed, a much smaller area of land could sustain a much better fed population, with substantial return of land to the natural environment.

7.5 African savanna

One of the most perceptive studies of African land degradation is the Save (pronounced Sahvee) study, carried out in eastern Zimbabwe by the University of Zimbabwe under UN sponsorship. The region chosen includes a communal or tribal area, an area occupied by small peasant farmers who hold land individually, and a resettlement area under a land reform program. In general, throughout the region the land is under heavy population stress, but in the part where individual land holdings exist, crop yield is roughly 3.5 times greater than on similar communal land of the same area. In the communal area, there is a significant shortage of draft animals and manure, and woodland is deeply degraded. In the individually farmed area woodland is in a better state, although decreasing and being cut for sale at a nearby mining town. Wood for building is scarce, as is thatching grass, and in the communal areas traditional huts are being replaced by brick structures with asbestos roofs. In general, the area is undergoing desertification and loss of soil fertility.

Local attitudes to the changes are interesting. The local people recognize the environmental destruction and its link to the falling crop yields and declining carrying capacity of animals, but "while people often accepted that their farming practices are degrading the land, they also believed that they should not be expected to do much about this since they are poor and short of labour and equipment: land exploitation appeared to them to be a right, and land conservation a luxury" (Campbell et al. 1989). As population increases and common land is degraded, there is demand for more land. One solution is resettlement into the undegraded commercial farming areas nearby, historically reserved for white farmers. Unfortunately, population growth is so rapid and the level of skills and capital is so low that resettlement areas too are rapidly devastated. The inference

YIELD
kg/ha

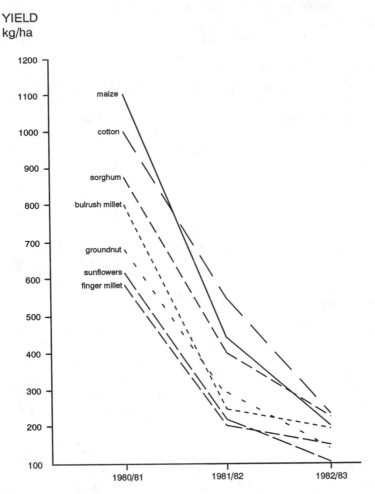

Figure 7.3. The effect of drought on crops in a degraded rural area in Africa. Crop yields, expressed in kilograms per hectare, are from the Chiweshe ward in the same study. The year 1980/81 was reasonably good, 1981/2 and 1982/3 were bad. As population rises, the inhabitants become increasingly vulnerable to drought because the overuse of cropland means there is less and less carryover in soil moisture and soil nutrients from the better years, and cultivation cannot shift to unutilized fallow land. (From Campbell et al. 1989.)

is that resettlement contributes little to the sustainable alleviation of human poverty and environmental degradation, as any brief improvement in the availability of land is rapidly cancelled by the increase in population. Eventually, all that is achieved is biological collapse over a wider area than before.

The prognosis for the area is grim. Initially, the part of the region that has individual tenure will do much better than the communal area, but eventually as the population–resource imbalance worsens, both regions will collapse, the communal region first, then the area of individual tenure. Eventually, this is the path to the state of degradation seen in the famine years of Ethiopia.

Yet potentially, the area can support many people. Commercial farms in comparable conditions grow about 2.45 tons of maize per hectare; in the study area yields vary from 1.1 tons per hectare in good years to 0.25 tons in bad years. Under wildlife, areas of similar soil and rainfall carry far more meat on the hoof than degraded pastureland. To improve conditions, substantial investment is needed, and changes in land use. The needs include training in farm methods and environmental management as well as roads, fences, and good water management and supply. It is important that marketing facilities are radically improved, so that the produce can easily be sold. Rural electrification is essential to reduce stress on wood reserves, which currently provide most of the energy supply. Most important of all, there should be some means of producing a profit: a surplus.

7.5.1 Restoring the African land

It is apparent from this and other studies that land reform should involve not resettlement onto less damaged land but a system of land-use rights, such as individual tenure and individual land rights. The communal system that is traditional in Africa works effectively for a light population, kept small by disease and war, that is living well within its resources. Communality has many advantages, but it can fail when the resources are limited, as each user is tempted to overexploit the commons. Eventually, the commons collapses biologically. Individual tenure, even if granted equitably at the outset, means eventually that unprofitable farmers will be bought up by profitable farmers. This may or may not bring good land management, depending on the forces shaping profit. It also often brings social distress.

Capital investment is needed to bring about restoration of the land, doubling or quadrupling the productivity of land that remains in production while marginal land is retired. Surprisingly, this capital investment is not necessarily large. In the period immediately after independence, Zimbabwe sharply and very inexpensively improved maize output from its communal land, partly by improving marketing and transport of grain. In the long term, though, sustained improvement requires sustained profitability through low transport and marketing costs, motivated farmers with individual tenure, adequate market prices, and above all, skill. As wealth increases, the use of livestock for plowing and transport drops, further reducing land pressure on pastures. If this occurs, productivity can rise sharply and marginal land can be retired while the population is fed better.

Rural electrification is a necessary part of sustained improvement. Electricity, if relatively inexpensive, can substitute for wood as the main energy supply for cooking and lighting energy. In some regions of Africa, eucalypt plantations are common, planted to supply a fast-growing substitute for indigenous hardwoods. Their wood makes excellent construction material but can be unpopular as cooking fuel because it burns too rapidly. Electric slow cookers are a good substitute, especially in areas where diets contain foods that need long cooking times.

If rural productivity rises and if land ownership is changed to an individual-tenure system, the corollary is large-scale urbanization. As good farmers buy up bad and population grows inexorably, people drift into towns. This took place in Europe in the last century and is occurring rapidly in parts of Africa. For example, in some districts of central Zimbabwe there has already been a decline in rural population because the degraded common land is unable to support even its present population. Urbanization is regarded as a serious problem by many economists, but it occurs because, quite simply, the towns are better places to live than collapsing rural areas. For the land, the more rapidly people leave and go to towns, the better. More generally, the best hope for Africa's remaining woodland and surviving soil may be rapid urbanization and urgent rural electrification in parallel with the deliberate creation of a rural rich-peasant class of Swiss-style small landholders exporting food to the villages and towns from reduced but more productive cropland. As a parallel, in Greece after the devastation of the Second World War, the economy of mountain villages was reestablished by food-for-work programs. Peasants were paid for their labor in creating infrastructure, such as roads and water-supply systems, or in planting trees. Relatively small amounts of capital injected in this way, together with access to markets and the ability to enter the cash economy, helped to trigger economic takeoff. With a rising standard of living and a rural surplus comes the ability to protect the land.

There is no reason for famine in Africa to continue, or for the continued destruction of African woodland and forest. Suitably directed aid, land reforms, and the provision of adequate markets can help in the restoration of the environment. At present, the opposite is happening.

As trees are cut the land becomes brighter, hotter, and barer. Soil erosion accelerates, rainfall may become more erratic, animal species disappear even if not hunted, and the woodland becomes a savanna or scrubland inhabited only by cattle and humans.

I possess an old geological map of Matabeleland, southwestern Zimbabwe, published in 1897. The map, which is remarkably accurate (at least in its geology), lists the area of land suitable for "native" cultivation as 30,000 square miles, with 6000 square miles suitable for European-style agriculture. The area "presently" cultivated (in 1897) is given as 300 square miles. Were that map to be drawn today, it would show almost complete utilization of all suitable arable land and the cultivation of much that ought not to be farmed. In 1950 most of Africa was still wild. Today, much of the land is occupied. As water boreholes are drilled, the population is able to spread further and further into the remaining bush, degrading more and more land. Yet the population, by the standards of Europe, China, or India, is still not high – perhaps today two million people in the area of my map, which is significantly larger than England and Wales. Africa is not yet densely populated in relation to the potential carrying capacity, if well-managed. In much of Africa the intensive cultivation of perhaps 10% of the presently farmed area could sustain the population.

Government policy has historically been to improve farming skills but also to extend cultivation to "unused" or wild land. The impact of aid agencies has been enormous, though not always in the intended direction, and typically not toward sustainable use of resources. The emphasis on developing exports of beef to Europe is a good example. In the rich nations of the late twentieth century there is strong demand for game meat, as New Zealand farmers have discovered. Shipping impala and kudu meat from Africa to Europe is more profitable than exporting beef to rich nations that have enough. There are problems, of course, especially in dealing with disease, but in general the production of game meat is more efficient and better for the environment than the production of cattle meat. It is also exceptionally profitable to operate a wildlife park, as Kenya has found.

Overseas aid and advice has been a substantial cause of the degradation of Africa, but it is still possible to redirect aid toward environmental improvement. Because of the interconnectedness of world climate, it is in the strong self-interest of the rich nations to sustain the African biosphere, as it is in the self-interest of the people of Africa themselves. Arguably, virtually all so-called development aid should be directed at the environment, for a healthy people can only exist in a healthy environment. The chief targets should be population control (discussed in the next chapter), improvement of markets and infrastructure so wealth can enter the rural areas, rural electrification, rural land reform, and environmental education.

7.5.2 Pilanesberg: an example

Land can be restored surprisingly quickly in the African bushland. An example comes from the small South African black homeland of Bophuthatswana. When this state was created in 1977 under the apartheid policy of South Africa, a decision was taken to set aside 70,000 hectares of the new homeland as natural parks, the largest being Pilanesberg. From these rather improbable beginnings has come a striking experiment in the recreation of natural habitat.

The Pilanesberg park is roughly circular, about 25 kilometers in diameter. The land had a complex history of tribal settlement, followed by white farming at a low level of skill, and was returned after 1936 to tribal use. By the 1970s the land was seriously degraded and carried little wildlife. In 1969 a university study suggested that the region could be turned into a recreation resort and nature reserve, and during the late 1970s this was carried out. The cost was not especially high: about 12 million South African rand (the rand's purchasing power was then about equal to $1) to fence in the park area, to purchase animals, and to build facilities. Nature foundations and volunteer labor helped. A total of 5957 mammals of 19 species was introduced, and exotic vegetation was cleared.

The park, though large, is much too small to be an unmanaged ecosystem. It needs direct, constant management, including grass burning and vegetation management, together with annual culling to maintain stable populations. In 1987 the natural increase in animal population was estimated to be worth 1.2 million rand.

Table 7.3. *Land conserved in natural parks and game reserves in selected countries*

	% conserved	Overall population density (per sq. km)
Botswana	18.16	2
Tanzania	12.11	29
Malawi	11.26	84
Namibia	8.2	1
Zambia	7.89	10
Kenya	7.5	43
Zimbabwe	7.39	26
United States	4.48	27
Australia	4.1	2
Swaziland	3.47	43
South Africa	3.23	29
West Germany	1.67	247
Canada	1.3	3
Brazil	1.2	17
France	0.7	101
USSR	0.58	13
Mexico	0.32	45
China	0.23	119

Source: Brett (1989).

For the displaced human population the park has had substantial impact. They have lost their land, to anyone a distress and for tribal people a brutalizing catastrophe. In exchange, however, the park directly employs considerable numbers of people as laborers, game guards, and managers, and probably produces substantially more food than the previous degraded farms. This food is wild venison, which is much in demand in local cities. The park is also the focus of a large recreation town that employs 4000 people and has annual sales of over 500 million rand, making a large contribution to the Bophuthatswana budget.

This example, despite its curious political background, demonstrates that the return of degraded land to wildlife can be profitable for the local inhabitants in several ways. Sustainable meat production is high, and probably maximized because wildlife is a much more effective user of vegetation resources than cattle. The tourist potential is also high. The Pilanesberg reserve attracts mostly local tourists, but Africa as a whole can potentially entice very large numbers of tourists from Europe. Pilanesberg is not an ideal example, because of its political history, but it demonstrates that land can successfully be returned to the wild state, with a gain both to the biosphere and also to the human inhabitants of the region.

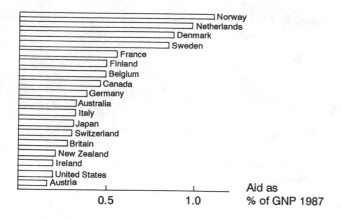

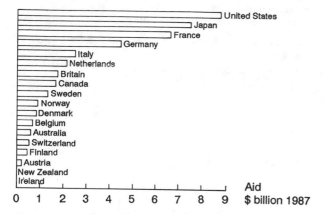

Figure 7.4. International aid expenditure for OECD nations, 1987. Since 1987, Japanese aid has increased substantially. (Data from OECD, Paris, as reported in media.)

7.6 The rainforest nations

Rainforest covers large parts of these nations and regions: Brazil, Peru, Ecuador, Colombia, the Guianas, and Venezuela in South America; Central America; central and west-central Africa (especially Zaire, Congo, Cameroon, Gabon, and Central African Republic); Madagascar; Southeast Asia (especially Burma, Malaysia, Brunei, and Indonesia); and New Zealand. Thailand was until recently also a rainforest nation. Apart from Brunei and Thailand, virtually all these nations are in a financial morass. With the exception of some of the African countries, the rainforest states have very strong nationalist aspirations with patriotism often verging on xenophobia.

The most powerful of the rainforest countries is Brazil, and Brazilian leaders have a strong vision of Brazil as a superpower of the future. In environmental

Table 7.4. *Demands on land in Brazilian Amazonia*

	Area (thousands of sq. km.)	% of total area
Already cleared[a]	340	10
To be flooded for hydroelectricity	150	4.4
Oil deposits	10	0.3
Forest reserves, maintained for timber production	500	15
Projeto Grande Carajas[b] (uncleared land to be affected by development in eastern Amazonia)	540	16
Softwood plantations (other than those associated with Projeto Grande Carajas)	130	4.8
Colonization (up to 1989)	430	13
Total officially earmarked for development	2100	63.5
Protected in national parks, etc. (at end of 1987)	116	3.4

Note: These figures do not take into account factors such as ongoing deforestation by established settlers, ranches, etc.
[a] Estimate. Other estimates vary from 3% to 20%.
[b] This project has been substantially supported by the European Community through its Coal and Steel Community.
Source: Johns (1988).

Table 7.5. *Southeast Asian forest statistics, selected countries, 1986*

	Total forest area (sq. km)	Protected forest (sq. km)	Production forest (sq. km)	Virgin forest (sq. km)	Concessions (sq. km)
Thailand	142,958	n.a.	n.a.	n.a.	194,188
Malaysia					
Peninsular	63,532	10,679	43,607	9,600	10,607
Sabah	44,869	6,000	34,064	7,815	64,000
Sarawak	94,384	24,200	70,184	50,387	20,042
Philippines	63,830	16,800	44,030	10,420	56,754
Papua New Guinea	359,900	n.a.	n.a.	n.a.	11,761
Indonesia	1,439,700	490,410	644,036	524,000	533,740

Note: Some categories overlap. Burgess (1989) draws attention, for example, to the concessions in Thailand over an area substantially greater than the total forest area. A significant part of Indonesia's production forest in Borneo has no trees.
Source: Burgess (1989).

terms, the terms of the true economy, it already is a superpower, one of the most important states on Earth. Brazil is well aware that the northern nations are equally destructive of the environment and that modern Canadian and Soviet forest policy is little different from Brazilian.

7.6.1 The tropical and subtropical forest

Protecting the tropical forest is the most urgent single priority in global environmental management. As the forest is destroyed, a critical part of the global air conditioner is being removed, and it will take centuries or millennia to restore the damage. One cannot easily speed up tree growth. No other single environmental problem is as urgent or as difficult to cure once the damage is done.

The causes of deforestation are complex. Many are connected with market forces, especially the demand for tropical timber in the richer nations. Initially, this demand was for specific species, such as teak for the British, and sandalwood which was widely distributed across the Pacific but cut almost to extinction for Australian traders. More recently, clear-cutting of forest for timber has supplied many markets, but especially Japan. For example, the JANT operation in Papua New Guinea cleared forest to produce hardwood chips for the manufacture of paper, board, and reconstituted wood. There is often little or no financial return on these operations for the owner of the forest. The jointly owned subsidiary in the JANT project operated for 10 years without declaring a profit for its owners, the Papua New Guinea government and a Japanese company. In general, demand for timber is strong worldwide; as long as timber owners charge little for felling, the market for cheap tropical hardwood products will be large.

Another cause of deforestation is conversion of forest land to agriculture. This is taking place worldwide, repeating in the tropics the older European history of forest clearance for cropland. The best-known modern example is in Brazil, but similar conversion is taking place in West Africa, Southeast Asia, and Madagascar. In some regions, the cause of the conversion is the individual actions of landless peasants, but in large areas the conversion has been initiated or supported by central governments or organizations such as the European Community, and often carried out by large companies or politically powerful groups. In Brazil, Tibet, and Indonesia the deforestation has in part been driven by colonization, in programs to settle colonists who are poor members of dominant cultural groups.

Deforestation has been sharply increased in some areas by the accidental side effects of projects directed or financed by rich nations. These include oil exploration and extraction, where communication lines have been cut, allowing the ingress of squatters into the forest, gold-mining ventures, and projects (especially roads) supported by overseas aid organizations or the World Bank. Often these aid projects, for instance to build hydroelectric power installations, are disguised mechanisms for subsidizing industries in the donor nations that build machinery for the aided project.

Table 7.6. *African forest statistics, selected countries*

	Original forest (sq. km)	Remaining forest (sq. km)	Annual loss (sq. km)
Cameroon	230,000	179,000	1500
Congo	250,000	213,000	1100
Ivory Coast	145,000	20,000	n.a.
Gabon	209,000	200,000	150
Ghana	81,000	16,000	n.a.
Liberia	96,000	46,000	500

Source: Rietbergen (1989).

7.7 The regional distribution of deforestation

7.7.1 Asia

Deforestation in southern and eastern Asia is progressing rapidly. Bangladesh, once a forest land, is now 6% forest. In Thailand, forest removal has now gone so far that the country is becoming a wood importer, not exporter, and the results have been significant soil erosion and massive environmental degradation. Peninsular Malaysia has also lost much of its virgin forest, and although forest reserves were created in the early 1900s, many reserves were converted to agriculture soon after independence. To some extent, forest has been replaced by plantations of rubber, palm oil, and cacao. In Sarawak and Sabah, despite resistance from local tribes, cutting has been savage. Papua New Guinea still retains much of its forest cover, but there and in the Solomon Islands the pressure to cut is strong. In Burma recent concessions have been granted to Thai logging interests eager to replace the now vanishing Thai resources. Indonesia is the dominant forest nation of Asia. In the early 1960s, Indonesia exported some timber, but exports rose sharply in the 1970s, reaching 15–20 million cubic meters per year. By 1985, log exports had fallen and been restricted, but sawnwood and plywood exports were growing. Large-scale colonization has accompanied the logging, in transmigration schemes to populate Kalimantan (Borneo). These schemes have been heavily supported by aid programs.

In China, reforestation was formerly a major objective of government policy, and large-scale tree planting was attempted, though with incomplete success. In the 1980s the domestic demand for timber reversed the situation, with heavy felling, especially in northeastern and southwestern China and Tibet, and China became a large timber importer. Cutting is generally wasteful, with low utilization rates.

7.7.2 Africa

African forests have had a similar history in the postcolonial era. The major inventories remaining are in Zaire, Congo, Gabon, Central African Republic,

Table 7.7. *Tropical moist forest, selected countries, late 1980s*

	Original forest cover (sq. km)	Current deforestation (% per year)	Present primary forest (sq. km)	Remaining primary forest (%)
Bolivia	90,000	2.1	45,000	50
Brazil	2,860,000	2.3	1,800,000	63
Burma	500,000	3.3	80,000	16
Cameroon	220,000	1.2	60,000	27
Central America	500,000	3.7	55,000	11
Colombia	700,000	2.3	180,000	26
Congo	100,000	0.8	80,000	80
Ecuador	132,000	4.0	44,000	33
Gabon	240,000	0.3	100,000	42
Guyanas[a]	500,000	0.12	370,000	74
India	1,600,000	2.4	70,000	4
Indonesia	1,220,000	1.4	530,000	43
Ivory Coast	160,000	15.6	4,000	2.5
Kampuchea	120,000	0.75	20,000	17
Laos	110,000	1.5	25,000	23
Madagascar	62,000	8.3	10,000	16
Malaysia	305,000	3.1	84,000	27.5
Mexico	400,000	4.2	110,000	27.5
Nigeria	72,000	14.3	10,000	14
Papua New Guinea	425,000	1.0	180,000	42
Peru	700,000	0.7	420,000	60
Philippines	250,000	5.4	8,000	3
Thailand	435,000	8.4	22,000	5
Venezuela	420,000	0.4	300,000	71
Vietnam	260,000	5.8	14,000	5
Zaire	1,245,000	0.4	700,000	56

Note: Present primary forest excludes second-growth forest and plantations. Some numbers differ significantly from those in tables 7.5 and 7.6 because of differing classification methods, varying interpretations, and lack of information. Burma has recently been renamed Myanmar. Kampuchea is also called Cambodia.
[a] Guyanas include Suriname, French Guiana, and Guyana.
Source: Myers (1990).

and Cameroon. In some nations, such as the Ivory Coast and Ghana, virtually all the original forest has gone. In others, for example Cameroon, exploitation is just beginning. Cameroon is attempting to link forest conservation with tourism, while allowing continuing cutting in other areas. In Liberia, the Ivory Coast, Nigeria, and Ghana, the policy has been one of simple removal and replacement by agriculture, although in some countries this is now ending as the small amount of residual forest is protected. In Madagascar, similar attempts are being made to preserve residual forest, though most is gone. In Zaire, exploitation is still limited by poor communications.

7.7.3 South and Central America

Deforestation in South and Central America has been widely publicized in the media recently, and there has been much discussion about the underlying causes. The extent of the problem in Brazil is shown in Table 7.4 and illustrated in Figure 7.5. In the broadest terms, the causes of Amazonian deforestation include deliberate government policy, including tax concessions and military support for the colonization of Amazonia; the activities of foreign aid organizations; mineral exploitation for oil and gold, including that by foreign companies from the United States, Canada, and Britain; hydroelectric schemes; demand for beef; and the desire of Americans and Europeans for mahogany.

The effects of deforestation on the human economy have been examined at length in many works, some of which are listed in the bibliography at the end of the book; it would be superfluous to detail them here. Less appreciated, though, is the impact of deforestation on the climate of South America as well as on the climate of North America and the globe as a whole. Figure 7.6 illustrates the regional effect of clearing the trees from Amazonia, as calculated in a numerical simulation. In this simulation, if Amazonia is cleared, there will be a significant surface temperature increase in the region and a major change in South America's rainfall, not just in Amazonia but also in Venezuela and the Brazilian Northeast. This change could be irreversible; once the forest is cleared, it could become too dry to sustain rainforest, so replanting would be impossible.

7.8 Conservation

Various methods have been used in the attempt to conserve forest. These include outright legal prohibition, attempts to devise sustainable exploitation regimes, replacement of markets for tropical hardwoods by substitutes, and regrowth of original forest. Attempts to preserve rainforest began with the Indian Forest Service over a century ago. As yet, the success, worldwide, has been limited.

Many environmental groups now consider that the removal of tropical forest is so serious that attempts should be made to inhibit cutting by law. It is very difficult to enforce prohibitions on the ground in the forest nations, because interest groups ranging from large companies to individual squatters are so strongly in favor of cutting. What is more feasible is the removal of market. This involves banning all hardwood imports into the rich nations from selected nations that do not demonstrate sustainable management of forest, or even an outright ban on the export of *all* tropical timber, in the same way that the ivory trade is banned.

Concerted political action of this nature would be difficult to achieve but would be very beneficial to the remaining forest, especially in Asia and Africa. If the major important blocs – the European Community, the United States, especially Japan – were to ban all imports, the economic effects on the rich nations would

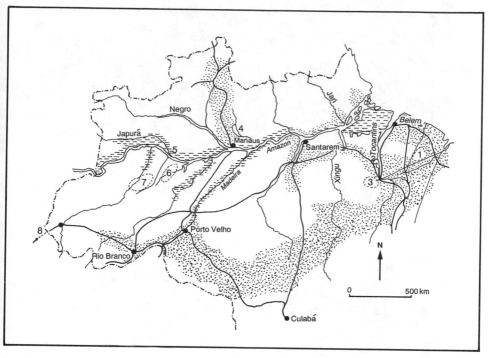

Figure 7.5. The extent of economic change in Brazilian Amazonia. The regions where significant forest cover has been lost are indicated by stippling. Seasonally flooded regions are also marked. Numbers refer to development or conservation sites: (1) Gorupi forest reserve; (2) Tucurui hydroelectric power scheme; (3) Serra dos Carajás ore deposits and railroad; (4) Balbina hydroelectric power scheme; (5) Várzea ecological station; (6) Urucú basin oil deposits; (7) Juruá River oil deposits; and (8) planned highway extension to Peru. Certain points are relevant. Hydroelectric lakes are typically shallow, inundating large areas that may subsequently become CH_4 sources. The Carajás iron project, supported by European funding, is associated with large-scale pig-iron production depending on charcoal, and will demand over 1.1 million tons of charcoal annually, to be derived from several million tons of wood (Fearnside 1989). Exploitation of oil and natural gas deposits is possible without permanent settlement or disruption of the local ecology, but in similar circumstances in East Asia the record is poor. A highway extension to Peru would allow export of wood to the Pacific and further colonization. The presence of conservation sites should also be noted. (From Johns 1988.)

be minor; the impact on the exporting nations of Asia would be large but not catastrophic. However, it is difficult to see such a ban being implemented in the near future. More likely, perhaps, is the use of public opinion against importers of tropical hardwoods in Europe and North America to discourage such things as mahogany kitchens and hardwood furniture, followed by bans on imports to Europe and the United States and selected boycotts of imported products from other rich nations that use wood products.

Japan is the most important rich country involved in Asian deforestation. If

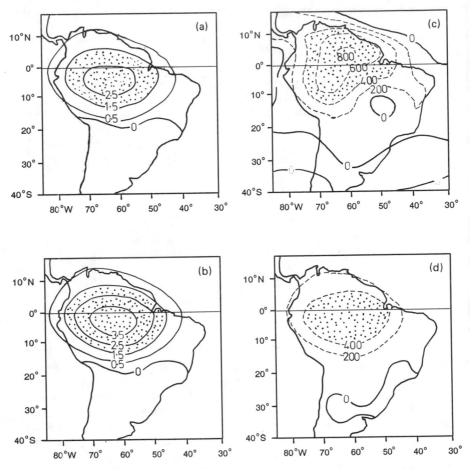

Figure 7.6. Calculated climate changes if Amazonia were to be deforested. (a) Mean surface temperature increase in degrees centigrade. (b) Mean deep-soil temperature increase in degrees centigrade. (c) Mean annual precipitation decrease in millimeters. (d) Mean annual evaporotranspiration decrease in millimeters. (From a computer simulation by Shukla et al. 1990. Copyright 1990 by the AAAS.)

the forest is to be saved, Japan must soon stop importing, either through its own choice or because of external pressure from diplomatic and market opinion. In the near future, it is likely that environmental groups in Europe and the United States will begin to target Japan and Japanese goods – cars and electronics – for economic action because of its poor environmental record in the forests and the seas, in the same way that pressure was brought to bear on South Africa for its political record.

In some cases, sustainable use of forest may be attempted. For instance, in Cameroon it may be possible to develop a limited and highly selective lumber industry, linked with a major tourist industry in the forest. Costa Rica has already

Figure 7.7. The effects of deforestation on soil erosion. Space photograph of the southern reaches of the Rio Grande o'Guapay in Bolivia (18.5°S, 63.5°W). The major tributary shown in the photograph carries a massive load of newly eroded material from the deforested foothills of the Cordillera Oriental. (From G. L. Wells [personal communication]; NASA photograph S84.29400.)

begun to accomplish this, and derives a substantial part of its foreign income from tourism. The natural seasonally flooded forests of Brazilian Amazonia may become the greatest tourist attraction in South America (if the boats can be screened from mosquitoes!). Tourism can be linked with highly selective, labor-intensive, small-scale wood harvesting of chosen trees for high-quality timber, used for example in furniture, to create forest industries that are much more profitable than clear-cutting for wood pulp.

The world needs wood, however, and replacement supplies must be found to meet the demand for pulp and board, as well as sawn logs. Or demand can be reduced. North American newspapers are often 100 pages, in contrast to British, which survive on 16–32 pages. In part, recycling can answer the demand for paper. At present, only a small amount of the world's paper is recycled. The rest goes to landfills to generate CH_4 to CO_2 or to incinerators. Effective recycling in the United States and Europe would substantially reduce demand. Japan already is a large-scale recycler.

Tree plantations are widespread and often profitable. In many areas of the Southern Hemisphere, large plantations of eucalypts and pines have been grown to supply wood needs. Brazil, for example, has an estimated 2 billion eucalypt trees. In Africa, eucalypts such as the saligna gum are widespread. Addis Ababa, so the legend goes, was chosen as Ethiopia's capital because eucalypts grew well there and could supply fuel in place of exhausted native wood. Conifers, such as the Monterey pine, which grows naturally only in a limited area of California, are also very widely planted. In New Zealand, extensive pine plantations in areas that had previously been deforested have created an extremely fecund silviculture industry. Japan is extensively forested, but uses little of its own wood, preferring to import. For similar reasons, land in Britain that was deforested centuries ago is not now reforested, because of local opposition: it is easier to import wood, much of it tropical, and to export the environmental problem. China could potentially become a massive exporter of wood from plantations, as could India and parts of Africa. The potential and mismanagement of the northern forest have already been discussed.

7.9 Financing an end to deforestation

Stopping deforestation will be difficult and expensive. It may be comparatively easy to deny a market, by boycott or legislation, but a more stable and long-lasting halt to deforestation will need the willing cooperation of the forest nations. This will be expensive. The most obvious sources of finance are the use of debt owed by forest nations to rich nations, the use of research funds and development aid money, and the imposition of regrowth targets on industries in the rich nations, such as coal-fired power stations, that emit CO_2, so that CO_2 emissions can be balanced with recapture.

Debt is a powerful tool. The external debt of the rainforest nations, much of it held by Western banks, is of the order of $400 billion. This is enough to buy the whole forest at present prices, although these debts are rapidly being written down by the lenders. More appropriately, because all nations are concerned about sovereignty, debt can be swapped or, more properly, used internally within the rainforest nations for the establishment of very extensive national parks under local control and a demonstrated attempt to protect those parks. Recently, Brazil and several other South American nations have endorsed such proposals. There are several ways this goal can be achieved even by the actions of modest powers. An example is the opportunity that the Canadian government lost in 1987 to make a gesture in defense of the forest. The sums quoted in the following discussion are in Canadian dollars. In mid to late 1987 Canadian banks, like banks in other rich nations, chose to write down several billion dollars that they had loaned with inadequate security to Latin America. This write-down took place with the approval and encouragement of the Canadian government. A significant

Table 7.8. *External debt of selected countries, 1987*

	External debt ($ million)	Long-term debt as % of GNP
Sub-Saharan Africa		
Botswana	518	35
Burkina Faso	861	42
Cameroon[a]	3,508	22
Congo[a]	4,636	195
Gabon[a]	2,071	54
Ivory Coast[a]	10,291	90
Madagascar[a]	3,377	162
Togo[a]	1,223	91
Zaire[a]	8,630	143
Zambia	6,400	244
Latin America and Caribbean		
Brazil[a]	109,497	29
Colombia[a]	15,482	41
Costa Rica[a]	4,437	89
Ecuador[a]	10,407	93
Guyana[a]	1,285	353
Peru[a]	16,625	29
Venezuela[a]	29,015	52
East Asia and Pacific		
Indonesia[a]	48,477	63
Philippines[a]	28,446	65
Solomon Islands[a]	89	77
Thailand[a]	17,603	30
South Asia		
Bangladesh	9,506	51
Burma[a]	4,348	47
India	42,928	15

[a]Rainforest nation.
Source: World Resources Institute (1990), from World Bank data.

part of the write-down was financed by the government, which forgave several billion dollars in taxes that the banks should have paid. This was done by allowing the banks to set against their revenues the losses incurred in the write-downs.

In the internal accounts of the banks written-down debt is worth less than its original value, but it is still valid at full price until sold back at low price in a secondary market to the original debtor. At the time of the write-downs, it was suggested that the Canadian government should have allowed the billions of dollars of tax benefits to the banks only on condition that some of the ownership of debt was transferred to the Canadian nation. Canada could then have entered into a bilateral treaty with Brazil to cancel the debt against the protection of forest assets. This was *not* done and, as it has taken place, the write-down in the debt is, in effect, a gift via the banks from the people of Canada (the ordinary domestic bank customers and the taxpayers) to the debtor nation, but without any strings. Similar write-downs have taken place, on a larger scale, in the United States, West Germany, and the United Kingdom.

Another Canadian example illustrates the availability of money for environmental research. In 1983 the Canadian government decided to finance scientific research in private industry by generous tax credits. By the time the program was terminated, at least $2.8 billion had been paid out. Of this, at least $900 million was clearly identified as waste, of which roughly $500 million was outright criminal fraud. Of the remainder, the research included $22 million given to a project studying the difference between red-and-white and black-and-white cows. Many research projects were not up to this level of excellence. Canadian provinces have similarly wasteful programs. One province alone, British Columbia, spent $1.5 billion on a disruptive and now near-bankrupt coal mine. As a rough average, between $1 billion and $3 billion a year are wasted on projects of this sort in Canada. The money spent on research tax credits, together with the tax credits allowed the banks for their written-off loans, totals $6 billion or more. Most of the rich nations have comparable programs. Many of these programs have environmental implications, including billions spent on energy projects.

Canadian International Development Aid amounts to about U.S. $2 billion a year. The Canadian aid program is generous, yet anecdotal evidence from observation in the field leads to the conclusion that some overseas assistance is of more value to industrial suppliers or to the conscience of the giver than to the welfare of the recipients. Part of the aid helps to subsidize Canadian companies to produce goods or services otherwise unwanted or unmarketable in Canada. There has until recently been virtually no assessment of the environmental consequences of the funds spent. With some exceptions, most other national programs are equally casual about the environment.

Canada is a modest, well-managed power. If such fiscal carelessness occurs in Canadian banks and government, one hesitates to estimate what is taking place in other, richer nations and in less well-organized countries. Furthermore, this discussion has not considered military spending. The originally planned U.S. expenditure on the Stealth bomber program alone could replace all China's coal-fired power stations with nuclear power, at a cost of $1 billion per GW. The benefits to the United States would be much improved environmental security and a gain in international goodwill too.

Private institutions and corporations have played a major role in damaging the tropical environment. They can also have an important part in reconstruction. Recently, some power utilities, especially in Holland and the United States, have begun to replant tropical forest in an attempt to balance their output of CO_2 by recapturing equal amounts of CO_2 in wood carbon. As a short-term device, this

Table 7.9. *Net transfer of resources to poor countries (in billions of U.S. dollars)*

Capital flow	1979	1980	1981	1982	1983	1984	1985
Net transfer from loans	30.7	30.6	27.7	0.8	−8.6	−22.0	−41.0
Net transfer from all resource flows	41.4	39.3	41.5	10.4	−0.3	−12.5	−31.0
Net transfer from all resource flows to Latin America	15.6	11.9	11.4	−16.7	−25.9	−23.2	−30.0

Source: World Commission on Environment and Development (1987), quoting UN, *World Economic Survey 1986* (New York: UN, 1986).

is laudable, and could be much encouraged by tax benefits or required by legislation. It will take some decades to eliminate CO_2 emissions even if a determined
policy of reduction is implemented as advocated in the last chapter. In the interim, a requirement that fossil fuel users should finance the replanting of forest
would have the double benefit of reducing greenhouse problems and simultaneously providing the capital needed for reforestation.

Even if capital is available, reforestation of tropical forest is very difficult.
The best approach may be to plant fast-growing species such as pines, and then
in the shaded understory beneath the pines to introduce indigenous slow-growing
species. In the pine plantations of the Eastern Cape, South Africa, and New
Zealand, this process is occurring naturally, with podocarps and tree ferns of the
indigenous forest growing as an understory under the shade of radiata pines,
despite the seemingly inhospitable pine-needle carpet.

This book is concerned with the state of the environment, not the wealth of
humanity. The two problems, though, are intricately linked. Neglect of one will
cause collapse of the other. The Brundtland Commission drew attention to the
transfer of wealth from the poor countries to the rich nations that is taking place
as a result of international debts. In 1985 there was a sizable net outflow of
money from the poor countries.

In the face of this transfer of capital, it is difficult or impossible to expect poor
countries to care about environmental issues. It is also difficult for them to become wealthy or retain skills. If the poor nations cannot keep skilled people from
emigrating, they cannot generate the numbers of educated, concerned people
each nation needs to take care of the environment. And if the environment collapses, the basis for wealth goes too. Recently there have been encouraging steps
to resolve the problem of the debt of poor states, especially in Africa. More help
is needed.

One relatively inexpensive way in which financial help can slow deforestation
is through the funding of scientific research into forests and wildlife. Most tropical countries have relatively small scientific communities, and scientists have
little voice in decision making. Financial aid to sponsor research, if directed to
local scientists in the forest nations, would create a local constituency within
each country that is concerned about the fate of the forest. This constituency
would then be able to argue in favor of forest at a local level and to devise ways
of making it in the interest of local people to protect forest. Ultimately, the fate
of the forests depends on the willingness of the people of each forest nation to
protect their forest, and this willingness will only come if there is a voluble and
well-reasoned local voice to speak in favor of the forest.

7.10 Hogsback: macroenvironmental policy and microenvironmental management

Good environmental management depends on skillful local administration of the
biosphere. Large-scale policies can be set by central authority using legislation,
inducements, and programs, but macroenvironmentalism can only be effective

if there is proper implementatioin on a microenvironmental scale. This needs local planning that is based on environmental grounds and an environmentally educated population. An example from South Africa illustrates many of the problems of poor central and local planning.

South African tourist literature often carries the slogan "a world in one country." The slogan refers primarily to the natural diversity of the country, but it also accurately describes the complexity of South African society. Nowhere else is the contact between rich and poor humanity so close. South Africa has all the problems of the wealthy nations – acid rain, overdependence on automobiles, chemical pollution of many sorts, the debate about nuclear power, and so on. It also has all the intractable difficulties of poor countries – poverty, a rapidly increasing population, degradation of agricultural land. The apartheid policy of the 1950s to 1970s sought to separate rich and poor by imposing strict controls on migration, by segregating all facilities, and especially by dividing the land. This policy of land segregation is closely comparable to the migration controls imposed at the borders of all rich nations to keep out citizens of poor countries. There is little difference between a policy of using passes to keep migrants out of the historically "rich" Cape Peninsula and a policy of using passports to keep migrants out of the historically rich Europe or North America. The immigration officers at Heathrow or Kennedy airports are implementing the same separation policy and pass laws as used to be implemented by the South African police. Both, in the long term, must fail, if only under sheer pressure of numbers, because it is not possible in an interconnected biosphere to cut off one isolated sector and to declare it to be specially advantaged.

The Hogsback Mountains lie in the Border district of the Cape Province of South Africa, inland from the towns of East London and Grahamstown. The Border district is so named because for a century or more it was the border between the predominantly European culture and mixed population of the Cape and the predominantly African, Xhosa culture to the northeast. The mountains form an escarpment 1500 to 1800 meters (5000–6000 feet) high, overlooking a plain. On the plain, former white farmland has recently been added by compulsory purchase to old tribal reserves; the land is now heavily populated with peasant farmers and planted with orange groves. Large dams on the plain trap water from the mountains.

The mountain escarpment rises in two steps, with natural grassland on the flat part of each step and cloud forest on the steep slopes between. The cloud forest is extremely rich in tree species but is of limited area, so there are few individuals in each species. For instance, this is one of the important natural reserves of a beautiful flowering tree known as the Cape chestnut, or calodendron, but there are probably at most a few hundred individual Cape chestnuts in the area. Much of the forest is yellowwood, which are podocarps, trees distantly related to the English yew. The grassland is also very rich in plant species, such as proteas, especially along the edges of the cloud forest. Many common commercial flowers, such as arums, lilies, and geraniums, grow wild in this area.

In the period from 1850 to 1950 the area was progressively occupied. British

Figure 7.8. The Hogsback escarpment, Cape Province, South Africa. Mountains in the distance (from right: First, Piglet, Second, and Third Hogsback) rise from high grassland, now seriously disrupted by the planting and subsequent burning of pine forest and by the invasion of exotic species after the fire. Escarpment slope (center-right of photo) has residual cloud forest vegetation. Lower plateau (extreme right) is dominated by exotic pines and Australian blackwoods.

troops established a fort high in the mountains to keep watch over the region, and probably introduced blackberries and some exotic trees. Farmers established fruit orchards and began a battle with fruit-eating parrots and monkeys from the cloud forest that continues to this day. Forestry began, and a variety of exotic trees was introduced, including Australian blackwoods and American pines as well as Californian redwoods, Himalayan deodars and blue pines, and various cedars. Today there are probably several hundred exotic tree species on the mountains. The early foresters, trained by the Indian Forest Service, rigorously protected the cloud forest. Their first concern was for the vegetation; people were few and were kept out.

With peace, population increased. On the plain, natural increase was augmented by the effects of influx control in the South African cities, and eventually the area became part of the Ciskei, a notionally independent and very poor black state. On the mountains, the early white settlers were mostly intellectuals and missionaries from the town of Alice, a center of higher education on the plain. Although not wealthy, they included very distinguished economists, social historians, and theologians, as well as physicists and botanists, in close touch with the main currents of liberal thought in the rich nations and often in conflict with the South African government. The boundary between the intellectually rich on the mountains and the financially poor on the plain was drawn mostly along the base of the cloud forest.

Over the last 40 years deliberate government policy has sharply changed the natural vegetation. On the plain, much of the natural grassland and bush has been removed or degraded, replaced by intense but poorly capitalized cultivation. The area is densely populated, poor, and ill-governed, sometimes violent. On the mountains, the grassland has been replaced by state-owned exotic pine plantations that have been grown to the edge of the cloud forest, eliminating many areas of outstanding genetic diversity, including the ecotones at the edges of the escarpments. Smaller oak and eucalypt plantations have also been grown. Large pine plantations were established on the lower step of the escarpment, between two belts of cloud forest. Later, pines were planted on the high grasslands, with massive ecological disturbance. Some years after planting, the natural fire ecology of the pines came into effect – a minor fire escaped and burnt much of the mountainside, stopping eventually at the edge of the cloud forest. Since then, naturally reseeded pines, together with exotic blackberries and other invaders, have turned the burnt-out plantations on former grassland into an ecological riot of regrowth, a catastrophic genetic mélange. The area was once a major wintering ground for thousands of European storks, but far fewer are now seen (seven in late 1990), confined to remaining meadows.

The white community on the mountains has settled in individual plots of land, each of many hectares, because of land regulations. These plots have become overgrown with pines and invading Australian blackwoods, which have replaced the natural vegetation. Indigenous species are little regarded by local authorities, and exotic silver birches have been planted on the roadside despite the rich genetic resources of local forests. Gardens typically have trees and plants from all continents. Relationships with the community of poor peasants on the plain below are distant, and recently a historic footpath used for well over a century by domestic workers to walk up from the plain was closed. Between the two communities, the cloud forest, although protected, appears to be in poor health, challenged by exotic species, by illegal wood gathering, and by hydrological changes induced by pine plantations and land use changes. The forest cliffs used, until recently, to be inhabited by bateleur eagles. These appear to have been killed for feathers used in tribal rituals. Other animals, such as leopards, may be in decline too.

This litany of disaster has been written by accident rather than design – poor forestry practice, population increase, inappropriate introduction and lack of control of exotic species, local effects of national policies on land use. The cloud forest, most significantly, *has* been protected – in many other poor countries it would have been destroyed – because of the value placed upon it by the community. Generally, however, the Hogsback region demonstrates with its bad vegetation management and population pressure the problems of both the rich and the poor sectors of the world.

Unchecked, the future of the region is likely to include the collapse of the cloud forest ecosystem to a few residual patches too small to survive genetically, and the destruction of much of the mountain community by a major fire through the pine and blackwood forests. In the plain below, increasing population pressure may lead to migration upslope, whether or not this is legally permitted, and degradation of the land until there is a major food supply problem, perhaps famine. Eventual migration into the mountain land is probable. Although the area is too cold for local staple crops, a community based on potatoes and pulses could survive there, with eventual complete loss of the mountain biosphere too.

Only detailed microenvironmental management, coupled with sophisticated national macroenvironmental policy, can save the region. On a national scale, direction needs to given to local government and especially to the forest service, to order them to establish environmental management as a high priority. In the case of local government, this would mean employing environmental officers, regulating encroachment by exotic species and plantations, and careful design of roads and water use. For the forest department, the environment, especially the protection and restoration of natural habitat, needs to be set as the prime good, not the production of wood.

South African wood management is poor. The country recycles little paper and is heavily littered. Wood demand could be substantially reduced by national recycling schemes. Forestry operations are typically wasteful, are often loss-making (e.g., after fires), and use poorly skilled labor and direction. More efficient forestry would much reduce the area of plantation for the same value of output. Local hardwoods could be planted for use as high quality, expensive timber, initially as an understory shaded by pines. More appropriate exotics that are less of a fire risk and better timber than pines could also be used. Forestry can still be profitable, but needs to be more skilled and less wasteful. In general, the forest service and local government need an influx of skill and finance for environmental research.

The problem on the plain is more complex, and is rooted in population growth, which is discussed in the next chapter. On a local scale, however, there can be no escape from degradation unless there is a surplus of wealth to improve farming methods and reduce the pressure on the land. This will require a significant initial infusion of capital as aid, either from central government or from overseas aid agencies.

One local microeconomic factor may prove helpful in microenvironmental management. The Hogsback area is very attractive as a retirement community and tourist center because of its extraordinary natural beauty. Retired people introduce capital to the community in exchange for the amenity of beauty. To date, virtually all of the retired are white, living on the mountains. The South African government now makes significant social payments to black pensioners, and a black retired class is being created. Retired people, rich or poor, value their surroundings, are typically moderate in their environmental demands and in the nuisance they create, and are often strongly interested in the environment. Tourists only come because they are attracted by beauty. For the Hogsback region, the most optimistic future scenario in a postapartheid South Africa would include a much reduced but much more skilled foresty industry, with the forest service more concerned about indigenous forest and grasslands; there would be local planning to develop an unobtrusive tourist industry, coupled with high-density, low-impact retirement communities in small houses and apartments for rich and poor, carefully set in environmentally suitable locations that are beautiful but biologically not invasive. The revenue brought in by pensioners to the community could be used to restore and sustain the natural environment.

The Hogsback illustration demonstrates in one small region the difficulties of the conflict between the rich and poor parts of the world and the need for microeconomic forces to sustain local environmental management. In each local community there must be profitable reasons for maintaining the ecosystem. In Hogsback, the income will probably be from pensioners and tourists. On a larger scale, in Costa Rica for example, the use of rainforest to draw tourists can influence national economic policy. The seasonally flooded forests of Amazonia have similar potential. In Switzerland, to take another example, national policy is matched by local advantage, so that the countryside is maintained with great care. On a planetary scale this marriage of macro- and micro-scale policy is essential, so that the global concerns of the rich nations can be translated into local opportunity to support good environmental management of each small community in the poor nations.

7.11 Government and vegetation

I began writing this book while visiting a small patch of reclaimed Zimbabwe highveld, a place where we watched the Intertropical Convergence Zone suddenly strengthen, to break years of drought, and where we rode horses amongst the white rhino. Those few acres of nature have been carefully guarded by government policy, but for the most part the vegetation and animals of the tropics are under assault from the policy of local governments and from misdirected aid from overseas nations.

The assault by government and the consequent extinctions began centuries ago. In Victorian times, my own ancestors were largely responsible for the deliberate introduction of Australian eucalypts into Africa. Today these are the dominant trees in many parts of what was once Gondwanaland: from Africa to the hills of Venezuela, they paint the monotonous landscape. There have been benefits and losses, but the vegetation is permanently changed. In Africa, change began even earlier when the government allowed or failed to prevent the extinction of the bloubok, arguably the most beautiful of all African antelope. It has become so obscure that even in southern Africa most educated people have never heard of this large and lovely animal. We will never know with what vegetation it coexisted, what species were lost when it died. The quagga, a variety of zebra, is better remembered, in the wild probably hunted to its end in a great shoot organized for Queen Victoria's son, and lingering briefly in a zoo and perhaps on an island in Auckland harbor, New Zealand. Attempts are now being made

Figure 7.9. The African veld. Photograph, taken from rail line, shows a maize field in the middle distance and the SASOL synfuel plant behind. Little more than a century ago, this area was open veld. Today, a substantial part of the arable land is under maize, and many of the trees are also exotics, especially eucalypts. The region is now heavily populated and economically based on coal. Several large coal-fired electric power stations have been built in the vicinity, and the SASOL plant is one of the world's largest coal-based chemical factories, making a significant part of South Africa's gasoline. The environmental impact on the remaining fragments of the natural ecosystem has been severe, and the health of the human population has also been impaired by the pollution. Historically, there are close analogies between South Africa and former East Germany, both of which had strong central direction of a coal-based economy in order to meet external political challenges. In both states, the need to protect the environment was not fully debated during the planning of major projects.

to retrieve it, with the help of molecular biology. As the animals passed away, the vegetation with which they coexisted changed too.

These are the effects of policies implemented by our modern structure of government. Vegetation change made modern government possible. In the East, the development of rice culture millennia ago lay behind both the organization of government and the population growth that made that organization essential. In the Mediterranean world, civilization arose from the organization of the fields of the Nile and Mesopotamia. Loss of soil fertility and poor crops, the consequences of environmental damage, also played their role. When the thin soils of Attica could no longer sustain Athens, the city turned outward and upward. Even in the New World, civilization began with maize: the history of that change is written in the record of carbon isotopes in food grains and the shift from C3 to

C4 food plants. Humanity, primitive or technological, has always sought to impose its will on the soil and to exploit nature.

We now occupy a subjugated world. Mother Nature no longer exists. What grows, grows because we ordain it. In the words of one of Bob Dylan's best songs, life is "hanging in the balance of the reality of Man, like every sparrow falling, like every grain of sand." We have left Eden and can never return. Our task now is somehow to manage the vegetation of the planet, to find a sustainable stable state of the world that will support us and our children.

The needs of the biosphere are clear, and it is urgent that government policy should address them. In the rich nations, agricultural programs should be designed not only to restore the environmental balance in the fields and forests of Europe, North America, and the Soviet Union, but also to influence land use in the poor nations toward better management of the biosphere. Destroying tropical forest to feed the pigs of Europe is a mistake that should never be repeated. The rich nations need also to examine their regulation of multinational companies that operate in the tropics, so that Amazonia is not threatened by British or Canadian gold exploration, or the rhinos of Sumatra and the Zambezi valley by U.S. oil exploration. In the poor nations substantial aid needs to be directed away from massive development projects and toward the environment. The environment in Africa is best protected by training rural cultivators and reforming land tenure systems, together with an injection of capital into marketing and infrastructure, especially electrification, so that a rural agricultural surplus can be achieved on a smaller area of cropland. Finally, dramatic action to protect the tropical moist forest is imperative, and should include strict control of timber imports into the rich nations together with the use of debt as an inducement to protect forest in national parks or reserves.

Throughout history, societies have been faced with the problems of land allocation and management. Ideas about land use and population lie at the root of modern economics. In some nations, mostly English-speaking, farming has become an industry in which short-term profit, via intense and competitive production, is the main goal. This system has served the urban consumer well, providing abundant cheap food, but has failed to respect the long-term needs of the environment. More catastrophic has been the collective system of the centrally planned nations, which devastated the environment while simultaneously failing to provide food. Historically, communal systems of poor societies have often worked well, in stable societies with populations limited by disease or war. However, as population has expanded, communal systems have failed. In general, the communal system of Africa has led to environmental damage or collapse, because under communal ownership there is no incentive or reward for skill. The most successful in the struggle to survive are those who mine the most out of the land, as quickly as possible, before a neighbor can extract it. In consequence, skills are easily lost and have to be relearned, and the land is degraded and the people become progressively poorer, while population rises to supply children as labor.

Perhaps the most stable of systems are those where relatively small-scale family farms manage the land. Swiss farms, for instance, are tiny, but the farmer is skilled, has a reasonable income, and cherishes the land. In such societies, urban consumers in effect pay the farmer to become the custodian of the land. The politics of agriculture are immense and complex, but throughout the world the environment would be better guarded if farmers were given not one but two tasks: to provide food and also to be custodians of the land. In practice, this would mean developing networks of subsidy to finance not crops, but skilled environmental management. Production agriculture should be constrained by environmental imperatives but entirely free of subsidy, except to maintain basic stocks of food for security. Instead of funding subsidies, urban consumers should direct their floods of money to nourish the environment.

In the early decades of the nineteenth century, the west of Ireland was a semi-feudal society of poor peasants and landlords far from the center of wealth and social conscience. It was some days' journey from London, rather further than the hills of Ethiopia or the scrub of Sudan are distant today from Washington. The population of Ireland had discovered the potato, and had bred rapidly, to the concern of the Malthusian economists of the day. In London, the government maintained strictly protectionist Corn Laws, which had various effects but were in some ways analogous to modern agricultural policies in Europe and North America.

Disaster struck. The potato crop, grown from a tiny genetic base, was devastated by disease. In London, the government was led by intelligent, fiscally responsible administrators. These were civilized, thoughtful men, compassionate, not monsters. Yet they were slow, too slow, to realize the scale of the famine and they failed to provide the help required. Eventually, when they did send help, they provided soup kitchens to feed 2 million, public works programs, and other relief measures. Half to three-quarters of a million people died and a million emigrated, many to America.

It was in Ireland, over a century before the great famine, that Jonathan Swift, Anglican Dean of Dublin, had penned his modest proposal. This was the suggestion, supposedly on the advice of an American, that the children of Ireland should be bred deliberately to supply meat for the English dinner table – perhaps the most vicious satire ever written. At least such a proposal would have provided a profitable reason to feed the children. Instead, they died. I began my geological career mapping in County Mayo, across fields marked on the map with the quiet note that they were children's burial grounds. The political repercussions of the various Irish famines continue to this day.

Africa, with its rapidly increasing millions and the genetic impoverishment of its environment, is in some ways a massive Ireland. The Common Agricultural Policy and the agricultural programs of the United States are comparable to the English Corn Laws. It is debatable whether the agricultural leaders in Brussels and Washington possess a stature or an insight greater than that of the British government 150 years ago. Perhaps they do; perhaps they may remember. But if they fail, for centuries to come there will be scars both on the human society and on the environment in the tropical lands.

Reading list

Body, R. (1982). *Agriculture: the triumph and the shame*. London: Temple Smith.

Brett, M. R. (1989). *Pilanesberg: jewel of Bophuthatswana*. Sandton, S. Africa: Frandsen Publishers.

Campbell, B. M., R. F. Du Toit, and C. A. M. Attwell (1989). *The Save study*. Harare: University of Zimbabwe Press, P.O. Box M.P. 45, Mount Pleasant, Harare.

Carson, R. (1962). *Silent spring*. New York: Houghton Mifflin. London: Penguin.

Cottrell, R. (1987). *The sacred cow*. London: Grafton (Collins).

Franklin, M. (1988). *Rich man's farming: the crisis in agriculture*. London: Routledge.

Kingdon, J. (1990). *Island Africa: the evolution of Africa's rare animals and plants*. London: Collins.

Knystantas, A. (1987). *The natural history of the U.S.S.R.* London: Century Hutchinson.

Organization for Economic Cooperation and Development. (1989). *Agricultural and environmental policies: opportunities for integration*. Paris: OECD.

Poore, D. (1989). *No timber without trees: sustainability in the tropical forest*. London: Earthscan.

Prance, G. T., and T. E. Lovejoy, eds. (1985). *Amazonia*. Oxford: Pergamon.

Visser, M. (1986). *Much depends on dinner*. Toronto: McClelland and Stewart, 481 University Ave.

Westoby, J. (1989). *Introduction to world forestry*. Oxford: Blackwell.

Wolman, M. G., and F. G. A. Fournier (1987). *Land transformation in agriculture*, SCOPE 32. Chichester: Wiley.

8

The management of man

You trample on the poor and force him to give you grain. Therefore, though
you have built stone mansions, You will not live in them.

Amos 5:11

8.1 Population

The growth in human population is the root of the problems discussed in this
book. In the previous chapters, the reasons for population growth have not been
considered in detail. This is because the symptoms need to be examined first:
population growth is the disease. It is the cause of most of the world's environmental ills. The figures are well known. In 1960 there were 3 billion people,
there are 5.4 billion today, almost inevitably there will be over 6 billion in the
year 2000. It is worth remembering that if crop yields in poor nations are raised
to the levels of the rich countries, the world could feed several times as many
people as at present. Alternatively, much of the land presently cultivated, especially in Africa, could be returned to a managed ''natural'' state if the population
stabilized. We can go either way. We can increase population, doubling our
numbers yet again, degrading more land and doubling the misery when famine
eventually strikes. Or we can return much agricultural land to ''nature'' by better
farming practice, not necessarily with much capital input.

Population growth can be stopped. Some decades ago, the Canadian province
of Quebec had one of the highest population growth rates in the developed world.

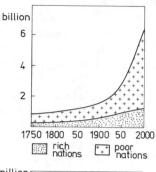

Figure 8.1. The population of the Earth. (From World Resources Institute 1987.)

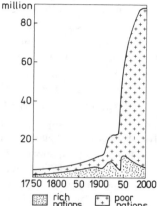

Figure 8.2. Increase in numbers of human beings, 1750–2000. Increase is expressed in population rise per decade, but shown as a smoothed curve. (Modified from World Resources Institute 1987.)

The province was fecund but poor. Quebec, which is predominantly Roman Catholic and therefore circumspect about birth control, now has one of the lowest fertility rates in the world. It is infecund and rich. Indeed, there is now serious concern in Quebec about a decline in the population, and financial incentives to give birth have been instituted. The major nation with the lowest reproduction rate in 1990 is Italy, where women are having, on average 1.29 children. As in Quebec, there is a likelihood that the population will contract markedly. Population growth can obviously be slowed and halted in a few decades, as the record of Quebec and Italy shows. Many wealthy nations, such as Germany, Denmark, Austria, Sweden, Switzerland, and Belgium, have static or declining populations, without compulsory birth control. So there is cause for optimism.

There are four fundamental factors that have caused population to stop growing in the richer nations. These factors have been understood for many years and have often been ably expounded, yet they are rarely acted upon. The factors are good health, good education, security in old age, and the rights of women. Other factors, such as the easy availability of contraception, are helpful in slowing population growth, but can only become important if the fundamental social structure is present.

The frequently repeated but rarely implemented prerequisites of a policy to halt population growth are: to make children expensive, so that parents recognize that they are a burden not lightly to be incurred; to make old age a secure time, so that aged parents depend not on their own individual offspring but rather on the collective promise of society; and to make it more attractive (to both women and men) for women to work in the economy than to bear children. Interest keeps peace: the only way to achieve these aims is to attempt large-scale social engineering.

Poor people have children because children are useful and profitable. Children look after aged parents; children work in the fields; children "keep women out of trouble" in male-dominated cultures. If children are likely to die, as in many poor nations, then the answer is to have yet more children. On the other hand, if old people are given state pensions, then there is less reliance on children. If children have to go to school, they are expensive and cannot work profitably in fields. If women are the equals of men in society, they can work and become money earners who do not need, in male opinion, to be "kept out of trouble" by pregnancy. If health care is good, children are expected to survive and parents do not have twice as many as are needed to ensure that enough survive.

More formally, the essentials of a stable population are social and economic forces that discourage large families. When children do not die as infants, when parents must bear the financial burden of raising them, when the parents can look for security to old-age pensions instead of relying on offspring, and when women and families are financially better off if they can take part in the nonreproductive activities of the society, then there is no reason to have large families.

Giving skill to women, including the skill to understand their own monthly reproductive cycle, is essential. Handing out condoms in male-dominated societies is perhaps good for the conscience of Western liberals, and it is amusing for small African children to inflate them. But condom distribution is a Western mechanical approach that is of minor or short-term impact on population control unless the father of the family actually uses them. In contrast to the sustained population growth in Africa, Quebec has no planned-parenthood missionaries from England and Scandinavia, remains Catholic, and does not overreproduce. Socioeconomic forces, not just the availability of contraception, determine population growth.

8.2 The Save study: population, resources, and attitudes

The Save study in Zimbabwe investigated local attitudes toward birth control and reproduction, and the results well illustrate the general problem in the poor nations. In the early 1980s the Chiweshe district, part of the study area, was capable, under present farming practice, of sustaining a population density of about 16 people per sq. km in average years, each person eating about 2500 kilocalories of food energy per day. In reality, the population density was about 29 people per sq. km in 1982, projected to rise to 36 in 1990. The food deficiencies are met largely through remittances from family members living in urban areas and through imports of food from other areas of Zimbabwe in which there is agricultural surplus. Nevertheless, the study found that over 20% of Chiweshe children were either moderately or severely undernourished according to weight-for-age criteria. In drought years, the stress on the population is much greater.

Commercial farming areas with similar land, also in Zimbabwe, obtain yields typical of technically sophisticated agricultural practice. If these yields were obtained in the Chiweshe area, with large-scale farming practice and an increase in the land actually cultivated to 25% of the total land (at present, much land is now derelict, after overuse), then the area could sustainably support 86 people per sq. km, far in excess of the present population. Surplus food could then be sold to town dwellers or used to produce meat. It is unlikely that food production could sustainably be raised much above this level with present technology, even if sophisticated.

The present annual population growth rate in the Chiweshe district is about 3.6%, with a total fertility rate of 6.32. Sixty percent of the labor force is under the age of 15. At this rate of population growth, even with sophisticated agriculture there will be a food deficit by roughly 2010. With present-day agricultural skills, on the other hand, there will have to be either massive importation of food enabling the population to survive into the next century, or sharply increased mortality in the next major drought.

The people of the Chiweshe area felt that human population pressure was becoming intolerable, but considered that this problem could readily be overcome by resettlement on commercial farming land. If carried out widely across Zimbabwe, resettlement would relieve the agricultural deficit in communal areas until about 2010, but at the price of removing the food source for the towns, which would have to depend on food imported from other countries. Foreign exchange to pay for imports would be limited, as the surplus agricultural production for export at present generated by the commerical sector would disappear. After about 2010, with a doubled population, the whole rural area, including the resettled area, would be in food deficit, and the country would be wholly dependent on food aid. Some African nations are already in this state.

Despite this, local attitudes on population measured by the Save study were robust and typical of poor regions worldwide. To quote one Ziyambe resident: "To have many children is to be rich." Fertility goals were high, with most women wanting six or eight children or more. The majority had access to a clinic during pregnancy, but most methods of contraception were viewed with caution or rejected. Hormonal contraceptives were rumored to cause serious side effects, including disease and deformity in children. Young couples were under much pressure from extended families to produce children.

In particular, the attitude of men is illustrative. They were totally opposed to contraception. One comment was that "a man without children has no power." Women who wished to use oral contraceptives had to do so secretly. Furthermore, many were threatened with divorce if their fertility was not maintained.

The likely future in the area is for steadily rising population and rapid degradation of the land, so that eventually population will be far in excess of carrying capacity. Even if resettlement occurs, the relief on the land will only be temporary, and the national food supply will be seriously endangered. Most probably,

Table 8.1. *Zimbabwean women's opinions of the "best" number of children to have, Chiweshe and Ziyambe areas, Save study*

Number of children thought "best"	Percentage of women so choosing	
	Chiweshe ($N = 110$)	Ziyambe ($N = 57$)
4–5	20	21
6–8	34	40
9 and over	15	18
"many"	31	21

Source: Campbell et al. (1989).

serious drought in the next one to two decades will cause mass mortality, even if alleviated by food aid from Europe and North America.

Alternatively, if population could be stabilized, say at 46 people per sq. km, and if food production could be increased to the levels of neighboring commercial farmland, the region could produce a surplus sufficient to sustain 40 people per sq. km over and above the needs of local inhabitants. Such a surplus, if exported to local towns and cities, would bring substantial revenue to the area and initiate an economic takeoff. Without a cessation in population growth, however, economic takeoff and environmental protection are virtually impossible.

A stable population is essential if Zimbabwe is to survive as an economically viable society. The pattern is the same in most of the poor nations. If the biosphere of the planet is to remain capable of processing the atmosphere, it cannot be allowed to collapse in the poor tropical nations. Any local collapse will damage or degrade the environment worldwide; regional or continental ecological collapse threatens the stability of the entire global air-conditioning system. For this reason, population control is not only essential to Zimbabwe but to all nations. It is in the strong interest of the rich nations to help stabilize the population of the poor countries.

The attitudes of the people of the Chiweshe area toward their own fertility are of the greatest worldwide importance. If these attitudes cannot be changed, the global prospects for a stable environment are poor. Yet the potential is there. The area can sustain a fertile and profitable agriculture and a protected environment if action is rapidly taken to reduce the rate of reproduction.

8.3 Education

In much of rural Africa, children are only educated for a few years. They spend much of their childhood herding cattle or tending crops. This labor is very valuable to the society, just as Victorian England exploited child labor. When child labor is abolished, wages rise. Because it is impossible to legislate against the use of child labor by parents, the only way to abolish it is to have full education from the age of 6 to 16.

Table 8.2. *Estimated size and growth of world population, 1955–2000*

Region	Estimated population (thousands)				Estimated annual % change				
	1960	1987	2000		1955–60	1965–70	1975–80	1985–90
World	3,018,878	4,997,609	6,121,813		1.86	2.04	1.75	1.63
Africa	280,051	589,208	871,817		2.31	2.62	2.97	3.02
Nigeria	42,305	101,922	161,930		2.63	3.23	3.49	3.49
North & Central America	268,579	411,933	487,379		2.03	1.60	1.53	1.38
Mexico	37,073	82,964	109,180		3.15	3.25	2.86	2.39
Cuba	7,029	10,214	11,718		1.79	1.87	0.84	0.98
United States	180,671	242,159	268,239		1.70	1.08	1.06	0.86
South America	146,839	279,384	356,351		2.73	2.45	2.27	2.08
Brazil	72,594	141,459	179,487		2.97	2.57	2.31	2.07
Asia[a]	1,668,165	2,913,191	3,548,994		1.95	2.43	1.86	1.63
Japan	94,096	122,053	129,725		0.93	1.07	0.93	0.51
Bangladesh	51,585	106,651	145,800		2.38	2.66	2.83	2.61
India	442,344	786,300	964,072		2.26	2.28	2.08	1.72
China	657,492	1,085,008	1,255,895		1.53	2.61	1.43	1.18
Europe	425,129	494,529	512,474		0.83	0.64	0.42	0.27
West Germany	55,433	60,628	59,484		1.13	0.56	−0.09	−0.18
USSR	214,335	283,993	314,736		1.77	0.91	0.93	0.93

Note: In Europe, Germany and several other countries have declining populations. In the rest of the world only Cambodia and Lebanon briefly suffered decline. In 1960, Nigeria had fewer people than England. It now has twice as many.

[a] Excludes USSR.

Sources: World Resources Institute (1987, 1988).

Table 8.3. *World population, 1950–85: key facts*

	1950	1960	1970	1980	1985
Total population (billions)					
World	2.5	3.0	3.7	4.4	4.8
Richer regions	0.83	0.94	1.05	1.14	1.17
Poor regions	1.68	2.07	2.65	3.31	3.66
Growth rate (%)[a]					
World		1.8	2.0	1.9	1.7
Richer regions		1.3	1.0	0.8	0.6
Poorer regions		2.1	2.5	2.3	2.0
Urban population (%)					
World	29	34	37	40	41
Richer regions	54	67	67	70	72
Poorer regions	17	22	25	29	31

[a] Annual rate over previous decade or, for last column, over previous five years.
Source: World Commission on Environment and Development (1987), quoting Department of International Economic and Social Affairs, *World population prospects: estimates and projections as assessed in 1984* (New York: UN, 1986).

Poor nations can afford education. Another way of phrasing this is to ask if the society can set aside from food gathering one adult for every 20 or 30 children, to teach them. The answer is that almost every human society has enough surplus to afford basic education. The surplus is demonstrated by the availability of money and labor for military purposes (Table 8.5). Diverting literate adults into education *is* possible. The standards may be low, but it should be possible within a decade to attain universal education. If this sounds optimistic, it is something that has nearly been achieved in a shorter period in countries as disparate as Zimbabwe and Cuba. There has been an explosive change in both these nations. In the period 1981–8 Zimbabwe has come close to universal education. When the children are in school, their parents come to understand that children who need to be fed and clothed to the age of 16 are expensive. In place of the present economic advantage to parents of children who can be used as labor, an economic disadvantage is substituted. This is true even if the children spend half their time playing soccer as physical education, or watching educational TV (if rural electrification can be achieved). Ideally, children should be taught literacy, agriculture and the management of the local environment, and the rudiments of technology. The next generation may aspire to general relativity.

Basic education demands very little in the way of foreign currency. At its simplest, universal education is a matter of setting aside some proportion of the trained adults of a state to teach the next generation. Seen in the context of defense spending in poor countries, the denial of universal education is one of the world's greater immoralities.

Table 8.4. *Population living in urban areas, 1950–2000*

Region	1950	1985	2000
	Percentage		
World	29.2	41.0	46.6
Wealthier regions	53.8	71.5	74.4
Poorer regions	17.0	31.2	39.3
Africa	15.7	29.7	39.0
Asia	16.4	28.1	35.0
China	11.0	20.6	25.1
India	17.3	25.5	34.2
Latin America	41.0	69.0	76.8
Temperate South America	64.8	84.3	88.6
Tropical South America	35.9	70.4	79.4
	Total (in millions)		
World	734.2	1,982.8	2,853,6
Wealthier regions	447.3	838.8	949.9
Poorer regions	286.8	1,144.0	1,903.7
Africa	35.2	164.5	340.0
Asia	225.8	791.1	1,242.4
Latin America	67.6	279.3	419.7

Source: World Commission on Environment and Development (1987), quoting UN Population Division, *Urban and rural population projections, 1984, unofficial assessment* (New York: UN).

There is a determination in some but not all poor states to attain universal education quickly. Other nations place a lower value on education, preferring instead to use resources for military purposes (Table 8.5). Aid from rich nations has traditionally been for both military and civilian purposes. Arguably, it is in the strong interests of the richer nations to ensure their own environmental security by directing as much aid as possible away from the military and into educational and environmental purposes. Furthermore, pressure can be brought on the governments of poor nations to increase their own educational and social spending. However, such policies can be attacked as paternalistic intervention.

Poor nations are independent and fiercely antipaternalistic. They ask not to be treated as children in the family of nations, and this should be honored. Western nations have their goals and so do the poor nations. Negotiation should be between equals, without paternalism. This statement has its reverse side too. Negotiations without paternalism, between equals, can be hard. Each side has specific goals and specific weapons. Western nations have certain beliefs, among them the belief in human rights. They are untrue to their own ideals when they fail to demand that poor nations shall place education before defense or corrup-

Table 8.5. *Military and educational expenditures, selected countries*

	Military expenditure (1987)		Educational expenditure (latest year obtainable; % of GNP or GDP)	Ratio of military to educational expenditures	Literacy rate (%)
	$ billion	% of GNP or GDP			
Australia	4.99	2.5	5.6	0.45	99.5
Bangladesh	0.32	1.8	2.1	0.86	33.1
Botswana	0.02	2.2	6.0	0.37	70.8
Bulgaria	6.66	10.3	7.1	1.45	95.5
Burkina Faso	0.05	3.1	2.5	1.24	13.2
Canada	8.84	2.2	7.4	0.30	95.6
China	20.66	4.4	2.7	1.63	72.6
Costa Rica	0.02	0.6	5.2	0.11	92.6
Cuba	1.60	5.4	6.3	0.86	96.0
Egypt	6.53	9.2	5.5	1.67	44.9
Ethiopia	0.44	8.5	3.9	2.18	3.7
France	34.83	4.0	6.1	0.66	98.8
Guyana	0.34	8.9	10.1	0.88	95.9
Iran	21.12	7.9	3.8	2.08	61.8
Iraq	16.70	30.7	3.8	8.08	45.9
Japan	24.32	1.0	5.1	0.20	100.0
Lesotho	0.01	2.3	3.5	0.66	73.6
Mauritania	0.04	4.2	7.9	0.53	28.0
Mozambique	0.10	8.4	1.2	7.00	16.6
Nigeria	0.18	0.8	1.8	0.44	42.4
Pakistan	2.23	6.5	2.1	3.10	25.6
Peru	2.20	4.9	2.9	1.69	87.0
Senegal	0.10	2.2	4.7	0.47	22.5
South Africa	3.40	4.4	3.8	1.15	79.3
Tanzania	0.08	3.3	4.3	0.77	85.0
USSR	303.00	12.3	7.0	1.76	99.0
U.K.	31.58	4.7	5.2	0.90	100.0
United States	296.20	6.5	7.5	0.87	95.5
Venezuela	1.38	3.6	6.8	0.53	89.6
Zambia	0.17	6.6	5.4	1.22	68.6
Zimbabwe	0.28	5.0	7.9	0.63	76.0

Note: Literacy is differently defined in various countries. Thus U.K. and Japanese figures reflect all considered capable of reading, Canadian figure excludes functionally illiterate. GNP, GDP = gross national/domestic product.
Source: Encyclopedia Britannica Yearbook 1990.

tion as a national priority. If a state, because of corruption, refuses to educate its people, it can be argued that the West should remove all aid and institute graduated sanctions, ranging from selective taxes on exports and imports to, if necessary, complete isolation. In the past three decades, aid to poor nations has

Table 8.6. *Change in gross domestic product (GDP) in low- and middle-income economies, selected regions*

	1986 GDP ($ billion)	Annual change in per capita GDP		
		1985–6	1987	1988
Sub-Saharan Africa	154.3	−2.6	−4.5	−0.4
East Asia	631.1	6.7	7.0	7.9
Europe, Middle East, North Africa	748.1	1.1	−0.5	0.3
Latin America, Caribbean	699.5	−0.7	−0.6	−0.6

Note: Figures for each region include only low- and middle-income economies and exclude high-income nations. For comparison, the 1986 GDP of sub-Saharan Africa, excluding South Africa whose gross national product was about $60 billion, was about $95 billion. The 1986 gross national product of Belgium was $91 billion; of Switzerland (pop. 5.4 million), $115 billion.
Source: World Bank Annual Report, 1989.

been given for strategic military or political reasons, not to support environmental policy. Often this has meant that corrupt regimes have been sustained. Environmental policy, in contrast, can only succeed in nations with an educated and aware population. Instituting sanctions against corrupt and dictatorial regimes may seem severe on the common people. But, as was pointed out during the independence debates in the 1960s, the people of Africa must govern themselves, be free to choose a bad government, and be prepared to accept and live with the consequences; this is a necessary corollary of independence. The poor nations, as adults in the family of nations, have the right of equal action and denunciation against Western polluters. Their best weapon is the opinion poll in the democracies.

Universal education must become one of the chief goals of Western foreign policy. The right to education is a basic human right. It is also an urgent goal in the struggle to halt population growth. Only in school do children become consumers of family income, not profitable family income earners.

8.4 Health

Good health is the next prerequisite. If children commonly die, families will have more children in order to compensate for the loss. If a parent wishes to have two children, but expects to lose half of all offspring, then there is an incentive to have five or six so that eventually at least two live.

The obvious answer is public health. Health, like education, is not by definition expensive, although it can become so. Doctors need not consume an excessive proportion of national income. Cuba in many ways has been a most unattractive society that has existed on the largesse of the Soviet Union, but Cubans

Table 8.7. *International food aid: selected recipient and donor countries*

Recipient country	Average annual receipt of cereals (in thousand tons)		Average annual population growth (%, 1975–80)
	1974–6	1985–7	
Africa	1799	6587	2.97
Chad	30	91	2.10
Egypt	620	1909	2.69
Ethiopia	79	746	2.32
Kenya	4	195	4.03
Madagascar	24	92	2.70
Mali	109	142	2.19
Morocco	100	424	2.27
Mozambique	31	328	4.42
Somalia	61	182	4.23
Sudan	50	881	3.08
Tanzania	89	82	3.42
Tunisia	73	223	2.61
Zaire	8	98	2.86
Zambia	3	105	3.08
Americas			
El Salvador	4	233	2.93
Haiti	20	108	2.38
Jamaica	4	253	1.24
Bolivia	44	207	2.59
Peru	32	211	2.63
Asia	4202	2899	1.86
Bangladesh	1308	1463	2.83
India	981	204	2.08
Indonesia	236	233	2.14
South Korea	261	0	1.87
Pakistan	585	417	2.84
Sri Lanka	175	309	1.71

Donor country	Average annual donations of cereals (in thousand tons)	
	1974–6	1985–7
Australia	271	393
Canada	770	1133
United States	4060	7357
Argentina	10	40
China	59	−172 (recipient)
Japan	188	413
France	164	202
West Germany	175	242
Italy	51	137
Netherlands	48	128
Sweden	143	77
United Kingdom	96	129

Source: World Resources Institute (1988, 1990).

Table 8.8. *Percentage of population over the age of 60, selected countries (most recent census)*

Australia	15.0
Bangladesh	5.7
Botswana	6.8
Bulgaria	17.6
Burkina Faso	5.5
Canada	14.8
China	7.6
Costa Rica	5.6
Cuba	10.9
Egypt	6.2
Ethiopia	6.2
France	17.6
Guyana	5.4
Iran	4.8
Japan	14.7
Lesotho	7.5
Mauritania	4.6
Mozambique	4.3
Nigeria	3.5
Pakistan	6.9
Peru	6.0
Senegal	6.1
South Africa	6.2
Tanzania	6.1
USSR	13.0
U.K.	20.2
United States	15.7

Note: These figures should be contrasted with figures in Table 8.5 for military expenditure as a percentage of gross national or domestic product, to judge whether old age care is affordable.
Source: Encyclopedia Britannica Yearbook 1988.

have a life expectancy at birth that is comparable to that in the United States. The effectiveness and quality of health care in the two societies are thus similar, but the Cuban medical system is much less expensive. Its high efficiency is achieved in part by preventive public health measures, which are much less costly than the curative individual health care practiced in the United States. It is cheaper to avoid becoming sick than to become sick and then be cured. There are, of course, many things wrong with Cuban society, but for a poor nation its health care system is impressive.

It is not a coincidence that the annual growth rate of the Cuban population was

Table 8.9. *Female literacy and birthrate in selected countries*

	Female literacy (%, 1985)	Crude birthrate (per thousand, 1985–90)
Algeria	37	40.2
Botswana	69	47.3
Burkina Faso	6	47.2
Cameroon	55	41.6
Egypt	30	36.0
Ethiopia	1–2?	43.7
Gabon	53	38.8
Kenya	49	53.9
Libya	50	43.9
Nigeria	31	49.8
Uganda	45	50.1
Zimbabwe	67	41.7
Canada	99	14.1
Costa Rica	93	28.3
Cuba	96	16.0
El Salvador	69	36.3
Haiti	35	34.3
Jamaica	93	26.0
Mexico	88	29.0
United States	99	15.1
Brazil	76	28.6
Chile	96	23.8
Paraguay	85	34.8
Peru	78	34.3
Uruguay	94	18.9
Afghanistan	8	49.3
Bangladesh	22	42.2
China	56	20.5
India	29	32.0
Indonesia	65	27.4
Iran	39	42.4
West Germany	99	10.4
Sweden	99	11.2

Note: For West Germany, Sweden, Canada, and the United States, literacy rate of 99% is assumed for nondisabled population. In recent years, birthrates have fallen significantly in the more prosperous African countries, such as Zimbabwe, but may be rising in many Islamic states, China, and India.

Source: World Resources Institute (1990).

1.87% in 1965–70 and is now below 1%. The crude birth rate was 32 per 1000 in 1965–70; it is now 18.2 per 1000 and the fertility rate has fallen from 4.3 to below 2.

The spread of AIDS in some poor nations has now become so wide that the disease may check population growth, both through direct mortality and by selective killing of young adults and small children. At present, slowed population growth is only being seen in some districts of central Africa, but it is likely that as AIDS infection rates increase in Latin America and Asia, the same effects will be seen there. Some planners have quietly welcomed this, as a means of controlling population. Such attitudes are high immorality, comparable to the views of Nazi eugenicists. AIDS must be vigorously fought in Africa and elsewhere by public education and by social programs, especially demilitarization (it is in Africa's armies that AIDS rates are highest). If it is not fought, it will selectively decimate the young skilled people in the poor nations, removing the only work force that can create an environmentally sound and sustainable economy. It will also, inevitably, spread to the heterosexual population of the rich nations. Both ethics and interest therefore demand that AIDS be strongly attacked in the poor nations, with support from the rich countries.

8.5 Old-age pensions

Most third-world nations have a pyramid-shaped population structure, with a very broad base of young people and a very small peak of elderly. Typically the young adults outnumber the elderly manyfold. In the area of the Save study, half the population was under 15, and only 3% over 65. In this context, the imposition of taxes on the working population to care for the elderly is not difficult. Old-age pensions represent a collective contribution from the younger, working population to care for the old. Any society can afford them. In many poor countries the population over 65 is so small that the cost is trivial to the society in general – quite the opposite of the situation in many richer nations.

The psychological impact of pensions is immense. A proper pension scheme that guarantees the security of the old (including especially old women in male-dominated cultures) removes much of the psychological need for children as supports of the elderly. The society as a whole – the children of other people – assumes the burden. Implementing pension schemes in Africa and and Latin America needs to become a prime goal of Western policy and of all negotiations to reduce the debt of the poorer nations. Fortunately, pensions are like public health and education. Their cost in foreign exchange is minimal. In Africa they will impose little internal fiscal strain because of the extraordinarily youthful structure of the population. Pension schemes there are cheap and easy to implement and cost far less than maintaining armies (see Tables 8.5 and 8.8). Pensions may be costly 50 years from now, but that is a problem for the next generation.

One common African custom, related to pensions, is that of the bride-price paid by the groom to compensate the parents of the bride for the effort of raising her. In some African cultures, generations ago, the bride-price was symbolic –

a few mice or birds – but today it has become equal to several years' income of the groom or many cattle. Typically, it is little more than the sale of a girl. The income from the sale of daughters is, of course, an incentive to raise daughters. It is also a substitute for a pension; daughters are the guarantee of a comfortable old age. The bride-price system is widespread and immoral as practiced today. It can be eliminated by the institution of state pension schemes in southern and central Africa, if the schemes are accompanied by vigorous efforts, both in law and in advertising, to persuade the population that state pensions are equivalent to the bride-price. Such a move would be welcome to young men, who would no longer need to labor to marry; it would be accepted by the elderly. Divorce, on grounds of infertility, is common ground for the request that the bride-price be repaid. This means that family pressure on a woman to produce children is strong, so that her parents need not return the bride-price. If the price were removed, in the short term there would be an increase in marriage and children. In the medium and long term, the incentive to have children, and the father's need to have salable daughters, would fade. Peasant societies have children for good reasons. If the reasons disappear, so will the children.

8.6 The status of women

This leads to the question of the status of women. The attitudes of a society are illustrated by the attitudes of its leaders. The title of this chapter is carefully chosen. Once on a flight in central Africa I sat next to a leader, a professor of politics in a local university. We discussed the plight of the inhabitants of South Africa. Then I raised the other great oppression of Africa, the treatment of women. His reaction was immediate and aggressive. It echoed strangely and exactly the most reactionary views about race. In his opinion, women were born inferior and had no need for liberation, because they would not benefit from it. They were best looked after by their superiors, men. Even in Zimbabwe, which is politically sophisticated, women were legally minors until 1982 and still have inferior status in many communal areas. This philosophy is shared by the leaders in much of Africa, southern Asia, and Latin America. Until recently, it was also the opinion of the West.

If the rich nations believe their own present opinion that women are the equals of men, then they have no honest course except to attempt by whatever means possible to change the opinions of the poorer countries. This is not unprecedented – societies have always imposed their opinions on others. Within the relatively recent past, the Canadian courts held that women were not persons. Many Canadian women now alive grew up as nonpersons. It took the British House of Lords to impose the belief on Canadian society that women really are people. The attitude of the American founding fathers and nineteenth-century British Liberals was that certain liberties were so important that they had to be told abroad. This view, which spread democracy through the world, is now unfashionable; we shrink from influencing other cultures. Yet, just as Canadian society is now pleased that women are people, so too may other women, in those poorer nations that are male-dominated, be grateful for learning this truth.

Among the rural poor of the Earth, especially in Africa, women are the main source of physical labor. In the area of the Save study, they hew wood, draw water, plant crops, and prepare food. Childbearing interrupts this process only in a constant, minor way, especially in the polygamous societies that are standard in Africa and parts of Asia. Cooking food, hours for each meal, occupies a substantial part of the day. In contrast, in an electrified society, where cooking can be done rapidly, or by electric slow-cookers that do not need tending, women can enter the money-earning labor force. Working affects attitudes on childbearing. It becomes more profitable for the mother to work than for her to produce children, especially when children are an economic drain, not an asset. Children become part, not all, of a woman's life, and husbands begin to participate in childrearing.

The emancipation of women is better achieved by subtle economic means than by direct propaganda. If women are allowed to abandon their wood and water gathering, are able to cook their food in less than two hours' labor, and are able to send their children to school, they will find or invent money-making work. Political leaders seem to forget that it is women who have children. If women are at money-making work then men will find it profitable not to force them to have children.

Most women, even in the richer nations, are ignorant of the detailed physiology of their own reproductive cycles. They are not taught or told when ovulation occurs, or when they are fertile. Consequently, for most women, contraception is an aggressively intrusive business, whether by means of the pill, condoms, or diaphragms (as in the West), or by abortion (as in most centrally planned countries and in parts of the East). For African women, mechanical methods are prohibitively expensive unless given as foreign aid. For many, especially in the strongly Roman Catholic or Protestant districts that are widespread in rural Africa, abortion is anathema because it makes the weakest die to help the weak.

One contraceptive method that is acceptable throughout the poorer nations is the Billings method, which is endorsed by the Roman Catholic Church. In this method, women are taught the simple facts of their ovulation cycle, and how to detect the days of fertility by understanding the pattern of cervical secretions expressed at the vulva. The method needs no technology. It is controlled by the woman, not the man (unlike contraception by condom), and it is secure in a stable relationship. The method, in consequence, is a powerful influence toward building long-term partnerships in which the woman has an equal, not a subordinate, role. Given an economic climate in which rearing children is financially disadvantageous, the Billings method, easily learned and costing nothing, could be the instrument for a massive reduction of fertility in the poor nations, while at the same time helping to construct stable monogamous family units as the foundations of a revitalized, skilled society in which women are treated as equals.

Slowing population growth is social engineering on a grand scale, but it will come about if the economic forces are right and the knowledge is available.

Women will have freedom; they will be needed in the economy and to teach. If societies are electrified the women will be able to work, and not just cook or gather fuel. If society is urbanized, the women and children will not be in the fields, nor will they be gathering water. The liberation of women will not easily be achieved, but when it occurs the population crisis will be over.

The population crisis *is* manageable. Poor people act just as rich people do when faced with the same social forces. The examples of Quebec and Cuba are relevant to Africa and Asia and have shown that in a decade the Earth's population can be stabilized. If we act quickly, we shall, early next century, have a planet of 7–8 billion people, but this will be a stable, perhaps slowly declining population. If we do not act soon, we shall have not an Earth, but a desolation – 10 or more billions. Ten billions surely can not be sustained; mostly likely, what will follow will be collapse, warfare, and diseases such as AIDS and malnutrition that thrive in an unstable society. There will be innumerable migrants pounding on the borders of the West. Many tens of millions will get through, to stress the social systems of once-complacent Europe, North America, and Japan.

Health, education, pensions, social reform – the remedies to the problems of population – are expensive but not impossible to implement in poor nations. They may cause drastic inflation in those third-world countries that choose also to maintain large armies of soldiers, but most of those states are already economically mismanaged. The example of Cuba – floating on subsidy from the Soviet Union – shows, ironically, that it is possible simultaneously to maintain large armies of soldiers, teachers, and doctors *and* to reduce population growth.

In the management of the planetary economy it is essential to replace a fast-growing, ill-educated, and unhealthy population with a stable, educated, long-lived humanity. This is an ethical goal. The poverty of the third world is intolerable. But it is also a matter of interest. If the goal is not attained, the security of the rich nations will be gravely threatened. The goal *is* attainable, and it must be attained very soon. The means must be the vigorous and uncompromising use of financial aid, debt, and advice as weapons in the war for humanity.

8.7 Implementation: the 1950 Earth model

Biological reactions are very curious. By the smallest of nudges they attain the most improbable results. Sometimes in the workings of a living cell there are tiny inputs of persuasion derived from farther down the chain, and on other occasions the kinetics of a reaction are exploited, so that simply because it is quicker to get started in one direction the system follows a path that is not necessarily the route one would expect. Small changes over a long chain of events can have great results. It is how people are made. This is the only way that it will be possible to shift the workings of society toward a sustainable economy of the globe.

The first, difficult step is to recognize the seriousness of the crisis, yet to acknowledge how little we know and to avoid action on the basis of inadequate knowledge, without using this as an excuse for inaction. In other words, we should not panic, but we should begin action and simultaneously improve our knowledge. All we firmly know is that there are too many of us, that we are adding rather too much to the atmosphere, and that we are drastically changing the Earth's vegetation. The only, but adequate, basis for action is the strong suspicion that this is not wise.

The year 1950 is a good reference point for the natural Earth. The state of the climate and vegetation in this year was moderately well recorded. Nineteen fifty is a datum used in the geochronological dating of carbon. It stands at the start of the four decades of industrial growth in the postwar period and represents perhaps the last time at which the global climate could be described as natural. Before 1950, although major change had already begun and the landscape of the temperate continents was man-made, human agriculture and the natural vegetation of the planet were still roughly adapted to the rainfall and climate patterns of the preceding centuries. Since 1950, change has been rapid and climate will shift. Should humanity ever decide collectively to manage the climate, 1950 could well become a reference point to which we may decide to adjust the workings of the biosphere. In fact, 1950 was a cool year, but the weather of the five years around 1950, on average, is representative of the conditions under which most of today's adult humanity grew up. Scientifically, we know that in 1950 the vegetation of the planet was still in reasonably good health although already damaged by deforestation, by the expansion of cropland, and by industrial emissions. As a conservative starting point, it is probably sensible *to devise actions that will tend to return the natural vegetation of the world toward its state in 1950*. We cannot, obviously, attain that state, simply because we now have too many people; we have left Eden. But it is a known state that supplies a reference standard.

An alternative would be to use computer models to design a best possible Earth. We may do this in some distant future, but there is the risk that the models may be wrong. Furthermore, people do not agree, and nations would be differently affected. The use of the 1950 Earth is much safer *as a basis for international negotiation*. We can agree on it since we are all still adapted to it. We should, however, also recognize that all our efforts must include a component of adapting to change, for we shall never fully reattain the 1950 state. Too much is lost.

8.8 Multilateral policy: GATT to GATE?

Over the past 40 years a wide variety of multilateral constraints on human behavior has been agreed upon. These include the General Agreement on Tariffs and Trade (GATT), the rules of the international monetary system, and regional alliances such as the Treaty of Rome, under which the European Community is constituted, the U.S.-Canada Free Trade agreement, and many others, such as the Lomé convention between poor nations and Europe. Each agreement has involved some loss of sovereignty of the contracting parties in exchange for the reward of increased national security. There is now no longer any country, except perhaps Burma (Myanmar) or Albania, that is fully independent. The agreements, surprisingly, have had a major influence on world behavior, not because they are in themselves especially powerful, but because they set standards that are nearly honored by most nations. There are of course many fouls in the game, and the referee is often blind or partisan, but on the whole the game continues because it is better than fighting.

There has been much talk about a "Law of the Atmosphere," an excellent idea and probably impractical. Biological organisms, in a chain of reactions, never go straight to an end point – it is too difficult, too costly, and the organism would die before it was achieved. We do not clearly know where we are going. Even if we did, a Law of Atmosphere would take a generation to achieve. It would probably eventually become like the Law of the Sea Convention, which is often a profound hindrance to cooperation and to scientific research, and has in some cases become an excuse for exercises in self-interest, national greed overriding environmental need.

More successful and more important is the General Agreement on Tariffs and Trade (GATT). GATT has often been broken, but over its life it has been a powerful force in regulating world trade and in moderating relationships between nations. If you know the laws, you tend to break them only a little. GATT provides for a negotiating body that has an effective and long-established tradition of managing world trade and removing barriers to the movement of goods. At present, GATT is not concerned in any way with environmental problems and has been very unsuccessful in dealing with agricultural issues. It is responsible for agricultural trade but has failed almost completely to regulate this sector, in contrast to its success in dealing with manufactured goods.

Effective regulation of international trade in agricultural products is potentially a very powerful tool in protecting the environment. Much of the business of GATT is concerned with the prevention of *dumping* – selling goods in foreign markets at prices below the cost of producing the goods, either to destroy competing producers in the country of sale, or as a result of excessive subsidy in the country of origin. Most agricultural trade today is, in effect dumping, destroying producers in poor countries by subsidizing those in rich countries. Between the rich countries acrimonious arguments have broken out that threaten the future stability of all international trade.

Environmental dumping also occurs. Examples already discussed include cutting forest to grow cassava or rear cows, and allowing industry freedom to pollute and thereby be more competitive. Often, such dumping is difficult to prove, but no more so than it is to prove cases of dumping in industrial trade, a problem that GATT has successfully handled for many years. If industrial dumping is proven, GATT allows graduated sanctions to penalize the dumper. Similarly, if environmental dumping were to be proved, trade sanctions could be applied to make the dumping unprofitable and to penalize the offender.

A General Agreement on Trade and the Environment (GATE) could become a powerful force for protecting the biosphere. In such an agreement, agricultural markets and subsidy would have to be fully decoupled. Subsidy would go to farmers as environmental managers, *not* to them as producers. Each farmer would be paid to manage the land, but crop production would be wholly free of subsidy. In the United States at present, moves are being made toward this type of agricultural management, with farmers being paid for "set aside" acreage that is not used for crops. In Europe, the typical small family holding is well adapted to becoming a land management unit. Similarly, in Africa the communally held peasant communities could rapidly evolve into European-style smallholdings where freeholders are paid according to their success at managing the environment. If land management and food production were to be decoupled worldwide in this way, it would then be possible to have a free market in agricultural products, with international trade watched over and regulated by a GATE.

A GATE agreement, as outlined here, would be complex and would take many years to become fully effective, but it would grow naturally out of GATT. Its logical basis would be clear and straightforward, with none of the Byzantine intricacy of the CAP and the harem intrigues of the debates about agriculture in Washington. The sums of money spent today on subsidizing agricultural production are adequate to restore much of the planetary ecosystem, if better used. A GATE agreement, based on environmental policy, could set targets for environmental reconstruction, especially in the major temperate land masses. It will not be an easy matter to determine reference points for environmental targets, which is why the state of the planet in 1950 could be a better goal than some theoretical model of a perfect Earth. Nevertheless, given careful thought it should be possible to direct agricultural subsidy away from international dumping and toward environmental management. Agricultural policy *does* manage the planet's vegetation. In each nation, policy needs to be changed.

A global energy policy may be made possible by a general agreement to set goals for the amount of CO_2 and CH_4 released into the atmosphere – say a 20% reduction soon, with a much faster cessation of the release of CFCs, leading to a greenhouse-neutral economy early in the next century, with no net emission of CO_2 and CH_4. Nations determined by GATE to be failing to meet their commitments would be penalized by allowing other countries to impose import taxes on goods sold by the offending nations, in proportion to the extent to which the offenders missed their goals. If the potential cost of the tax penalty were greater than the cost of reducing CO_2 and CH_4 emission, this would impose a great incentive to behave. The taxes need not be great – previous GATT experience, for instance in Canada, and similar U.S. experience has shown that a small tax can remove a competitive edge and destroy a market.

Virtually every aspect of trade involves the environment, and any revision of the GATT agreement must recognize this. Does one trade cotton – what does this do to the flow of the Nile River? Does one import cassava? – what is the impact on forest? Does one import wood? – what is the effect on the latent-heat transfer in the atmosphere? Does one build cars? – what is the impact of their emissions? Since each question is complex, the 1950 standard is useful; we simply ask if the effect is toward or away from the natural vegetation and atmosphere of the Earth as it was then. If it is away from the 1950 standard, a determination of unfair environmental practice can be made.

It will take years to negotiate a GATE, but the task is not impossible. Even the attempt to negotiate, however, in itself will mean that countries begin to synchronize national policy. Once implemented, it will be a steady force toward a better world.

8.9 The G-7 meetings: G-12?

A recent development in international affairs has been the annual G-7 meetings, involving the heads of government of seven major industrial powers: the United States, Japan, Germany, France, and the United Kingdom, Italy, and Canada. Judging from media reports, not much, apparently, is done in these meetings; nevertheless, the meetings ensure coordination of policy. National policies are gently shifted by a network of consultations at lower levels, in an awareness of the international trend. Collectively, the outcome is the beginning of a rudimentary system of planetary government.

The environment, on a planetary scale, is very much the concern of the G-7 group, which coordinates policy on the debt of the poor nations and on energy. The responsibility of the G-7 group is to consider the proposal that debt be linked with environment, to consider the global impact of aid programs to the poor, and to consider global energy policy.

An initial step would be to link the G-7 meeting to an expanded G-12 meeting of the environmental superpowers, including the Soviet Union, China, India, Brazil, and Australia. It is often counterproductive to have too large a group, or too many meetings, but there would be many merits in a G-12 group. It would have the collective power to impose any decision. More likely, though, the exchange of views would serve not to decide but to coordinate, advise, and reduce disputes. It might even reduce international tensions so that part of the money that is at present spent to defend humanity from itself may instead be spent to improve the security of humanity. Scientific programs such as the International Geosphere-Biosphere Program, would come under the general aegis of a G-12 group. So would space exploration, fuel policy, vegetation policy, and the implementation of a General Agreement on Trade and the Environment – all these would represent a shift to a new global economy.

8.10 The Commonwealth and the Francophonie

The Commonwealth of Nations is in decay: perhaps it should be restored. At present it seems to be merely an organization of faction, dispute, and games. It has, however, the peculiar advantage of including many of the world's poorest countries, linked by the world's richest culture. Poor nations remain poor for a reason: the route to wealth is not so much via technology, but via the market organization that allows technology to prosper. Organization comes out of education, on the whole, and it may be the Commonwealth's best role to spread education through a Commonwealth-run chain of universities and technical colleges in the poor countries, possibly on extraterritorial Commonwealth-ruled land. The Commonwealth could also undertake the electrification of rural Africa, using power stations similarly sited on extraterrestrial land under Commonwealth control.

Nationalism has not served Africa and Asia well, nor South America. It may be time to create supranational institutions, but to base them in the poor countries. The bulk of Africa's external debt will have to be forgiven; the best way to do this is to extract repayment in internal currency. The money, drawn away from African defense spending, would be used to improve both the environment and education. Security could be provided by a small Commonwealth standing army, charged with maintaining peace in those countries that have chosen to reduce their own armies and to redirect the labor to the tasks of improving their schools, hospitals, and social security and protecting their natural vegetation.

The same is true in the Francophonie, the French-speaking equivalent of the Commonwealth, including France, Canada, Belgium, and former colonies. Perhaps both the Commonwealth and the Francophonie should become the joint financial responsibility of the European Community. Both the Commonwealth and the Francophonie are well-suited to the task of electrifying Africa. With European help, they could do this rapidly.

8.11 The European Community

History is rather like a good football match. For long periods, there is little apparent action. Pressures build up, subtle change takes place, but on the surface all is pushing and shoving, with no obvious movement. Suddenly, the break comes. Decisions have to be made, quickly. The right decision brings success; wrong decisions can spoil opportunity or even initiate disaster.

This book is appearing at a moment of breakout. In Eastern Europe, all is flux. Nationalism, suppressed under the brutal Stalinist system, is reemerging; economies are in decay. Western Europe is distracted by internal political events in the major nations and is reacting to, not acting so as to influence, the changes in the East. In environmental terms, these events are of great global importance. The coal-based economies of the post-Communist nations in Eastern Europe and the Soviet Union are among the chief sources of global pollution. The countryside of these nations is under severe environmental stress and is locally in biological collapse. Western financiers are urging the leaders of Eastern Europe to rescue their ailing, debt-ridden economies by stoking up the fires of industry. If this is done, using present factories and industrial plant, the pollution that will result will severely stress the plant life of much of Europe and will have global consequences to CO_2 levels, Arctic haze, and other forms of pollution.

For environmental reasons, it is urgent that the industry of Eastern Europe be recapitalized, to enable it to become fuel-efficient and much less polluting. The only source of funds to carry this out is the European Community to the west, which suffers so much from acid rain generated by Eastern European industry. This is, perhaps, the victim-pays principle rather than polluter-pays, but there is no other source of finance.

For political reasons, it is also urgent that the European Community involve itself directly in Eastern European affairs. Eastern European politics in the 1920s and 1930s, filled with innumerable conflicts arising from local nationalism and degenerating into a political climate fit to be exploited by the barbarism of Hitler, is an example not to be forgotten. The nationalism of the region can only be stably and safely accommodated if it is held in the protective web of a broader democratic federation that has environmental as well as economic goals. This applies also to the internal problems of the Soviet Union, the empire that now has no valid name, which can only evolve securely in a broader federal framework, perhaps to become a set of states within a larger European federation.

Both the political and the environmental dynamics of the region demand rapid action. The football match is now on the move: correct decisions will lead to great things; inaction or incorrect decisions will produce conflict and disaster. One obvious solution to both problems would be to extend the European Community eastward as rapidly as possible by offering conditional membership immediately to those nations that are prepared to accept the democratic values forming the constitutional soul of the Community. Transitional arrangements might take a decade or more, but there is an opportunity to extend the Community as far as Moscow, which means that a single decision-making confederation would extend from Lisbon to Vladivostok, possibly also including Turkey and Morocco. Such an entity, once integrated, would become the dominant manager of the planetary environment. The core of opinion and wealth in the Western European democracies would be powerful enough to remake our Earth. Looking further, a political confederation of this magnitude would be attractive to other nations. Vladivostok is not far from Tokyo, Ottawa is culturally descended from Paris and London: here are the seeds of an economic union spanning the entire Northern Hemisphere.

Political reality changes rapidly. The writing of this book reflects the viewpoint of 1990. By the end of 1991, all may be stasis again, or chaos, or a welter of nationalism, difficult to integrate. But a chance currently exists to manage much of the planet.

8.12 Nation-states, bilateral arrangements, and development aid

The decisions made in G-7, Commonwealth, and Francophonie meetings can become reality only if implemented by member nations. It is at the level of national policy that things get done; the grander meetings tend to coordinate, not to implement. Moreover, it is a good general rule in all management that decisions should always be made at the lowest level that is competent and informed, because it is generally that level which will make the best decision.

Bilateral environment policy agreements are needed. They should be based as much as possible on existing relationships between rich and poor nations. Each rich nation has skills to contribute to global management and each poor nation has its own field of responsibility. It is worth illustrating these generalities with specific examples.

A common criticism voiced by intellectuals in poor nations is that most "development aid" programs are not purely altruistic. Aid programs often spread political largesse around the donor nation. This is done by subsidizing a factory or a product or an agricultural sector in the donor nation, or simply diverting money to a lobby group of "consultants" in the giving country. There are exceptions but the criticism is often valid. Moreover, such aid fosters corruption in the recipient nation, as those who stand to gain in the donor nations (e.g., equipment manufacturers) buy support in the recipient country for their product.

An anecdotal account illustrates the general but undocumented problem of aid-generated corruption. An electrical engineering group from a small neutral country put in a bid to supply equipment to a major project in Africa. This bid, though technically excellent and competitively priced, was rejected in favor of a bid from a supplier from a larger nation, at much higher cost. The cost, apparently, was borne by a soft loan, and a substantial kickback appeared to have gone from the successful bidder to the decision maker in the poor African nation. The net outcome was that the people of the African country took on a much higher debt than necessary, even if on soft terms; the supplier received the money recycled from its own government's aid budget; the African decision maker became rich, in hard currency; and the representative of the unsuccessful group, though with wide experience of Africa and love for the continent, decided he was unable to participate in such bribery and withdrew home to Europe, leaving the market to the more corrupt supplier. This anecdote though generalized, reflects reality, and the results can be seen in the large cars of Africa, the aid budgets of the rich nations, and the profits of certain suppliers. It is time for a more honest approach: we should admit that much aid is spent for the internal purposes of the donor nations, and recognize the corrupting effect it has on the poor states. Political decisions should be made openly, not under the table.

Two examples discussed earlier illustrate the argument. Canada is skilled at making comparatively safe and efficient nuclear reactors. If its present aid budget of about $2.5 billion (Canadian) a year were mostly poured into the electrification of Africa, virtually all the money would be spent building equipment within Canada, yet the world as a whole would benefit, as would the security of Canada. The efficiency of the open spending in Canada under proper accounting procedure would be high, in contrast to the present situation where much aid money fails to reach its target, diverted by corruption and self-interest.

The United Kingdom, to take another example, has great expertise in its universities, which include some of the best in the world. It could staff new British or Commonwealth colleges in the poor nations simply by using academics on sabbatical, those who have chosen early retirement, or young postdoctoral researchers, and especially by extending its Open University, using satellites to teach at a distance. (Information travels more easily than people.) The salaries would return to or be spent in the United Kingdom, and Britain would benefit from a more stable academic employment structure. But there is a risk to this form of assistance. At present, aid to universities in poor nations is having a negative effect on local skills. By providing as aid cheap academic labor, the aid agencies have made it possible for local governments to reduce in real terms the salaries paid to their own academics. The result has been a general loss of the best people, either to the rich nations or to other sections of the local economy. In consequence, poor countries have lost skill. This has been particularly true in the sciences, which are a very low priority for most leaders of poor countries, with the result that the voice of local environmental concern in many of these poor countries is now very weak. Together, development aid and unsympathetic local politicians have destroyed the local scientific community in many tropical states. Substantial reassessment of aid to education, as well as an increase, is needed if the environment is to be safeguarded.

On a larger scale, the United States, Germany, and especially Japan, now the world's largest donor of aid, need to redirect their aid to address environmental problems. The debt of the poor nations must be reduced, but this is best done in exchange for the equity that is represented by a secure environment. Each of the rich nations should, bilaterally or in a multilateral framework, bargain in self-interest for environmental ends. The most important of these is the implementation of social programs, such as education and health care, to restrain population growth.

The cost of this mix of aid and bargaining need not be excessive. It would be somewhat greater than the present cost of development aid but not necessarily grossly so. The total world debt of the poor nations at the close of 1988 was around a trillion U.S. dollars, with Brazil, Mexico, and Argentina alone owing $290 billion. At the end of 1987, Africa owed $218.1 billion, of which $128.8 billion was owned by the sub-Saharan nations. The enormous principal behind this debt was transferred to the poor nations without much obvious strain on the economies of the richer nations (into whose armaments factories some of the money was recycled). Coordinated bilateral arrangements could be devised to manage this debt slowly, using repayment *within* the debtor nations to implement social reforms in order to stabilize population growth. The risk in this is rampant inflation in the debtor nations, but not if they were constrained to reduce their armies of soldiers as rapidly as they expanded their armies of teachers and doctors. In the longer run, a stable population in a stable environment is a formula for prosperity, to the benefit of rich and poor nations alike.

In 1833 the British freed all the slaves in their empire. To do this, their parliament voted 20 million pounds sterling to buy and set free every slave. The sum of money was fabulous. One way of expressing this amount in modern terms is to consider military budgets. In the 1830s the active cost (excluding pensions, etc.) of the British Royal Navy was about 2.5 million pounds a year. At the time, the Royal Navy was the most powerful military organization on Earth, responsible worldwide for keeping the Pax Britannica. Today, the Pentagon has a similar role. The modern equivalent of the cost of freeing the slaves, expressed in multiples (say 3 or 4) of the Pentagon's budget, would be over $1 trillion. The

British decision to free the slaves of their empire was made by a free and altruistic electorate acting on the judgment that slavery was morally wrong and was corrupting British society. Slavery had to be stopped.

Today we face a similar crisis. The misery of the poor nations need not continue. Whatever the historical causes and the incompetent leadership that have produced that misery, they can be stopped. The poor will always be with us, but at least their suffering can be alleviated. When Britain was the richest nation on Earth, it did, on the occasion of the freeing of the slaves, use that wealth to bring freedom. The United States did the same in the Marshall Plan. Both nations recognized, even in their imperialism, that wealth corrupts unless it is used to help the poor. It is time for a third act of generosity, in altruism and self-interest.

One curious property of the need to manage the planet is that it leads to peace. We cannot build a new global economy unless we cooperate, both on a multilateral and on a bilateral level. The present system of financial management already spans half the world. If a broader structure becomes stable, it will enmesh its participants. Just as the European Community has made war between France and Germany a laughable notion, so will nourishing the planetary environment defuse the competition between nations by weaving a framework of rules that sustain peace.

Reading list

Billings, E. L., and A. Westmore (1981). *The Billings method.* New York: Random House.

Campbell, B. M., R. F. Du Toit, and C. A. M. Attwell, eds. (1989). *The Save study.* Harare: University of Zimbabwe Press, P.O. Box M.P. 45, Mount Pleasant.

Goudie, A. (1990). *The human impact on the natural environment.* Oxford: Blackwell.

Jones, H. (1990). *Population geography.* London: Paul Chapman Publishing, 144 Liverpool Road, London, N1 1LA. 321 pp.

Redclift, M. (1987). *Sustainable development: exploring the contradictions.* London: Routledge.

9
Summary: a new global economy

It is one of the greatest tragedies of today that man, having made himself autonomous, free of dependence on God as he thinks, has separated himself from Nature, and the more he knows, the greater the curse he can become to the Nature he was intended to rule. . . . It is a Christian task . . . to treat whatever part of the world God may place us in as part of God's creation to which we owe a duty.

> H. L. Ellison (1967), *Comment on Psalm 148*

But who should begin? Who should break this vicious circle? . . . responsibility cannot be preached, only borne, and the only possible place to begin is with oneself. . . . Whether all is really lost or not depends entirely on whether or not I am lost.

> Václav Havel, *Letters to Olga*, No.142 (written from prison)

Is there no balm in Gilead?

> Jeremiah 8:22

The word *economy* is derived from the Greek root *oikos,* meaning home. An economist is a person who is entrusted with the stewardship of the home. We now have a global economy, and we have become responsible for the stewardship of the planet. In a culture that can reach the limits of the solar system, there is no limit to growth in well-being except that imposed by stupidity, but despite our technical success we have not yet come to terms either with our power or with our responsibilities. The thesis of this book is simple: humanity now controls the Earth and has assumed the management of the planetary economy. Economics is no longer a matter of resources, exploitation, and production; global economics must now be based on the sound stewardship of our home. Only if we become guardians of the planet can the life of humanity improve.

9.1 The control of climate

For 4 billion years the Earth's surface environment has been self-managing. The natural controls on the environment have operated since the beginning of the geological record and have maintained the temperature so that the surface has neither frozen over nor boiled. As the solar input has grown and varied, the natural warming effect (the greenhouse effect) of the Earth's atmosphere has changed, so that the planet's surface temperature has stayed roughly constant. There have been warm periods and cold intervals, but the planet as a whole has never frozen, nor has it become so warm that life has died out.

Today, the human economy is of such scale that it has taken over the control of the globe, and humanity is beginning to determine the course of the global climate. Our influence is so profound that we are now powerful enough deliberately to change that climate. We are able, if we choose, easily to warm the planet, or, with somewhat more difficulty, to cool it. We can, if we wish, melt the polar ice; we can probably, if we like, create a semidesert in central North America; we may soon be able to impose rainfall in the Peruvian desert or drought in southern Africa. Climate, which is a long-term matter, is now technically under our control, although we cannot yet influence day-to-day weather on any significant scale.

In several ways human behavior now determines the climate. The products of our industrial activity are now important in terms of the total bulk of the atmosphere. Each year, we add trace gases to the atmosphere in such amounts that we have changed the composition of the air on a planetary scale. The addition is no longer simply local pollution; it constitutes an important part of the workings of the atmosphere, and we are rapidly losing track of what pristine nature was. Old air is a prized commodity, because we are now less and less able to characterize the natural environment. The atmosphere, from now on, will be what we make it to be.

The second major way in which humanity determines the planetary climate is in our control of the surface vegetation, the animal population, and the color of the Earth. Vegetation – everything from rainforest to plankton – is crucial to the global air-conditioning system. Plants help to cleanse, hydrate, and circulate the air. The processing system is enormously complex and subtle. The effect of vegetation on the water cycle and on the circulation of the air is poorly understood. Plants transpire water and in so doing they transfer energy into the air. The transfer of heat is so large that it is an important contribution to the energy driving the global air circulation. The northern forest and the tropical and subtropical dry forest are also important in helping to control the oxygen and carbon content of the air.

Clouds are immensely important in controlling the climate. Their impact on temperature is far more important than the direct effects of greenhouse gases or of vegetation. Small changes in greenhouse gases or in vegetation can change cloud distribution, and have extreme impact on the planetary climate. This is one of the nightmares of global change, that some small change in cloud patterns may induce a much larger catastrophe that transforms the world climate. Other nightmares include a runaway emission of methane from the Arctic hydrates, collapse of atmospheric hydroxyl concentration, a runaway increase in pollutants, and the danger of change in sea circulation.

Humanity now controls around a third to a half of the primary biological production of the land. Large parts of all the continents except Antarctica are now our gardens in which we plant crops, cut forests, graze cattle, or burn grass. Aeschylus no longer speaks the truth: the air is heaven's protectorate no more. In taking over the direction of the continental surface, we have also become responsible for the atmosphere: the decision to remove forest in Borneo alters the weather in South Asia, the burning of the Amazon forest influences the jet stream over the United States. As yet, we do not understand the extent and significance of these effects, and we are so rapidly losing the natural world that it may be gone before we understand it. It is extremely important to the future of humanity that *deforestation be stopped, and that agricultural practice, worldwide, be reformed so that good environmental management becomes a chief goal of human use of the land and seas.*

9.2 The response to challenge

A Luddite reaction to these problems would be to deny our duty of managing the planet, to smash our industries, and to return to a better, natural world. This would have the immediate consequence of consigning the poorer parts of the world to an existence without hope, in perpetual poverty, starvation, and disease, until enough people had died to restore the old pretechnological balance of nasty, brutish, and short lives. The ''natural'' Earth (if such a thing could be restored) would probably sustain less than a billion of the 5 billion people who now live. Fortunately for the poorer peoples of the Earth, this option is not possible since the most powerful nations are democracies, and the average electors will not decide to execute the poor. Nor will they cheerfully accept a government that orders them to go without fast foods, plastic packaging, cars, television, soft drinks, baseball caps, and the other marks of civilization. Any government which demands that these essentials be abandoned will be swiftly voted out of office. Persuasion, by subtle forces, is the only option. We cannot demolish our present industry *unless* we can replace it with a more attractive substitute.

The duty of economists as stewards of society is therefore to find alternatives to our present pattern of behavior in the richer and also in the poorer parts of the world, alternatives so attractive that they will actually be implemented. Any new global economic system must be capable of supplying energy to the richer nations and of providing food and some hope of an improvement in living standards to the poorer nations.

The generation of useful energy and the expending of it dominates the economic needs of the rich nations and is at the root of most atmospheric pollution. Given an adequate supply of energy, there is virtually no limit to the increase in wealth, if the energy is atmospherically clean and wealth is defined as the well-being of humanity in a stable biological environment. If energy is not clean, the increase in wealth is limited, and the environment is dangerously threatened. Some economists feel that the wealthy nations can simply carry on regardless, as we can always buy food even if rich farmers struggle and those of the poor nations fail. This is a valid, if amoral, point of view, but it ignores the strains on the social fabric of the rich nations that will come when they are faced with major internal agrarian changes; it also ignores the problem of innumerable starving migrants, legal and illegal, pounding on the gates closed at borders.

Our energy at present is generated from coal, oil, natural gas, and hydroelectric and nuclear sources. Coal is unacceptable in atmospheric terms and must be abandoned. Oil and natural gas are slightly less noxious, but the carbon dioxide and methane they produce are unacceptable unless waste gas can be eliminated or reinjected into the ground. Hydroelectric power is attractive in moderation but is not a viable way of producing enough electricity to satisfy the planet. It is environmentally damaging and it is limited in its resources.

There are several sources of energy that do not alter the air. Solar energy is the most obvious possibility. At present it is very expensive and inefficient to produce, but it is likely that it will in future become much cheaper and more efficient. Much research is needed, urgently. Biomass burning is another form of solar power that, at first glance, is atmospherically neutral, as biologically fixed CO_2 is then burnt for energy. However, a very large area of land would be needed for biomass production if a significant part of humanity's energy came from this source. This land would have to be taken from agricultural land or from forest where it would otherwise support plants that would fix CO_2 on a rather longer term. The obvious short-term option, which is favored in this work, is nuclear energy. Nuclear power is the only major source of energy that is able to supply the planet for the next few decades without a massive reduction in the availability of electricity to the poorer nations. Over the longer term, solar energy will probably eventually be the chief support of the economy.

Finally, and most important, there is the option of energy conservation, which is laudable in the rich nations but anathema to poor countries that desperately need abundant electricity. In the short term, for the rich nations, conservation is the best and least expensive way of reducing emissions of greenhouse gases.

The generation of electrical power without the simultaneous generation of CO_2 or loss of CH_4 is not the whole answer to the problem of global change. Public transport is vitally important, especially when electrically powered. Cities ought to be reshaped so that individual transport is less needed. Endless suburbia and interminable rush hours are not necessary features of paradise. Neither is gasoline. We can use an alternative currency of energy, such as electricity or hydrogen produced from water by electricity.

A successful global economy of the period 2000–2050 may not be based on fossil fuels, as today. Instead it may be a nuclear-electric-hydrogen economy. In such an economy the electricity generated in nuclear power stations would be used to power industry and to produce hydrogen to power independent transport.

Eventually, there would be a transition to solar power. In contrast, a global economy that continues to use fossil fuel is likely to be unsuccessful and will eventually fall to social unrest induced by climate change. Whatever the shape of the future, *it is important that the industrial economy should make no net emissions of CO_2, CH_4, or other pollutants* and that this state should be brought about soon.

9.3 The implementation of change

Can the global economy change? Do we possess the institutional structures to plan and to implement change? The answer must be optimistic: the alternative is a new Dark Age. We now possess a rudimentary network of global economic management. We have a growing web of international meetings and organizations such as the G-7 group, the International Monetary Fund and World Bank, the General Agreement on Tariffs and Trade, the Organization for Economic Cooperation and Development, and so on. Competition between nations is moderated by a wide variety of formal and informal agreements to regulate world trade and currencies. It is through this web of international cooperation that a new global economy can be developed, in which formal treaties, such as the 1990 London agreement on atmospheric ozone, will be supplemented by bilateral and multilateral agreements on reducing greenhouse emissions, on using the debt of poor nations as an instrument for protecting tropical vegetation, and on research programs for understanding the biosphere. A new, environmentally sound global economy *can* be implemented, without great disruption, and without limiting the growth in wealth of the rich or poor nations. The alternative is global catastrophe.

At the source of the flood of disruption that is changing the Earth is the fountain of human births. The planetary ecosystem can only be sustained if the human population is stabilized. An ill-educated, poor, and fecund humanity can only destroy the fabric of nature. Poor people are not stupid people; they have many children for the good reason that it is advantageous to them. Children provide labor, economic benefit, and hope for a secure old age. Those nations that have halted the growth in their populations all have universal education, good health care, social security in old age, and women who are treated as the equals of men. *Population growth can be stopped, but only if there is a strong social will to remove the underlying causes of the growth.*

As a global society, humanity can choose a policy of little action now, followed by a massive forced change in our civilization in the early decades of the next century. Alternatively, we can change our economy in a series of relatively small shifts of behavior, beginning now, in the hope that over the next few decades we attain a balanced planet.

Our science may not be advanced enough to manage the Earth with success. For most of this century the earth sciences and biology have been reductionist. There have been triumphs – plate tectonics, DNA – but also there has been blindness. We study the geochemistry of ytterbium in great detail, but have no time to discover why the air is one-fifth oxygen. J. E. Lovelock, the most thoughtful planetographer of our age, has pointed this out: our scientific funding committees and our training drive us to minutiae, but not to knowledge. We do not yet understand the physiology of the Earth. Consequently, we cannot manage it. It is urgent that we study the biosphere, not just with computers but on the ground, in the air and from space.

But already we control our home; blind, humanity steers the globe. A crash may extinguish us. Without knowledge or vision to go on, we must stop and restore. If we learn to manage, we can drive the chariot of our civilization across the solar system. Should we not learn, what is left of the biosphere may simply excrete our remains from the wreckage of our planet.

9.4 A note of optimism

Some years ago I was clearing the accumulated debris of a century from my family's cottage as part of the sad task of selling the home. Under the eaves was a large closet, filled with abandoned clothes and shoes. One of the boxes was heavier; fortunately I stayed its journey to the refuse dumpster outside. In it were some books. Months later, I looked through them. One was a translation in four volumes of Homer's *Iliad,* published by an assortment of interesting people including a whole row of our intellectual fathers – "A. Horace, P. Virgil and T. Cicero, in Paternoster-Row, J. Milton in St. Paul's Churchyard, D. Plato and A. Pope in the Strand": Pope's Homer.

The book is filled with wonder and footnotes of wonder about the world. It encapsulates the wisdom and, more important, the spirit of both the dawn of civilization and of the birth of our modern age. Our civilization was built on this spirit. In the past decades we have come close to denying that heritage in our quest for wealth.

Alongside Homer in that box destined for the refuse was another volume, older, though a "new edition, augmented by the author." This volume was even more forgotten than Pope's Homer. It was in French, the works of Fontenelle. In the volume is his essay "A Digression on the Ancients and the Moderns," written in 1688. In England, this was the year of the Glorious Revolution that marks the start of the modern world order, and the year after the publication of Newton's *Principia.* The date three centuries ago marks the beginning of a society based on science. Fontenelle was a universal man. He was perhaps the first modern optimist. He looked into the future, believing that humanity could learn from the past, could improve, and could meet the challenge of the human condition. His thesis was that we have now gone further than the ancient intellectuals. We are in unknown territory, and we should go forward with hope.

Fontenelle died in his 100th year, in 1757. Between his birth and death the world had, superficially, changed little, though Newton's work had produced its results: the scientific revolution was beginning. In 1759 my four volumes of Pope's *Iliad* were published. This was the *annus mirabilis,* the year of victories, in which Pitt, the Great Commoner, triumphed. Parliamentary democracy assumed the responsibility of leading the planet. By the time of the French Revolution of 1789, a century after Fontenelle's essay, the modern world order, begun in 1688, was fully established.

In the next century, the changes were beginning to have impact, though much of the world was still the domain of nature. My grandfather was born in Africa, in the year 1887, in Tokai, a lovely home in a Cape forest. He died during the writing of this book, in 1988, having lived like Fontenelle a century, enough time to encounter a planet with no frontier. His forest birthplace had become a remnant enclave in a metropolis. It was during his single lifetime that the changes implicit in Fontenelle's life became explicit. The wondrous world of the *Iliad*, so filled with mystery, has vanished. We have set foot everywhere. We are now responsible for the world's state. My grandfather shared a friend with Napoleon; Napoleon would have known someone who knew Fontenelle. It is a measure of the rapidity of the change that it has been only three centuries, three such human lifetimes since Fontenelle's essay in 1688, which first pointed out that we had surpassed the ancient civilizations, and that we were reaching into the unknown.

During a break in a meeting arranged by the American Geophysical Union to discuss the Gaia hypothesis – and, incidentally, the state of the planet in 1988 – two of us escaped to the San Diego Zoo. The zoo should really be called the San Diego Ark. It is a treasure house of nature. We found ourselves, an atmospheric physicist and a geologist, both ignorant of zoology, facing a reconstructed African koppie, complete with animals, plants, and even Bushmen paintings. The koppie was very beautiful, and it magnificently captured the flavor of Africa. But we had both grown up in the true Africa. There is a remote chance that I have inherited a gene or two from those distant painters. We could see the flaws and the artificiality in the structure. It was a fine counterfeit, not a koppie. Ours is the last generation of humanity to see wild nature. We must eventually be responsible for our planet, but we cannot restore it to its original state. Too much is lost. We have left Eden, and there is no way back.

There have been other critical moments in the histories of nations that closely parallel the present crisis. In some nations, the right actions have been taken, in others the warnings were not heeded and disruption or revolution followed. One of the closer parallels is the debate during the passage of the Reform Bill in the United Kingdom in the 1830s. This bill, which gave new life to British democracy, was passed by the narrowest margin. It probably averted a revolution. Had it failed, the evolution of freedom would have been stunted not just in Britain, but throughout the world. Macaulay's speech, which helped to win the impassioned debate in Parliament, applies as well today as then. The language is the flowery prose of the period; the truths it expresses remain valid. From this excerpt I have only deleted brief contemporary references and substituted the word "Earth" for "England":

Turn where we may, within, around, the voice of great events is proclaiming to us, reform, that you may preserve. Now, therefore, while everything at home and abroad forebodes ruin to those who persist in a hopeless struggle against the spirit of the age, now, while we see on every side ancient institutions subverted, and great societies dissolved, now, while the heart of Earth is still sound, now, while old feelings and old associations retain a power and a charm which may too soon pass away, now, in this your accepted time, now, in this your day of salvation, take counsel, not of prejudice, not of party spirit, not of the ignominious pride of a fatal consistency, but of history, of reason, of the ages which are past, of the signs of this most portentous time. Pronounce in a manner worthy of the expectation with which this great debate has been anticipated, and of the long remembrance which it will leave behind. Renew the youth of the State. Save property, divided against itself. Save the multitude, endangered by its own ungovernable passions. Save the greatest, and fairest, and most highly civilised community that ever existed, from calamities which may in a few days sweep away all the rich heritage of so many ages of wisdom and glory. The danger is terrible. The time is short.

Western society, which through its demands and its inventions is the principal cause of the attack on nature, is founded on Judeo-Christian faith, which asserts the unity of creation, under God. The Bible describes Adam and Eve before the Fall as monarchs of the garden of Eden, not as tyrants. We are those monarchs, ruling creation. In Hebrew tradition, a monarch is a shepherd who rules but is responsible for protecting and defending the subjects. Kings, such as Saul, who fail to remember their station, have their divine right to rule withdrawn. The biblical covenant with Noah explicitly included the wild animals, the living things that inhabit the Earth. The ox that labors should not be muzzled; the crop should not be picked to the last grain; the whales and dolphins are not to be eaten; the poor should be nourished and the land should be rested. Job teaches us to listen to the Earth and to nature. The Christian ethic reemphasizes this. Again and again, Christ retires to a garden or to the wilderness; Saint Paul talks of all creation in Romans 8:22. The choice of meat is widened, but the covenant with Noah and nature remains in force, sealed by the rainbow. Only in the past two centuries has this covenant been challenged by the power and self-assertion of humanity. Even the rainbow itself, the symbol of promise, has now been dimmed in many countries by air pollution that filters the colors from the sunlight.

Despite our losses, we are intellectually and physically richer than any other generation of humanity. Our poverty is spiritual. It is well within our power to be optimists, if we can dispel the cynicism of the past decades. If we are optimists, most things are possible. The challenge to cherish the planet, to construct a new global economy, is far less than the challenge, in 1940, to defeat the last threat against human hope. We are strong, capable, and wealthy. We should welcome the task. The path to a new global economy will be long and difficult, but on the horizon for the first time it is possible to see the faint image of humanity at peace with itself, and at peace with its environment. Perhaps that vision is not a mirage.

Appendix
Atmospheric chemistry

Atmospheric chemistry is extremely complex and not easily summarized. Each gaseous species affects the others, reaction series compete, small changes in the abundance of trace chemicals can have major influence. Brief accounts of important processes are to be found in *The changing atmosphere,* edited by F. S. Rowland and I. S. A. Isaksen (Chichester: Wiley, 1988). A good summary is in the article by J. S. Levine, "Photochemistry of biogenic gases," in *Global ecology,* edited by M. B. Rambler, L. Margulis, and R. Fester (Boston: Academic Press, 1989). Much more detailed is Peter Warneck's excellent work, *Chemistry of the natural atmosphere* (San Diego: Academic Press, 1988). In an appendix this work lists the parameters of 250 important reactions. The specific problems of urban air pollution are addressed by J. H. Seinfeld in a review article in *Science,* 10 February 1989. The brief summary that follows is derived from the above works.

A.1 Some short-term processes

The atmosphere, in its present state, is a biogenic product. All the components of the air, except the rare gases, are biologically cycled. For most gases, a nested hierarchy of cycles can be identified. Short-term processing, both chemical and biological, occurs close to the air–surface boundary or under the influence of sunlight. Medium-term exchange occurs between different parts of the atmosphere and between the air and the sea or ground. Longer-term cycling is geological, via the system of plate tectonics, and involves the Earth's crust and mantle. Each process affects the others. For humanity it is the short term that concerns us most. A few of these short-term processes are outlined here.

For humans, oxygen is the most important species, produced during photosynthesis.

$$n H_2O + m CO_2 + h\nu \xrightarrow{\text{chlorophyll}} C_m (H_2O)_n + m O_2 \tag{1}$$

In this equation, $C_m(H_2O)_n$ is carbohydrate produced in the plant cell, h is Planck's constant, and ν is the frequency of visible light from the Sun.

As a result of photosynthesis, oxygen (O_2) is sustained in the atmosphere. In the early part of the Earth's history, before life, oxygen may have been produced abundantly by photodissociation of water; today, it is generated by green plants.

In an oxygen-rich atmosphere, ozone (O_3) can be produced:

$$O + O_2 + M \rightarrow O_3 + M \tag{2}$$

In this reaction, oxygen atoms react with oxygen molecules to make ozone. M represents any third body, for example nitrogen (N_2) or O_2, that removes the energy of the reaction and stabilizes O_3. Where do the oxygen atoms come from? At high altitude, in the stratosphere, energetic deep-ultraviolet light is available that can split up oxygen molecules. In the troposphere, there is much less ultraviolet, but oxygen atoms are produced by the photodissociation of nitrogen dioxide. For example,

$$NO_2 + h\nu \rightarrow NO + O \tag{3}$$

where ν is light with a wavelength between 280 and 430 nm. The nitric oxide (NO) that is produced reacts rapidly with ozone to restore NO_2.

$$NO + O_3 \rightarrow NO_2 + O_2 \tag{4}$$

The net effect of these reactions is to establish a steady concentration of ozone in the tropospheric air, around 20–60 parts per billion by volume at mid-latitudes in the Northern Hemisphere, 10–20 in the Southern Hemisphere near the surface.

In the presence of light of wavelength 310 nm or less, ozone photodissociates.

$$O_3 + h\nu \rightarrow O(^1D) + O_2 \tag{5}$$

The $O(^1D)$ is an excited oxygen atom. Some of the excited oxygen atoms produced in this process are deactivated by collision with N_2 or O_2 molecules, but some react with water vapor to form hydroxyl (OH) radicals.

$$O(^1D) + H_2O \rightarrow 2\, OH \tag{6}$$

Hydroxyl is the chief scavenger of the lower atmosphere. Its production is highest at low latitudes at noon (most sunlight), in parts of the troposphere where H_2O vapor is abundant. This is especially true in the equatorial regions, where the solar ultraviolet flux is least reduced by stratospheric ozone. Despite its importance, OH abundance is not well known, because it is difficult to measure. It is highly reactive, and concentrations are low. An example of the role OH plays as a major oxidant in the lower atmosphere is the methane oxidation cycle.

A.1.1 Methane and carbon monoxide

Methane (CH_4) is attacked by OH:

$$OH + CH_4 \rightarrow H_2O + CH_3 \tag{7}$$

$$CH_3 + O_2 + M \rightarrow CH_3O_2 + M \tag{8}$$

$$CH_3O_2 + NO \rightarrow CH_3O + NO_2 \tag{9}$$

$$CH_3O + O_2 \rightarrow CH_2O + HO_2 \tag{10}$$

Formaldehyde (CH_2O) is the product of this sequence. By several pathways, it is oxidized to carbon monoxide (CO); for example:

$$CH_2O + h\nu \rightarrow CO + H_2 \tag{11}$$

These reactions produce ozone (e.g., via reaction 9, then 3 and 2) and OH. For every methane molecule destroyed, 3–4 ozone molecules and 0.5 OH radicals are produced, in the presence of adequate NO_x.

Carbon monoxide is oxidized by OH:

$$OH + CO \rightarrow CO_2 + H \tag{12}$$

Also,

$$H + O_2 + M \rightarrow HO_2 + M \tag{13}$$

$$HO_2 + NO \rightarrow NO_2 + OH \tag{14}$$

The methane oxidation processes outlined above depend in part on NO (reactions 9 and 14). The reactions given here are only part of what in reality is a much more complex chain of cycles. In clean air, where NO is present at less than about 10 parts per trillion, methane oxidation consumes OH in producing CO_2 and H_2O. Ozone is also consumed in clean air: in polluted air, in contrast, it is produced. The various steps in the methane oxidation process are outlined by Cicerone and Oremland (1988), given in the reference list for Chapter 3. These authors point out the instability in the CH_4, CO, and OH system. Increases in CH_4 or CO can decrease OH, thereby further increasing CO or CH_4. The role of nitrogen oxides is also significant: pollution has complex and contradictory effects. In the circumstances, it is a matter of concern that OH concentrations are so poorly known.

A.1.2 Polluted air

In polluted air, OH is produced by photochemical dissociation of ozone. Ozone is produced by reactions 2, 3, and 4, and in urban air can reach levels as high as 400 parts per billion. OH is also made from carbonyl compounds and nitrous acid. In the photodissociation of aldehydes:

$$RCHO + h\nu \rightarrow R + HCO \tag{15}$$

$$HCO + O_2 \rightarrow HO_2 + CO \tag{16}$$

$$HO_2 + NO \rightarrow NO_2 + OH \tag{17}$$

Nitrous acid (HONO) occurs in urban air:

$$HONO + h\nu \rightarrow OH + NO \tag{18}$$

On the other hand, OH is removed by NO_2 to make nitric acid (HNO_3). NO_2 occurs in urban and rural environments in the Northern Hemisphere.

$$OH + NO_2 + M \rightarrow HNO_3 + M \tag{19}$$

This is a major sink for nitric oxides. Another complication of pollution is the presence in urban air and in the free troposphere of peroxyacetyl radical, which reacts with NO_2 to produce peroxyacetylnitrate, commonly called PAN:

$$CH_3\,C(O)O_2 + NO_2 \rightleftharpoons CH_3C\,(O)\,O_2\,NO_2 \tag{20}$$

This is a reversible reaction, and PAN is a temporary reservoir for NO_2 with a lifetime of about 45 minutes in low-level air, though much longer in the cold upper troposphere. PAN concentrations in city air can be as high as 5–10 ppb in California.

Urban aerosols constitute a significant hazard to the health of animals and plants. They include a mixture of primary emissions, such as ash and soot, and also the species that are the sinks of atmospheric chemical processes. Sulfate forms after the atmospheric oxidation of SO_2 to sulfuric acid (H_2SO_4), which also reacts with ammonia. Nitrate salts are common, though unlike H_2SO_4, nitric acid has a high vapor pressure and does not condense on aerosols. Aerosol carbon is common – in Los Angeles, it makes up 40% of fine-particle mass. This carbon is derived from gasoline and diesel engines and from fuel burning. Black elemental carbon emissions, part of the total carbon aerosol, are mainly from diesel exhaust.

A.1.3 Ammonia

Another reaction of interest is the destruction of ammonia (NH_3):

$$NH + OH \rightarrow NH_2 + H_2O \tag{21}$$

The result is the formation of the amine radical (NH_2). This can be further reacted with nitrogen oxides or ozone:

$$NH_2 + O_3 \rightarrow NO_x + products \tag{22}$$

$$NH_2 + NO \rightarrow N_2 + H_2O \tag{23}$$

$$NH_2 + NO_2 \rightarrow N_2O + H_2O \tag{24}$$

or with atmospheric sulfuric acid. NH_3 is also rained out.

A.1.4 Reduced sulfur species

Hydrogen sulfide (H_2S) and dimethyl sulfide ((CH_3)$_2$S) are organically produced sulfur compounds. Dimethyl sulfide (DMS) may be particularly important in the natural return of sulfur from the sea to the land. It is emitted by marine algae. DMS is probably oxidized by OH, producing methane-sulfonic acid, which becomes associated in aerosol particles that help to nucleate clouds. Once in the aerosol, the methane-sulfonic acid is rapidly oxidized to sulfuric acid, some of which falls in rain on land, returning sulfur. Curiously, the European practice of spreading manure and sewage on fields, which are then washed to the North Sea, appears to favor algal growth and DMS production in the sea, the DMS eventually returning to land as acid rain. Ammonia from the manure, on the other hand, produces alkaline rain.

A.2 Composition of the lower atmosphere (troposphere)

Gas	Concentration
Nitrogen (N_2)	78.08%
Oxygen (O_2)	20.95%
Argon (Ar)	0.93%
Water vapor (H_2O)	variable
Carbon dioxide (CO_2)	0.035–0.036%

Trace gases	
Ozone (O_3)	10–100 ppbv
Hydrogen (H_2)	0.5 ppmv
Nitrous oxide (N_2O)	305 ppbv
Ammonia (NH_3)	0.1–1 ppbv
Nitric acid (HNO_3)	50–1000 pptv
Hydrogen cyanide (HCN)	about 200 pptv
Nitrogen dioxide (NO_2)	10–300 pptv
Nitric oxide (NO)	5–100 pptv
Nitrogen trioxide (NO_3)	100 pptv
PAN (see A.1.2)	50 pptv
Dinitrogen pentoxide (N_2O_5)	1–10 pptv
Methane (CH_4)	1.75 ppmv
Carbon monoxide (CO)	40–200 ppbv
Formaldehyde (CH_2O)	0.1 ppbv

Carbonyl sulfide (COS)	0.5 ppbv
Dimethyl sulfide ((CH$_3$)$_2$S)	0.4 ppbv
Hydrogen sulfide (H$_2$S)	0.2 ppbv
Sulfur dioxide (SO$_2$)	0.2 ppbv
Hydrogen chloride (HCl)	10–60 ppbv
Methyl chloride (CH$_3$Cl)	0.5 ppbv
Methyl bromide (CH$_3$Br)	10 pptv
Methyl iodide (CH$_3$I)	1 pptv
Neon (Ne)	18 ppmv
Helium (He)	5.2 ppmv
Krypton (Kr)	1 ppmv
Xenon (Xe)	90 ppbv

ppmv – part per million by volume (10^{-6})
ppbv – part per billion by volume (10^{-9})
pptv – part per trillion by volume (10^{-12})

A.3 Mass of the atmosphere

Mass of whole atmosphere:	5.137×10^{18} kg
Mass of troposphere:	$4.22 \ \times 10^{18}$ kg
Mass of stratosphere:	$0.91 \ \times 10^{18}$ kg
Mass of uppermost air:	

A.4 Solar irradiance

There has been much discussion of the effects on the Earth's climate of variation in the Sun's output. Recent work suggests that there may be a correlation between sea surface temperature and solar irradiance. As the Sun's radiation fluctuates, so does the temperature of the top of the oceans. However, land-based temperature records do not show such strong correlation. Overall, it is possible that solar variations have made a contribution to climatic fluctuation in the past few centuries. Nevertheless, the conclusion that the greenhouse warming will soon outweigh other natural climatic fluctuations is very strong indeed, unless some very unusual natural event occurs. The topic is discussed further by G. C. Reid (1991), *Journal of Geophysical Research*, vol. 96, pp. 2835–44; T. M. L. Wigley and S. C. B. Raper (1991), *Geophysical Research Letters*, Vol. 17, pp. 2169–72; and D. J. Gorney (1991), *Reviews of Geophysics*, vol. 28, pp. 315–36.

Further reading

The following books and journal articles were consulted in the preparation of this work. Full titles of all sources are given so that readers can investigate specific topics or order books from or for libraries. Books are preceded by an asterisk.

1: Introduction

* Goudie, A. (1990). *The human impact on the natural environment*. 3d ed. Oxford: Blackwell. 388 pp.
* McKibben, B. (1990). *The end of nature*. New York: Viking Penguin. 212 pp.
* Mungall, C., and D. J. McLaren, eds. (1990). *Planet under stress*. Toronto: Oxford University Press. 344 pp.
* Scientific American (1990). *Managing planet earth: readings from Scientific American*. New York: Freeman. 146 pp.
* Tickell, C. C. (1986). *Climate change and world affairs*. Lanham, MD: University Press. 76 pp.
* World Commission on Environment and Development (Chair: G. H. Brundtland) (1987). *Our common future*. Oxford: Oxford University Press. 400 pp.

* World Resources Institute (1987). *World resources 1987.* New York: Basic Books. 369 pp.

(1988). *World resources 1988–89.* New York: Basic Books. 372 pp.

(1990). *World resources 1990–91.* New York: Oxford University Press. 383 pp.

2: The natural Earth

Andreae, M., H. Berresheim, M. Bingemer, D. J. Jacob, B. L. Lewis, S. M. Li, and R. W. Talbot (1990). The atmospheric sulphur cycle over the Amazon basin. 2. The wet season. *Journal of Geophysical Research, 95,* 16813–24.

Angell, J. K. (1988). Impact of El Niño on the delineation of tropospheric cooling due to volcanic eruptions. *Journal of Geophysical Research, 93:* 3697–704.

Arnold, F., T. Buhrke, and S. Qiu (1990). Evidence for stratospheric ozone-depleting heterogeneous chemistry on volcanic aerosols from El Chichón. *Nature, 348:* 49–50.

Barber, R. T. (1988). Ocean primary productivity. *EOS, 69:* 1045.

Barnola, J. M., D. Raynaud, Y. S. Korotkovich, and C. Lorius (1987). Vostok ice core provides 160,000 year record of atmospheric CO_2. *Nature, 329:* 408–13.

* Barry, R. G., and R. J. Chorley (1987). *Atmosphere, weather and climate.* London: Methuen. 460 pp.

Bartlett, K. B., P. M. Crill, J. A. Bonassi, J. E. Richey, and R. C. Harriss (1990). Methane flux from the Amazon river floodplain: emissions during rising water. *Journal of Geophysical Research, 95:* 16773–88.

Berner, R. A., A. C. Lasaga, and R. M. Garrels (1983). The carbonate-silicate geochemical cycle and its effect on atmospheric carbon dioxide over the past 100 million years. *American Journal of Science, 283:* 641–83. (See also updated paper, Sundquist and Broecker 1985.)

Bigg, E. K. (1986a). Discrepancy between observation and prediction of concentrations of cloud condensation nuclei. *Atmospheric Research, 20:* 81–7.

(1986b). Technique for studying the chemistry of cloud concentration nuclei. *Atmospheric Research, 20:* 75–80.

Blanchard, D. C., and R. J. Cipriano (1987). Biological regulation of climate – discussion. *Nature, 330:* 526.

Bonnefille, R. (1987). Evolution forestière et climatique en Burundi durant les quarante derniers milliers d'années. *Comptes Rendues, Académie des Sciences, Paris,* 305, II: 1021–6.

Boyle, E. A., and L. Keigwin (1987). North Atlantic thermohaline circulation during the past 20,000 years linked to high-latitude surface temperature. *Nature, 330:* 35–40.

* Brimblecombe, P., and A. Y. Lein, eds. (1989). *Evolution of the global biogeochemical sulphur cycle,* SCOPE 39. Chichester: Wiley. 241 pp.

* Broecker, W. S. (1985). *How to build a habitable planet.* Palisades, NY: Eldigio Press. 291 pp.

Broecker, W. S., M. Andree, M. Klas, G. Bonani, W. Wolfli, and H. Oeschger (1988). New evidence from the South China Sea for an abrupt termination of the last glacial period. *Nature, 333:* 156–8.

Broecker, W. S., and G. H. Denton (1989). The role of ocean–atmosphere reorganizations in glacial cycles. *Geochimica et Cosmochimica Acta, 53:* 2465–502.

Broecker, W. S., J. P. Kennett, B. P. Flower, J. T. Teller, S. Trumbore, G. Bonani, and W. Wolfli (1989). Routing of meltwater from the Laurentide Ice Sheet during the Younger Dryas cold episode. *Nature, 341:* 318–21.

Brunig, E. F. (1987). The forest ecosystem: tropical and boreal. *Ambio, 16:* 68–79.

* Budyko, M. I. (1986). *The evolution of the biosphere.* Dordrecht: D. Riedel. 423 pp.

Burgermeister, S., R. L. Zimmermann, H. W. Georgii, H. G. Bingemer, G. O. Kirst, M. Janssen, and W. Ernst (1990). On the biogenic origin of dimethylsulfide: relation between chlorophyll, ATP, organismic DMSP, phytoplankton species, and DMS distribution in Atlantic surface water and atmosphere. *Journal of Geophysical Research, 95:* 20607–15.

Charlson, R. J., and S. G. Warren (1987). Biological regulation of climate-reply. *Nature, 330:* 526.

* Climate Data Programme (1987). *The global climate system, autumn 1984–spring 1986.* Geneva: World Meteorological Organization. 87 pp.

* Cloud, P. C. (1988). *Oasis in space – Earth history from the beginning.* New York: Norton. 508 pp.

Cockroft, M. J., M. J. Wilkinson, and P. D. Tyson (1987). The application of a present-day climatic model to the late Quaternary in Southern Africa. *Climatic Change, 10:* 161–81.

Cofer, W. R., J. S. Levine, E. L. Winstead, and B. J. Stocks (1990). Gaseous emissions from Canadian boreal forest fires. *Atmospheric Environment, 24a:* 1653–9.

COHMAP (1988). Climatic changes of the last 18,000 years: observations and model simulations. *Science, 241:* 1043–52.

Colinvaux, P. A. (1989). Ice Age Amazon revisited. *Nature, 340:* 188–9.

Crutzen, P. J. (1987). Role of the tropics in atmospheric chemistry. In *The geophysiology of Amazonia,* ed. R. E. Dickinson. New York: Wiley. 107–32.

(1988). Variability in atmospheric-chemical systems. In *Scales and global change,* SCOPE 35, ed. T. Rosswall, R. G. Woodmansee, and P. G. Risser. Chichester: Wiley. 81–108.

Currie, D. J., and V. Paquin (1987). Large scale biogeographical patterns of species richness in trees. *Nature, 329:* 326–7.

D'Arrigo, R., G. C. Jacoby, and I. Y. Fung (1987). Boreal forests and atmosphere–biosphere exchange of carbon dioxide. *Nature, 329:* 321–3.

* Degens, E. T. (1989). *Perspectives on biogeochemistry.* Berlin: Springer-Verlag.

* Deutscher Bundestag (1988). *Schutz der Erdatmosphar.* Report of Commission of Inquiry of the Eleventh West German Parliament, B. Schmidbauer, chair. Dt. Bundestag, Referat Offentlichkeitsarbeit, Zur Sache; 88, 5, Bonn.

Dickinson, R. E. (1987). Introduction to vegetation and climate interactions in the humid tropics. In *The geophysiology of Amazonia,* ed. R. E. Dickinson. New York: Wiley. 3–10.

* Dickinson, R. E., ed. (1987). *The geophysiology of Amazonia.* New York: Wiley. 526 pp.

Dickinson, R. E., and H. Virji (1987). Climate change in the humid tropics, especially Amazonia, over the last twenty thousand years. In *The geophysiology of Amazonia,* ed. R. E. Dickinson. New York: Wiley. 91–106.

Donahue, T. M., J. H. Hoffman, R. R. Hodges, Jr., and A. J. Watson (1982). Venus was wet: a measurement of the ratio of deuterium to hydrogen. *Science, 216:* 630–3.

dos Santos, J. M. (1987). Climate, natural vegetation and soils in Amazonia: an overview. In *The geophysiology of Amazonia,* ed. R. E. Dickinson. New York: Wiley. 25–36

* Ehrlich, P. R. (1986). *The machinery of nature*. New York: Simon and Schuster. 320 pp.

Falkowski, P. (1988). Ocean productivity from space. *Nature, 335:* 205.

Fan, S. M., S. C. Wofsy, P. S. Bakwin, D. J. Jacob, and D. R. Fitzjarrald (1990). Atmosphere–biosphere exchange of CO_2 and O_3 in the central Amazon forest. *Journal of Geophysical Research, 95:* 16851–64.

Fischer, G., D. Futterer, R. Gersande, S. Honjo, D. Ostermann, and G. Wefer (1988). Seasonal variability of particle-flux in the Weddell Sea and its relation to ice cover. *Nature, 335:* 426–8.

Fitzjarrald, D. R., K. E. Moore, O. M. R. Cabral, J. Scolar, A. O. Manzi, and L. D. de A. Sa (1990). Daytime turbulent exchange between the Amazon forest and the atmosphere. *Journal of Geophysical Research, 95:* 16825–38.

Foucalt, A., and D. J. Stanley (1989). Late Quarternary palaeoclimatic oscillations in East Africa recorded by heavy minerals in the Nile delta. *Nature, 339:* 44–6.

* Fowler, C. M. R. (1990). *The solid Earth*. Cambridge: Cambridge University Press. 472 pp.

Frakes, L. A., and J. E. Francis (1988). A guide to Phanerozoic cold polar climates from high-latitude ice-rafting in the Cretaceous. *Nature, 333:* 547–9.

Frederiksen, J. S., and P. J. Webster (1988). Alternative theories of atmospheric teleconnections and low-frequency fluctuations. *Reviews of Geophysics, 26:* 459–94.

Genthon, C., J. M. Barnola, D. Raynaud, C. Lorius, J. Jouzel, N. I. Barkov, Y. S. Korotkevich, and V. M. Kotlyakov (1987). Vostok ice core: climatic response and orbital forcing changes over the last climatic cycle. *Nature, 329:* 414–18.

* Grace, J. (1983). *Plant–atmosphere relationships*. London: Chapman and Hall.

Gras, J. L. (1990). Cloud condensation nuclei over the Southern Ocean. *Geophysical Research Letters, 17:* 1565–7.

Gregory, G. L., E. V. Browell, L. S. Warren, and C. H. Hudgsin (1990). Amazon basin ozone and aerosol: wet season observations. *Journal of Geophysical Research, 95:* 16903–12.

Guiot, J. (1987). Reconstruction of seasonal temperatures in central Canada since A.D. 1700 and detection of the 18.6 and 22 year signals. *Climatic Change, 10:* 249–68.

Guiot, J., A. Pons, J. L. de Beaulieu, and M. Reille (1989). A 140,000 year continental climate reconstruction from Europe pollen records. *Nature, 338:* 309–13.

Hamburg, S. P., and C. V. Cogbill (1988). Historical decline of red spruce populations and climatic warming. *Nature, 331:* 428–31.

Hanel, R. A., B. Schlachman, D. Rodgers, and D. Varnous (1971). Nimbus 4 Michelson interferometer. *Applied Optics, 10:* 1376–82.

Hare, F. K. (1985). Climatic variability and change. In *Climate impact assessment*, SCOPE 27, ed. R. W. Kates, J. H. Ausubel, and M. Berbenan. Chichester: Wiley. 37–68.

* Harland, W. B., R. L. Armstrong, A. V. Cox, L. E. Craig, A. G. Smith, and D. G. Smith (1989). *A geologic time scale 1989*. Cambridge: Cambridge University Press.

Harriss, R. C. (1987). Influence of tropical forest on air chemistry. In *The geophysiology of Amazonia*, ed. R. E. Dickinson. New York: Wiley. 163–74.

Harriss, R. C., and 13 others (1990). The Amazon boundary layer experiment: wet season 1987. *Journal of Geophysical Research, 95:* 16721–36.

Hartington, D. L., and D. Doelling (1991). On the net radiative effectiveness of clouds. *Journal of Geophysical Research, 96:* 869–91.

Harvey, L. D. D. (1988). Climatic impact of ice-age aerosols. *Nature, 334:* 333–5.

* Henderson-Sellers, A. (1983). *The origin and evolution of planetary atmospheres*. Philadelphia: A. Hilger. Bristol: Heyden and Son. 240 pp.

Heusser, C. J., L. E. Heusser, and D. M. Petet (1985). Late Quaternary climatic change on the American North Pacific Coast. *Nature, 315:* 485–7.

Hofmann, D. J. (1987). Perturbations to the global atmosphere associated with the El Chichón volcanic eruption of 1982. *Reviews of Geophysics, 25:* 743–59.

* Holland, H. D. (1984). *The chemical evolution of the atmosphere and oceans.* Princeton: Princeton University Press. 582 pp.

Horel, J. D., V. E. Kousky, and M. T. Kagano (1986). Atmospheric conditions in the Atlantic sector during 1983 and 1984. *Nature, 322:* 248–51.

Huntley, B., and I. C. Prentice (1988). July temperatures in Europe from pollen data 6000 years before present. *Science, 241:* 687–90.

Jablonski, D. (1986). Mass extinctions: new answers, new questions. In *The last extinction,* ed. L. Kaufman and K. Mallory. Cambridge, MA: MIT Press. 43–62.

Jouzel, J., C. Lorius, J. R. Petit, C. Genthon, N. I. Barkov, V. M. Kotlyakov, and V. M. Petrov (1987). Vostok ice core: a continuous isotope temperature record over the last climatic cycle (160,000 years). *Nature, 329:* 403–7.

Kasting, J. F., O. B. Toon, and J. B. Pollack (1988). How climate evolved on the terrestrial planets. *Scientific American, 258* (February): 90–7.

Kirchner, J. W. (1989). The Gaia hypothesis: can it be tested? *Reviews of Geophysics, 27:* 223–35.

Lacis, A. A., and J. E. Hansen (1974). A parameterization for the absorption of solar radiation in the Earth's atmosphere. *Journal of Atmospheric Science, 31:* 118–33.

Legrand, M., R. J. Delmes, and R. J. Charlson (1988). Climate forcing implications from Vostok ice-core sulphate data. *Nature, 334:* 418–20.

Legrand, M., and C. Saigne (1987). Measurements of methane sulphonic acid in Antarctic ice. *Nature, 330:* 240–1.

Lewin, R. (1988). Recount on Amazon trees. *Science, 239:* 563.

* Lewis, J. S., and R. G. Prinn (1984). *Planets and their atmospheres.* Orlando, FL: Academic Press. 470 pp.

Lindberg, S. E., and C. T. Garter (1988). Sources of sulphur in forest canopy through fall. *Nature, 336:* 148–50.

Lohrenz, S. E., R. A. Arnone, D. A. Wiesenburg, I. P. de Palma (1988). Satellite detection of transient enhanced primary production in the western Mediterranean Sea. *Nature, 335:* 245–7.

Lorius, C., N. I. Barkov, J. Jouzel, Y. S. Korotkevich, V. M. Kotlyakov, and D. Raynaud (1988). Antarctic ice core: CO_2 and climatic change over the last climatic cycle. *EOS, 69:* 681–3.

Lorius, C., J. Jouzel, D. Raynaud, J. Hansen, and M. le Treut (1990). The ice-core record: climate sensitivity and future greenhouse warming. *Nature, 347:* 139–45.

* Lovelock, J. E. (1987a). *Gaia – a new look at life on Earth.* New York: Oxford University Press. 157 pp.

   (1987b). Geophysiology: a new look at Earth Sciences. In *The geophysiology of Amazonia,* ed. R. E. Dickinson. New York: Wiley. 11–24.

* (1988). *Ages of Gaia.* New York: Norton. 252 pp.

   (1989). Geophysiology, the science of Gaia. *Reviews of Geophysics, 27:* 215–22.

Lovelock, J. E., and A. J. Watson (1982). The regulation of carbon dioxide and climate: Gaia or geochemistry? *Planetary and Space Science, 30:* 795–802.

* Lutgens, F. K., and E. J. Tarbuck (1986). *The atmosphere.* Englewood Cliffs, NJ: Prentice Hall. 492 pp.

Lyle, M. (1988). Climatically forced organic carbon burial in equatorial Atlantic and Pacific Oceans. *Nature, 335:* 529–32.

Manabe, S., and R. J. Stouffer (1988). Two stable equilibria for coupled ocean-atmosphere model. *Journal of Climate, 1:* 841–66.

Martin, D. W., B. Goodman, T. J. Schmit, and E. C. Cutrim (1990). Estimates of daily rainfall over the Amazon basin. *Journal of Geophysical Research, 95:* 17043–50.

* Martin, P. S., and R. G. Klein, eds. (1984). *Quaternary extinctions: a prehistoric revolution.* Tucson: University of Arizona Press.

Meehl, G. A. (1987). The tropics and their role in the global climate system. *Geographical Journal, 153:* 21–36.

* Mitchell, A. W. (1987). *The enchanted canopy: secrets from the rainforest roof.* Glasgow: Fontana/Collins. 255 pp.

Molison, L. C. B. (1987a). Micrometeorology of an Amazonian rain forest. In *The geophysiology of Amazonia,* ed. R. E. Dickinson. New York: Wiley. 255–72.

   (1987b). On the dynamic climatology of the Amazon basin and associated rain-producing mechanisms. In *The geophysiology of Amazonia,* ed. R. E. Dickinson. New York: Wiley. 391–408.

Myers, N. (1989). The future of forests. In *The fragile environment,* ed. L. Friday and R. Laskey. Cambridge: Cambridge University Press. 22–40.

Neftel, A., H. Oeschger, T. Staffelbach, and B. Stauffer (1988). CO_2 record in the Byrd ice core 50,000–5,000 years B. P. *Nature, 331:* 609–11.

* Nisbet, E. G. (1987). *The young Earth.* London: Allen and Unwin. 402 pp.

   (1990a). Climate change and methane. *Nature, 347:* 23.

   (1990b). The end of the ice age. *Canadian Journal of Earth Sciences, 27:* 148–57.

* Oeschger, H., and C. C. Langway (1989). *The environmental record in glaciers and ice sheets.* Chichester: Wiley.

Paegle, J. (1987). Interactions between convective and large-scale motions over Amazonia. In *The geophysiology of Amazonia,* ed. R. E. Dickinson. New York: Wiley. 347–90.

Palmer, T. N. (1986). Influence of the Atlantic, Pacific and Indian Oceans on Sahel rainfall. *Nature, 322:* 251–3.

* Pearce, F. (1989). *Climate and man.* London: Vision Books, 176 pp.

* Philander, S. G. (1990). *El Niño, La Niña and the Southern Oscillation.* San Diego: Academic Press. 291 pp.

Pimm, S. L. (1984). The complexity and stability of ecosystems. *Nature, 307:* 321–5.

* Prance, G. T., and T. E. Lovejoy, eds. (1985). *Amazonia.* Oxford: Pergamon. 442 pp.

Prell, W. L., and J. E. Kutzback (1987). Monsoon variability over the past 150,000 years. *Journal of Geophysical Research, 92:* 8411–25.

Ramanathan, V. (1988). The radiative and climatic consequences of changing atmospheric composition of trace gases. In *The changing atmosphere,* ed. F. S. Rowland and I. S. A. Isaksen. Chichester: Wiley. 159–86.

Ramanathan, V., B. R. Barkstrom, and E. F. Harrison (1989). Climate and the earth's radiation budget. *Physics Today, 42* (May 1989): 22–33.

Ramanathan, V., R. D. Cess, E. F. Harrison, P. Minnis, B. R. Barkstrom, E. Ahmad, and D. Hartman (1989). Cloud-radiative forcing and climate: results from the Earth Radiation Budget Experiment. *Science, 243:* 57–63.

* Rambler, M. B., L. Margulis, and R. Fester, eds. (1989). *Global ecology: towards a science of the bisophere.* San Diego: Academic Press. 204 pp.

Raynaud, D., J. Chappellaz, J. M. Barnola, Y. S. Korotkevich, and C. Lorius (1988).

Climatic and CH_4 cycle implications of glacial-interglacial CH_4 change in the Vostok ice core. *Nature, 333:* 655–6.

Reid, G. C. (1987). Influence of solar variability on global sea surface temperature. *Nature, 329:* 142–3.

Ritchie, J. C., and C. V. Haynes (1987). Holocene vegetation zonation in the eastern Sahara. *Nature, 330:* 645–7.

Salati, E. (1987). The forest and the hydrological cycle. In *The geophysiology of Amazonia*, ed. R. E. Dickinson. New York: Wiley. 273–96.

Salo, J., R. Kallida, I. Hakkinnen, Y. Makinen, P. Niemela, M. Puhakka, and P. D. Coley (1986). River dynamics and the diversity of Amazon lowland forest. *Nature, 322:* 254–8.

Sarmiento, J. L. (1988). Models of carbon cycling in the oceans. CDIAC communications, Winter 1988. Oak Ridge National Laboratory, Oak Ridge, TN: CDIAC. 6–9.

Satheyendranath, S., A. D. Gouveia, S. R. Shetye, P. Ravindran, and T. Platt (1991). Biological control of surface temperature in the Arabian Sea. *Nature, 349,* 54–6.

Schatten, K. H. (1990). Climate impact of solar variability. *EOS, 71:* 1103.

* Schneider, S. H., and R. Londer (1984). *The co-evolution of climate and life.* San Francisco: Sierra Books.

Sear, C. B., P. M. Kelly, P. D. Jones, and C. M. Gooden (1987). Global surface temperature responses to major volcanic eruptions. *Nature, 330:* 365–7.

Seiler, W., and R. Conrad (1987). Contribution of tropical ecosystems to the global budgets of trace gases, especially CH_4, H_2, CO and NO. In *The geophysiology of Amazonia*, ed. R. E. Dickinson. New York: Wiley. 163–74.

Sellers, P. J. (1987). Modelling effects of vegetation on climate. In *The geophysiology of Amazonia*, ed. R. E. Dickinson. New York: Wiley. 297–344.

Shackleton, N. J., J.-C. Duplessy, M. Arnold, P. Maurice, M. A. Hall, and J. Cartlidge (1988). Radiocarbon age of last glacial Pacific deep water. *Nature, 335:* 708–11.

Shukla, J. (1987). General circulation modelling and the tropics. In *The geophysiology of Amazonia*, ed. R. E. Dickinson. New York: Wiley. 409–62.

Smith, G. L., D. Rutan, T. P. Charlock, and T. D. Bess (1990). Annual and interannual variations of absorbed solar radiation based on a 10-year data set. *Journal of Geophysical Research, 95:* 16639–52.

Smith, J. L., and E. A. Paul (1986). The role of soil type and vegetation on microbial biomass and activity. *Proceedings, IV, ISME.* 460–6.

Smith, L. D., and T. H. von der Haar (1991). Clouds-radiation interactions in a general circulation model: impact upon the planetary radiation balance. *Journal of Geophysical Research, 96:* 893–914.

Sperber, K. R., S. Hameed, W. L. Gates, and G. L. Potter (1987). Southern oscillation simulated in a global climate model. *Nature, 329:* 140–2.

Stauffer, B., E. Lochbronner, H. Oeschger, and J. Schwander (1988). Methane concentration in the glacial atmosphere was only half that of the pre-industrial Holocene. *Nature, 332:* 812–14.

Street-Perrott, F. A., and R. A. Perrott (1990). Abrupt climate fluctuations in the tropics: the influence of Atlantic Ocean circulation. *Nature, 343:* 607–12.

Sundquist, E. T. (1987). Ice core links CO_2 to climate. *Nature, 329:* 389–90.

* Sundquist, E. T., and W. S. Broecker, eds. (1985). *The carbon cycle and atmospheric CO_2: natural variations, Archean to present.* Geophysical Monograph 32. Washington, DC: American Geophysical Union. 627 pp.

Symonds, R. B., W. I. Rose, and M. H. Reed (1988). Contribution of Cl- and F- bearing gases to the atmosphere from volcanoes. *Nature, 334:* 415–18.

Thompson, A. M., W. E. Esaias, and R. L. Iverson (1990). Two approaches to determining the sea-to-air flux of dimethyl sulfide: satellite ocean color and a photochemical model with atmospheric measurements. *Journal of Geophysical Research, 95:* 20551–8.

Tucker, C. J., I. Y. Fung, C. D. Keeling, and R. H. Gammon (1986). Relationship between atmospheric CO_2 variations and a satellite derived vegetation index. *Nature, 319:* 195–9.

Turner, T. R. G., J. J. Lennon, and J. A. Lawrenson (1988). British bird species distributions and the energy theory. *Nature, 335:* 539–41.

* Walker, J. C. G. (1977). *Evolution of the atmosphere.* New York: Macmillan.

Walker, J. C. G., P. D. Hays, and J. F. Kasting (1981). A negative feedback mechanism for the long term stabilization of the Earth's surface temperature. *Journal of Geophysical Research, 86:* 9776–82.

* Warneck, P. (1988). *Chemistry of the natural atmosphere.* San Diego: Academic Press. 753 pp.

Watson, A. J., and J. E. Lovelock (1983). Biological homeostasis of the global environment: the parable of the 'daisy' world. *Tellus, 35b:* 284–9.

Wells, G. (1989). Observing earth's environment from space. In *The fragile environment,* ed. L. Friday and R. A. Laskey. Cambridge: Cambridge University Press. 148–92.

Wetherald, R. T., and S. Manabe (1988). Cloud feedback processes in a general circulation model. *Journal of Atmospheric Sciences, 45:* 1397–1414.

Williamson, P. G. (1985). Evidence for an early Plio-Pleistocene rainforest expansion in East Africa. *Nature, 315:* 487–9.

Winograd, I. J., B. J. Szabo, T. B. Coplen, and A. C. Riggs (1988). A 250,000 year climatic record from Great Basin vein calcite: implications for Milankovitch theory. *Science, 242:* 1275–80.

Wolback, W. S., I. Gilmour, E. Anders, C. J. Orth, and R. R. Brooks (1988). Global fire at the Cretaceous-Tertiary boundary. *Nature, 334:* 665–9.

* Woodward, F. I. (1987). *Climate and plant distribution.* Cambridge: Cambridge University Press. 174 pp.

Zeng, X., R. A. Pielke, and R. Eykholt (1990). Chaos in daisyworld. *Tellus, 42b:* 309–18.

3: The causes of change

Abelson, P. H. (1989). The Arctic: a key to world climate. *Science, 222:* 377.

* Allaby, M. (1990). *Living in the greenhouse.* Wellingborough, NN8 2RQ, England: Thorsons. 192 pp.

Anderson, I. C., J. S. Levine, M. A. Poth, and P. J. Riggan (1988). Enhanced biogenic emissions of nitric oxide and nitrous oxide following surface biomass burning. *Journal of Geophysical Research, 93:* 3893–98.

Angell, J. K. (1988). Impact of El Niño on the delineation of tropospheric cooling due to volcanic eruptions. *Journal of Geophysical Research, 93:* 3697–3704.

  (1990). Influence of equatorial QBO and SST on polar total ozone, and the 1990 Antarctic ozone hole. *Geophysical Research Letters, 17:* 1569–72.

Ayers, G. P., and R. W. Gillert (1990). Tropospheric chemical composition: overview of experimental methods in measurement. *Reviews of Geophysics, 28:* 297–314.

Barnett, T. P. (1988). Recent changes in the world's oceans: how well are they known? *EOS, 69:* 1045.

Barnett, T. P., and M. E. Schlesinger (1987). Detecting changes in global climate induced by greenhouse gases. *Journal of Geophysical Research, 92:* 14772–80.

* Barr, B. M., and K. E. Braden (1988). *The disappearing Russian forest: a dilemma in Soviet resource management.* Totowa, NJ: Rowman and Littlefield. London: Hutchinson. 252 pp.

Baskerville, G. L. (1988). Redevelopment of a degrading forest system. *Ambio, 17:* 314–22.

* Berger, A., R. E. Dickinson, and J. W. Kidson, eds. (1989). *Understanding climate change.* Washington, D.C.: American Geophysical Union. 187 pp.

Blake, D. R., and F. S. Rowland (1988). Continuing world-wide increase in tropospheric methane, 1978–1987. *Science, 239:* 1129–31.

* Boden, T. A., P. Kanciruk, and M. P. Farrell (1990). *Trends '90: A compendium of data on global change.* ONRL/CDIAC-36, Oak Ridge, TN: Carbon Dioxide Information Analysis Center, Oak Ridge National Laboratory. 286 pp.

Bodhaine, B. A., and R. M. Rosson, eds. (1988). *Geophysical monitoring for climatic change,* No. 16. Summary Report, 1987. National Oceanic and Atmospheric Administration, Air Resources Laboratory. Boulder, CO: U.S. Department of Commerce.

Bojkov, R., L. Bishop, W. J. Hill, G. C. Reinsel, and G. C. Tiao (1990). A statistical trend analysis of revised Dobson total ozone data over the Northern Hemisphere. *Journal of Geophysical Research, 95:* 9785–9807.

Bolin, B. (1986). How much CO_2 will remain in the atmosphere? In *The greenhouse effect, climatic change and ecosystems,* SCOPE 29, ed. B. Bolin et al. Chichester: Wiley. 93–155.

(1989). Changing climates. In *The fragile environment,* ed. L. Friday and R. Laskey. Cambridge: Cambridge University Press. 127–47.

* Bolin, B., B. R. Doos, J. Jager, and R. A. Warrick, eds. (1986). *The greenhouse effect, climatic change and ecosystems,* SCOPE 29. Chichester: Wiley. 541 pp.

Bolin, B., J. Jager, and B. R. Doos (1986). The greenhouse effect, climatic change and ecosystems: a synthesis of present knowledge. In *The greenhouse effect, climatic change and ecosystems,* SCOPE 29, ed. B. Bolin et al. Chichester: Wiley. 1–32.

Bolle, H. J., W. Seiler, and B. Bolin (1986). Other greenhouse gases and aerosols. In *The greenhouse effect, climatic change and ecosystems,* SCOPE 29, B. Bolin et al. Chichester: Wiley. 157–204.

* Boyle, S., and J. Ardill (1989). *The greenhouse effect.* London: Hodder and Stoughton. 298 pp.

Brasseur, G., and M. H. Hitchman (1988). Stratospheric response to trace gas perturbations: changes in ozone and temperature distributions. *Science, 240:* 634–7.

* Brimblecombe, P. (1986). *Air: composition and chemistry.* Cambridge: Cambridge University Press. 224 pp.

* Bureau of Meteorology, Australia (annual). *Baseline atmospheric program (Australia).* Canberra.

Butler, J. H., J. W. Elkins, and T. M. Thompson (1989). Tropospheric and dissolved

N₂O of the West Pacific and East Indian Oceans during the El Niño Southern Oscillation Event of 1987. *Journal of Geophysical Research, 94:* 14865–77.

* Carbon Dioxide Information Analysis Center (1990a). *Global change research and data.* CDIAC communications, Spring 1990. Oak Ridge National Laboratory, Oak Ridge, TN: CDIAC.

* (1990b). *Trends '90: a compendium of data on global change.* ORNL/CDIAC-36, Environmental Services Division. Oak Ridge National Laboratory, Oak Ridge, TN: CDIAC.

Chappelaz, J., J. M. Barnola, D. Raynaud, Y. S. Korotkevich, and C. Lorius (1990). Ice-core record of atmospheric methane over the past 160,000 years. *Nature, 345:* 127–31.

Charlson, R. J., J. Langner, and H. Rodhe (1990). Sulphate aerosol and climate. *Nature, 348:* 22.

Charlson, R. J., J. E. Lovelock, M. O. Andreae, and S. G. Warren (1987). Oceanic phytoplankton, atmospheric sulphur, cloud albedo, and climate. *Nature, 326:* 655–61.

Chatfield, R. B., and A. C. Delany (1990). Convection links biomass burning to increased tropical ozone; however, models will tend to overpredict O₃. *Journal of Geophysical Research, 95:* 18473–88.

Cicerone, R. J. (1988a). How has the atmospheric concentration of CO changed? In *The changing atmosphere,* ed. F. S. Rowland and I. S. A. Isaksen. Chichester: Wiley. 49–63.

(1988b). Some difficult problems in atmospheric chemical trend detection. *EOS,* 69: 1045.

(1989). Analysis of sources and sinks of atmospheric nitrous oxide (N₂O). *Journal of Geophysical Research, 94:* 18265–71.

Cicerone, R. J., and R. S. Oremland (1988). Biogeochemical aspects of atmospheric methane. *Global Biogeochemical Cycles, 2:* 299–328.

Clark, J. S. (1988). Effect of climate change on fire regimes in northwestern Minnesota. *Nature, 334:* 233–5.

Cofer, W. R., J. S. Levine, E. L. Winstead, and B. J. Stocks (1991). New estimates of nitrous oxide emissions from biomass burning. *Nature, 349:* 689–91.

* Crosby, A. W. (1986). *Ecological imperialism: the biological expansion of Europe, 900–1900.* Cambridge: Cambridge University Press. 368 pp.

Crutzen, P. J., and L. T. Gidel (1983). A two-dimensional photochemical model of the atmosphere, 2. *Journal of Geophysical Research, 88:* 6641–61.

Crutzen, P. J., and T. E. Gradel (1986). The role of atmospheric chemistry in environment-development interactions. In *Sustainable development of the biosphere,* ed. W. C. Clark and R. E. Munn. Cambridge: Cambridge University Press. 213–29.

Derwent, R. G. (1990). Evaluation of a number of chemical mechanisms for their application in models describing the formation of photochemical ozone in Europe. *Atmospheric environment, 24a:* 2615–24.

* Deutscher Bundestag (1988). *Schutz der Erdatmosphar.* Report of Commission of Inquiry of the Eleventh West German Parliament, B. Schmidbauer, chair. Dt. Bundestag, Referat Offentlichkeitsarbeit, Zur Sache; 88, 5, Bonn.

Devol, A. H., J. E. Richey, B. R. Forsberg, and L. A. Martinelli (1990). Seasonal dynamics in methane emissions from the Amazon river floodplain to the troposphere. *Journal of Geophysical Research, 95:* 16417–26.

Dickinson, R. E. (1986). Impact of human activities on climate – a framework. In *Sustainable development of the biosphere*, ed. W. C. Clark and R. E. Munn. Cambridge: Cambridge University Press. 252–88.

* Durrell, L. (1986). *State of the ark*. New York: Doubleday. 224 pp.

* Earth System Science Committee (1988). *Earth System Science: a program for global change*. Washington, DC: NASA. Boulder, CO: University Corporation for Atmospheric Research. 208 pp.

Ehhalt, D. H. (1988). How has the atmospheric concentration of CH_4 changed? In *The changing atmosphere*, ed. F. S. Rowland and I. S. A. Isaksen. Chichester: Wiley. 25–32.

Elkins, J. W., and R. M. Rosson, eds. (1989). *Geophysical monitoring for climatic change*, No. 17. Summary Report, 1988. Boulder, CO: National Oceanic and Atmospheric Administration, Air Resources Laboratory, U.S. Department of Commerce.

Elwell, H. A., and M. A. Stocking (1988). Loss of soil nutrients by sheet erosion is a major hidden farming cost. *Zimbabwe Science News, 22:* 79–82.

Evans, W. F. J. (1988). A measurement of the altitude variation of greenhouse radiation from CFC-12. *Nature, 333:* 750–2.

Farman, J. C., B. G. Gardiner, and J. D. Shanklin (1985). Large losses of total ozone in Antarctica reveal seasonal ClO_x/NO_x interaction. *Nature, 315:* 207–10.

Finlayson-Pitts, B. L., F. E. Livingston, and H. N. Berko (1990). Ozone destruction and bromine photochemistry at ground level in the Arctic spring. *Nature, 343:* 622–5.

Fisher, D. A., M. Ko, D. Wuebbles, and I. Isaksen (1990). Evaluating ozone depletion potentials – reply. *Nature, 348:* 203–4.

Fishman, J., C. E. Watson, J. C. Larson, and J. A. Logan (1990). Distribution of tropospheric ozone determined from satellite data. *Journal of Geophysical Research, 95:* 3599–3618.

* Forgan, B. W., and G. P. Ayers, eds. (1989). *Baseline atmospheric program (Australia)* 1987. CSIRO Division of Atmospheric Research, P.O. Box 346, Smithton, Tasmania 7330, Australia: Bureau of Meteorology. 71 pp.

Fraser, P. J., P. Hyson, R. A. Rasmussen, A. J. Crawford, and M. A. K. Khalil (1986). Methane, carbon dioxide and methyl chloroform in the Southern Hemisphere. *Journal of Atmospheric Chemistry, 4:* 3–42.

* Friday, L., and R. Laskey, eds. (1989). *The fragile environment*. Cambridge: Cambridge University Press. 198 pp.

Goldan, P. D., R. Fall, W. C. Kuster, and F. C. Fehsenfeld (1988). Uptake of COS by growing vegetation: a major tropospheric sink. *Journal of Geophysical Research, 93:* 14186–92.

Goreau, T. J., and W. Z. de Mello (1988). Tropical deforestation: some effects on atmospheric chemistry. *Ambio, 17:* 275–81.

* Gribbin, J. (1988). *The hole in the Sky*. London: Bantam. 192 pp.

* (1990). *Hothouse Earth*. London: Bantam. 273 pp.

Gunn, J. M., and Keller, W. (1990). Biological recovery of an acid lake after reductions in industrial emissions of sulphur. *Nature, 345:* 431–3.

Hammitt, J. K., F. Camm, P. S. Connell, W. E. Mooz, K. A. Wolf, D. J. Wuebbles, and A. Bamezi (1987). Future emission scenarios for chemicals that may deplete stratospheric ozone. *Nature, 330:* 711–16.

Hansen, J., and A. A. Lacis (1990). Sun and dust versus greenhouse gases: an assessment of their relative roles in global climate change. *Nature, 346:* 713–19.

Harrington, J. B. (1987). Climatic change: a review of causes. *Canadian Journal of Forest Research, 17:* 1313–39.

* Harrison, R. M., ed. (1990). *Pollution: causes, effects and control.* Royal Society of Chemistry. Cambridge: Thomas Graham House. CB4 4WF.

Heath, D. F. (1988). Non-seasonal changes in total column ozone from satellite observations, 1970–86. *Nature, 332:* 219–27.

Heil, G. W., M. J. A. Werger, W. de Mol, D. van Dam, and B. Heijne (1988). Capture of atmospheric ammonium by grassland canopies. *Science, 239:* 764–5.

Heintzenberg, J., and E. K. Bigg (1990). Tropospheric transport of trace substances in the southern hemisphere. *Tellus, 42b:* 355–63.

Hofmann, D. J., and T. Deshler (1990). Balloonborne measurements of polar stratospheric clouds and ozone at $-93°C$ in the Arctic in February 1990. *Geophysical Research Letters, 17:* 2185–8.

Hofmann, D. J., T. L. Deshler, P. Aimedieu, W. A. Matthews, P. V. Johnston, Y. Kondo, W. R. Sheldon, G. J. Byrne, and J. R. Benbrook (1989). Stratospheric clouds and ozone depletion in the Arctic during January 1989. *Nature, 340:* 117–21.

Hough, A. M., and Derwent, R. G. (1990). Changes in the global concentration of tropospheric ozone due to human activities. *Nature, 344:* 645–8.

* Houghton, J. T., G. J. Jenkins, and J. J. Ephraums (1990). *Climate change: the IPCC scientific assessment.* Cambridge: Cambridge University Press. Geneva: World Meteorological Office.

Houghton, R. A. (1990). The future role of tropical forests in affecting the carbon dioxide concentration of the atmosphere. *Ambio, 19:* 204–9.

* Huntley, B. J., ed. (1989). *Biotic diversity in southern Africa.* Cape Town: Oxford University Press. 380 pp.

* International Geosphere-Biosphere Program (1990). *The land–atmosphere interface.* Stockholm: IGBP Secretariat, Royal Swedish Academy of Sciences. 39 pp.

Jahnke, R. A. (1990). Ocean flux studies: a status report. *Reviews of Geophysics, 28:* 381–98.

Janach, W. E. (1989). Surface ozone: trend details, seasonal variations, and interpretation. *Journal of Geophysical Research, 94:* 18289–95.

Johansson, C., and E. Sanhueza (1988). Emission of NO from savanna soils during rainy season. *Journal of Geophysical Research, 93:* 14193–8.

Judge, A. (1982). Natural gas hydrates in Canada. In *Proceedings, Fourth Canadian Permafrost Conference,* ed. H. M. French. Ottawa: Natural Research Council of Canada. 320–9.

Karl, T. R., J. D. Tarpley, R. G. Quayle, H. F. Diaz, D. A. Robinson, and R. S. Bradley (1989). The recent climate record: what it can and cannot tell us. *Review of Geophysics, 27:* 405–30.

Kaufman, L. (1986). Why the ark is sinking. In *The last extinction,* ed. L. Kaufman and K. Mallory. Cambridge, MA: M.I.T. Press. 1–43.

* Kaufman, L., and K. Mallory, eds. (1986). *The last extinction.* Cambridge, MA: MIT Press. 208 pp.

Keepin, W., I. Mintzer, and L. Kristoferson (1986). Emission of CO_2 into the atmosphere. In *The greenhouse effect, climatic change and ecosystems,* SCOPE 29, ed. B. Bolin et al. Chichester: Wiley. 35–91.

Kelly, P. M., and T. M. L. Wigley (1990). The influence of solar forcing trends on global mean temperature since 1851. *Nature, 347:* 460–2.

Kerr, R. A. (1990). Another deep Antarctic ozone hole. *Science, 250:* 370.

Khalil, M. A. K., and R. A. Rasmussen (1983). Sources, sinks and seasonal cycles of atmospheric methane. *Journal of Geophysical Research, 88:* 5131–44.

(1984). Carbon monoxide in the Earth's atmosphere, increasing trend. *Science, 224:* 54–6.

(1988). Carbon monoxide in the Earth's atmosphere; indications of a global increase. *Nature, 332:* 242–5.

(1990). Atmospheric carbon monoxide: Latitudinal distribution of sources. *Geophysical Research Letters, 17:* 1913–16.

Kinnison, D., H. Johnston, and D. Wuebbles (1988). Ozone calculations with large nitrous oxide and chlorine changes. *Journal of Geophysical Research, 93:* 14165–75.

Kinnison, D., and D. Wuebbles (1989). Nitrogen oxides from high altitude aircraft: an update of potential effects on ozone. *Journal of Geophysical Research, 94:* 16351–63.

Kirchhoff, V. W. J. H., A. W. Setzer, and M. C. Pereira (1989). Biomass burning in Amazonia: seasonal effects on atmospheric O_3 and CO. *Geophysical Research Letters, 16:* 469–72.

Klein, R. M., and T. D. Perkins (1987). Cascades of causes and effects of forest decline. *Ambio, 16:* 86–93.

* Knystantas, A. (1987). *The natural history of the U.S.S.R.* London: Century. 224 pp.

Koerner, R. M. (1989). Ice core evidence for extensive melting of the Greenland ice sheet in the last interglacial. *Science, 243:* 964–8.

Krueger, A. J., R. S. Stolarski, and M. R. Schoeberl (1989). Formation of the 1988 ozone hole. *Geophysical Research Letters, 16:* 381–4.

Kvenvolden, K. A. (1982). Origin and occurrence of marine gas hydrates. In *Proceedings, Fourth Canadian Permafrost Conference,* ed. H. M. French. Ottawa: National Research Council of Canada. 305–11.

(1983). Marine Gas Hydrates – I. Geochemical evidence. In *Natural gas hydrates: properties occurrence and recovery,* ed. J. L. Cox. Boston: Butterworth. 63–72.

LeLieveld, J., and P. J. Crutzen (1990). Inflluences of cloud photochemical processes on tropospheric ozone. *Nature, 343:* 227–33.

Levi, B. G. (1988). Ozone depletion at the poles: the hole story emerges. *Physics Today, 41* (July): 17–21.

Levine, J. S. (1990). Global biomass burning: atmospheric climatic and biospheric implications. *EOS, 71:* 1075–7.

Likens, G. E., F. H. Bormann, L. O. Hedin, C. T. Driscoll, and J. S. Eaton (1990). Dry deposition of sulfur: a 23-year record for the Hubbard Brook Forest ecosystem. *Tellus, 42b:* 319–29.

* Linklater, E. (1972). *The voyage of the Challenger.* London: John Murray. 288 pp.

* Liss, P. S., and A. J. Crane (1983). *Man-made carbon dioxide and climatic change.* Norwich: Geo Books.

Lobert, J. M., D. H. Scharffe, W. M. Hao, and P. J. Crutzen (1990). Importance of biomass burning in the atmospheric budgets of nitrogen-containing gases. *Nature, 346:* 552–4.

Lowe, D. C., C. A. M. Brenninkmeijer, M. R. Manning, R. Sparlis, and G. Wallace (1988). Radiocarbon determination of atmospheric methane at Baring Head, New Zealand. *Nature, 332:* 522–5.

MacGarry, B. (1988). Measuring urban and family fuel consumption. *Zimbabwe Science News, 22:* 93–4.

Makogon, Y. F. (1982). Perspectives for the development of gas-hydrate deposits. In *Proceedings, Fourth Canadian Permafrost Conference,* ed. H. M. French. Ottawa: National Research Council of Canada. 299–304.

Malingreau, J.-P., and C. J. Tucker (1988). Large-scale deforestation in the south-east Amazon basin of Brazil. *Ambio, 17:* 49–55.

Marland, G., R. M. Rotty, and N. L. Treat (1985). CO_2 from fossil fuel burning: global distribution of emissions. *Tellus, 37b:* 243–8.

* Martin, P. S., and R. G. Klein, eds. (1984). *Quaternary extinctions: a prehistoric revolution.* Tucson: University of Arizona Press. 892 pp.

Mayewski, P. A., W. B. Lyons, M. J. Spencer, M. S. Twickler, C. F. Buck, and S. Whitlow (1990). An ice-core record of atmospheric response to anthropogenic sulphate and nitrate. *Nature, 346:* 554–6.

McBride, J. P., R. E. Moore, J. P. Witherspoon, and R. E. Blanco (1978). Radiological impact of airborne effluents of coal and nuclear plants. *Science, 202:* 1045–50.

McElroy, M. B., and R. J. Salawitch (1989). Changing composition of the global stratosphere. *Science, 243:* 763–70.

Mossman, B. T., J. Bignon, M. Corn, A. Seaton, and J. B. L. Gee (1990). Asbestos: scientific developments and implications for public policy. *Science, 247:* 294–301.

Neftel, A., E. Moor, H. Oeschger, and B. Stauffer (1985). Evidence from polar ice cores for the increase in atmospheric CO_2 in the past two centuries. *Nature, 315:* 45–7.

Newman, P., R. Stolarski, M. Schoeberl, L. R. Lait, and A. Krueger (1990). Total ozone during the 88–89 Northern Hemisphere winter. *Geophysical Research Letters, 17:* 317–20.

Nisbet, E. G. (1989). Some northern sources of atmospheric methane: production, history and future implications. *Canadian Journal of Earth Sciences, 26:* 1603–11.

Nriagu, J. O., D. A. Holdway, and R. D. Coker (1987). Biogenic sulfur and the acidity of rainfall in remote areas of Canada. *Science, 237:* 1189–91.

Oeschger, H., and W. Siegenthaler (1988). How has the atmospheric concentration of CO_2 changed? In *The Changing Atmosphere,* ed. F. S. Rowland and I. S. A. Isaksen. Chichester: Wiley. 5–24.

Pandis, S. N., J. H. Seinfeld, and C. Pilinis (1990). The smog-fog-smog cycle and acid deposition. *Journal of Geophysical Research, 95:* 18489–500.

* Pearce, F. (1989). *Climate and man.* London: Vision Books. 176 pp.

Pearman, G. I., and P. J. Fraser (1988). Sources of increased methane. *Nature, 332:* 489–90.

Penner, J. E., C. S. Atherton, J. Dignon, S. J. Ghan, and J. J. Walton (1991). Tropospheric nitrogen: a three-demensional study of sources, distributions, and deposition. *Journal of Geophysical Research, 96:* 959–90.

Perry, H. (1983). Coal in the United States: a status report. *Science, 222:* 377.

Prather, M. J., and R. T. Watson (1990). Stratospheric ozone depletion and future levels of atmospheric chlorine and bromine. *Nature, 344:* 729–34.

Prinn, R. (1988). How have the atmospheric concentrations of halocarbons changed? In *The changing atmosphere,* ed. F. S. Rowland and I. S. A. Isaksen. Chichester: Wiley. 33–48.

Prinn, R., D. Cunnold, R. Rasmussen, P. Simmonds, F. Alyea, A. Crawford, P. Fraser, and R. Rosen (1990). Atmospheric emissions and trends of nitrous oxide deduced from 10 years of ALE-GAGE data. *Journal of Geophysical Research, 95:* 18369–85.

Ramanathan, V., L. Callis, R. Cess, J, Hansen, I. S. A. Isaksen, W. Kuhn, A. Lacis, F. Liether, J. Mahlman, R. Reck, and M. Schlesinger (1987). Climate–chemical interaction and effects of changing atmospheric trace gases. *Reviews of Geophysics, 25:* 1441–81.

Ramanathan, V., R. J. Cicerone, H. B. Singh, and J. T. Kiehl (1985). Trace gas trends and their potential role in climatic change. *Journal of Geophysical Research, 90:* 5547–66.

Rasmussen, R. A., and M. A. K. Khalil (1984). Atmospheric methane in the recent and ancient atmospheres: concentrations, trends and inter-hemispheric gradient. *Journal of Geophysical Research, 89:* 11599–605.

Richey, J. E., and M. de N. G. Ribeiro (1987). Element cycling in the Amazon Basin: a riverine perspective. In *The geophysiology of Amazonia,* ed. R. E. Dickinson. New York: Wiley. 245–54.

Rind, D., E-W. Chiou, W. Chu, J. Larsen, S. Oltmans, J. Lerner, M. P. McCormick, and L. McMaster (1991). Positive water vapour feedback in climate models confirmed by satellite data. *Nature, 349:* 500–3.

Rodgers, C. (1988). Global ozone trends reassessed. *Nature, 332:* 201.

Rodhe, H., E. Cowling, I. Galbally, J. Galloway, and R. Herrera (1988). Acidification and regional air pollution in the tropics. In *Acidification in tropical countries,* SCOPE 36, ed. H. Rodhe and R. Herrera. Chichester: Wiley. 3–39.

* Rodhe, H., and R. Herrera, eds. (1988). *Acidification in tropical countries,* SCOPE 36, Chichester: Wiley. 405 pp.

Rosemaria, A., ed. (1987). Forests today. *Ambio, 16,* no. 2–3.

Rotty, R. M. (1986). Estimates of CO_2 from wood fuel based on forest harvest data. *Climatic Change, 9:* 311–25.

Rowland, F. S., D. R. Blake, S. C. Tyler, and Y. Makide (1985). Increasing concentrations of perhalocarbons, methylchloroform and methane in the atmosphere. *Pontificae Academiae Scientiarum Scripta Vaia, 56:* 305–39.

* Rowland, F. S., and I. S. A. Isaksen, eds. (1988). *The changing atmosphere.* Chichester: Wiley. 282 pp.

Schimel, D. (1990). Biogeochemical feedbacks in the Earth System. In *Global warming: the Greenpeace report,* ed. J. Leggett. Oxford: Oxford University Press. 68–82.

Schindler, D. W. (1988). Effects of acid rain on freshwater ecosystems. *Science, 150:* 149–57.

Schnell, R. C. (1987). A major Arctic haze event north of Point Barrow, April 1986. In *Geophysical monitoring for climate change,* No. 15, ed. R. C. Schnell and M. Rossen. U.S. NOAA. Boulder, CO: U.S. Department of Commerce. 94–9.

Schnell, R. C., and M. Rosson, eds. (1987). *Geophysical monitoring for climatic change,* No. 15. U.S. NOAA. Boulder, CO: U.S. Department of Commerce.

Schoeberl, M. R. (1988). Dynamics weaken the polar hole. *Nature, 336:* 420–1.

Schoeberl, M. R., R. S. Stolarski, and A. J. Krueger (1989). The 1989 Antarctic ozone depletion: comparison with previous year depletions. *Geophysical Research Letters, 16:* 377–80.

Schutz, H., A. Holzapfel-Pschorn, R. Conrad, H. Rennenberg, and W. Seiler (1989). A 3-year continuous record on the influence of daytime, season and fertilizer treatment on methane emission rates from an Italian rice paddy. *Journal of Geophysical Research, 94:* 16405–16.

Schwartz, S. E. (1988). Are global cloud albedo and climate controlled by marine phytoplankton? *Nature, 336:* 441–5.

(1989). Acid deposition: unravelling a regional phenomenon. *Science, 243:* 753–62.

Sear, C. B., P. M. Kelly, P. D. Jones, and C. M. Goodess (1987). Global surface temperature responses to major volcanic eruptions. *Nature, 330:* 365–7.

Seiler, W., and R. Conrad (1987). Contributions of tropical ecosystems to the global budgets of trace gases, especially CH_4, H_2, CO and N_2O. In *The geophysiology of Amazonia*, ed. R. E. Dickinson. New York: Wiley. 133–62.

Seinfeld, J. H. (1989). Urban air pollution: state of the science. *Science, 243:* 745–52.

Simmons, A. H. (1988). Extinct pygmy hippopotamus and early man in Cyprus. *Nature, 333:* 554–7.

Solomon, S. (1988). The mystery of the Antarctic ozone "hole." *Reviews of Geophysics, 26:* 131–48.

(1990). Progress towards a quantitative understanding of Antarctic ozone depletion. *Nature, 347:* 347–53.

Solomon, S., and A. Tuck (1990). Evaluating ozone depletion potentials. *Nature, 348:* 203.

Spencer, R. W., and J. R. Christy (1990). Precise monitoring of global temperature trends from satellites. *Science, 247:* 1558–62.

Spivakovsky, C. M., S. C. Wofsy, and M. J. Prather (1990). A numerical model for parameterization of atmospheric chemistry: computation of tropospheric OH. *Journal of Geophysical Research, 95:* 18433–9.

Stauffer, B., E. Lochbronner, H. Oeschger, and J. Schwander (1988). Methane concentration in the glacial atmosphere was only half that of the pre-industrial Holocene. *Nature, 332:* 812–14.

Steudler, P. A., R. D. Bowden, J. M. Melito, and J. D. Aber (1989). Influence of nitrogen fertilization on methane uptake in temperate forest soils. *Nature, 341:* 314–16.

Stolarski, R. S. (1988a). The Antarctic ozone hole. *Scientific American, 258* (January): 30–7.

(1988b). Changes in ozone over the Antarctic. In *The changing atmosphere*, ed. F. S. Rowland and I. S. A. Isaksen. Chichester: Wiley. 105–20.

Stolarski, R. S., A. J. Krueger, M. R. Schoeberl, R. D. McPeters, P. A. Newman, and J. C. Alpert (1986). Numbus 7 satellite measurements of the springtime Antarctic ozone decrease. *Nature, 322:* 808–10.

Stolarski, R. S., M. R. Schoeberl, P. A. Newman, R. A. McPeters, and A. J. Krueger (1990). The 1989 Antarctic ozone hole as observed by TOMS. *Geophysical Research Letters, 17:* 1267–70.

Stringer, C. (1988). The dates of Eden. *Nature, 331:* 565–6.

Sullivan, T. J., J. M. Eilers, M. R. Church, D. J. Blick, K. N. Eshleman, D. H. Landers, and M. S. DeHaan (1988). Atmospheric wet sulphate deposition and lakewater chemistry. *Nature, 331:* 607–9.

Taylor, J. A., G. P. Brasseur, P. R. Zimmerman, and R. J. Cicerone (1991). A study of the sources and sinks of methane and methyl chloroform, using a global three-dimensional Lagrangian tropospheric tracer transport model. *Journal of Geophysical Research, 96:* 3013–44.

Thiemens, M. H., and W. C. Trogler (1991). Nylon production: an unknown source of atmospheric nitrous oxide. *Science, 251:* 932–4.

Thoning, K. W., P. P. Tans, and W. D. Komhyr (1989). Atmospheric carbon dioxide at

Mauna Loa Observatory, 2, Analysis of the NOAA GMCC Data, 1974–1985. *Journal of Geophysical Research, 94:* 8549–66.

Toggweiler, J. R. (1988). Deep sea carbon, a burning issue. *Nature, 334:* 467.

Twomey, W. (1988). Earth's radiation balance. *EOS, 69:* 1045.

U.S. National Oceanic and Atmospheric Administration (annual). *Summary Report, Geophysical Monitoring for Climate Change Program.* Boulder, CO: U.S. Department of Commerce.

* Velichko, A. A. (1984). *Late Quaternary environments of the Soviet Union.* Minneapolis: University of Minnesota Press. 327 pp.

Volz, A., and D. Kley (1988). Evaluation of the Montsouris series of ozone measurements made in the nineteenth century. *Nature, 332:* 240–2.

* Warneck, P. (1988). *Chemistry of the natural atmosphere.* San Diego: Academic Press.

Warren, S. G., and A. D. Clarke (1990). Soot in the atmosphere and snow surface of Antarctica. *Journal of Geophysical Research, 95:* 1811–16.

* Washington, W. M., and C. L. Parkinson (1986). *An introduction to three-dimensional climate modelling.* Oxford: Oxford University Press. 422 pp.

Waterman, L. S., D. W. Nelson, W. D. Komhyr, T. B. Harris, K. W. Thoning, and P. P. Tans (1989). Atmospheric carbon dioxide measurements at Cape Matatala, American Samoa, 1976–87. *Journal of Geophysical Research, 94:* 14817–29.

Waterman, R., and P. Cross (1988). Does rural sanitation promote deforestation in Zimbabwe? *Zimbabwe Science News, 22:* 88–92.

Watson, A. J., C. Robinson, J. E. Robinson, P. J. le B. Williams, and M. J. R. Fasham (1991). Spatial variability in the sink for atmospheric carbon dioxide in the North Atlantic. *Nature, 350:* 50–3.

Watson, C. E., J. Fishman, and H. G. Reichle (1990). The significance of biomass burning as a source of carbon monoxide and ozone in the Southern Hemisphere tropics: a satellite analysis. *Journal of Geophysical Research, 95:* 16443–50.

* Wayne, R. P. (1988). *Principles and applications of photochemistry.* Oxford: Oxford University Press. 268 pp.

* Wells, G. (1989). Observing the Earth's environment from space. In *The fragile environment,* ed. L. Friday and R. Laskey. Cambridge: Cambridge University Press. 148–92.

Whelan, T. (1988). Central American environmentalists call for action on the environment. *Ambio, 17:* 72–5.

Wigley, T. M. L., P. D. Jones, and P. M. Kelly (1986). Empirical climate studies. In *The greenhouse effect, climatic change and ecosystems,* SCOPE 29, ed. B. Bolin et al. Chichester: Wiley. 271–322.

Wigley, T. M. L., and S. C. B. Raper (1990). Climatic change due to solar irradiance changes. *Geophysical Research Letters, 17:* 2169–72.

Williams, J. D., and R. M. Nowak (1986). Vanishing species in our own backyard: extinct fish and wildlife of the United States and Canada. In *The last extinction,* ed. L. Kaufman and K. Mallory. Cambridge, MA: MIT Press. 107–40.

* Wilson, S. R., and G. P. Ayers (1990). *Baseline atmospheric program (Australia) 1988.* CSIRO Division of Atmospheric Research, P.O. Box 346, Smithton, Tasmania 7330, Australia: Bureau of Meteorology. 80 pp.

Woodwell, G. M., R. A. Houghton, T. A. Stone, R. F. Nelson, and W. Kovalick (1987). Deforestation in the tropics: new measurements in the Amazon Basin using Landsat and NOAA advanced very high resolution radiometer imagery. *Journal of Geophysical Research, 92:* 2157–63.

* World Resources Institute (1987). *World resources 1987.* New York: Basic Books. 369 pp.

(1988). *World resources 1988–89.* New York: Basic Books. 372 pp.

Yavitt, J. B., G. E. Lang, and A. J. Sexstone (1990). Methane fluxes in wetland and forest soils, beaver ponds and low-order streams of a temperate forest ecosystem. *Journal of Geophysical Research, 95:* 22463–74.

Zwally, H. J. (1988). Measuring global change in polar regions. *EOS, 69:* 1045.

4: The consequences of change

Adams, R. M., C. Rosenzweig, R. M. Peart, J. T. Ritchie, B. A. McCarl, J. D. Glyer, R. B. Curry, J. W. Jones, K. J. Boote, and L. H. Allen, Jr. (1990). Global climate change and U.S. agriculture. *Nature, 345:* 219–24.

Angell, J. K. (1990). Variation in global tropospheric temperature after adjustment for the El Niño influence, 1958–89. *Geophysical Research Letters, 17:* 1093–6.

* Berger, A., S. Schneider, and J. C. L. Duplessy, eds. (1989). *Climate and geo-sciences: a challenge for science and society in the 21st century.* Dordrecht: Kluwer. 724 pp.

Bethoux, J. P., Gentili, B., Raunet, J., and Taillez, D. (1990). Warming trend in the western Mediterranean deep water. *Nature, 347:* 660–2.

Bilham, R. (1991). Earthquakes and sea level: space and terrestrial metrology on a changing planet. *Reviews of Geophysics, 29:* 1–31.

Bolin, B. (1986). How much CO_2 will remain in the atmosphere? In *The greenhouse effect, climatic change and ecosystems,* SCOPE 29, ed. B. Bolin, B. R. Doos, J. Jager, and R. A. Warrick. Chichester: Wiley. 93–155.

(1989). Changing climates. In *The fragile environment,* ed. L. E. Friday and R. A. Laskey. Cambridge: Cambridge University Press. 127–47.

* Borgese, E. M. (1986). *The future of the oceans.* Montreal: Harvest House. 144 pp.

* Botkin, D. B., M. F. Caswell, J. E. Estes, and A. A. Orio (1989). *Changing the global environment.* San Diego: Academic Press. 459 pp.

Briffa, K. R., T. S. Bartholin, D. Eckstein, P. D. Jones, W. Karlen, F. H. Schweingruber, and P. Zetterberg. (1990). A 1400-year tree-ring record of summer temperatures in Fennoscandia. *Nature, 346:* 434–9.

Broccoli, A. J., and S. Manabe (1990). Can existing climate models be used to study anthropogenic changes in tropical cyclone climate? *Geophysical Research Letters, 17:* 1917–20.

Bromwich, D. H. (1990). Estimates of Antarctic precipitation. *Nature, 343:* 627–30.

Bryan, K., S. Manabe, and M. J. Spelman. (1988). Interhemispheric asymmetry in the transient response of a coupled ocean-atmosphere model to a CO_2-forcing. *Journal of Physical Oceanography, 18:* 851–67.

Cess, R. D., and 31 others (1990). Intercomparison and interpretation of climate feedback processes in 19 atmospheric general circulation models. *Journal of Geophysical research, 95:* 16601–15.

Cicerone, R. J. (1990). Atmospheric chemistry: greenhouse cooling up high. *Nature, 344:* 104–5.

Cook, E. R., M. A. Kablack, and G. C. Jacoby (1988). The 1986 drought in the southeast United States: how rare an event was it? *Journal of Geophysical Research, 93:* 14257–60.

Diaz, H. F. (1990). A comparison of "global" temperature estimates from satellite and instrumental data, 1979–88. *Geophysical Research Letters, 17:* 2373–6.

Dickinson, R. E. (1986). How will climate change? In *The greenhouse effect, climatic change and ecosystems,* SCOPE 29, ed. B. Bolin, B. R. Doos, J. Jager, and R. A. Warrick. Chichester: Wiley. 206–70.

Dickinson, R. E., and R.J. Cicerone (1986). Future global warming from atmospheric trace gases. *Nature, 319:* 109–15.

Dickinson, R. E., G. A. Meehl, and W. M. Washington (1987). Ice-albedo feedback in a CO_2-doubling simulation. *Climatic Change, 10:* 241–8.

Fairbanks, R. G. (1989). A 17,000 year glacio-eustatic sea-level record: influence of glacial melting rates on the Younger Dryas event and deep-ocean circulation. *Nature, 342:* 637–42.

* Fergusson, J. E. (1990). *The heavy elements: chemistry, environmental impact and health effects.* Oxford: Pergamon.

Fisher, D. A., C. H. Hales, D. L. Filkin, M. K. W. Ko, N. D. Sze, P. S. Connell, D. J. Wuebbles, I. S. A. Isaksen, and F. Stordal. (1990). Model calculations of the relative effects of CFCs and their replacements on stratospheric ozone. *Nature, 344:* 508–12.

Ghuman, B. S., and R. Lal (1987). Effects of deforestation on soil properties and micro-climate of a high rainforest in southern Nigeria. In *The geophysiology of Amazonia,* ed. R. E. Dickinson. New York: Wiley. 225–44.

Hansen, J., I. Fury, A. Lacis, D. Rind, S. Lebedoff, R. Ruedy, and G. Russell (1988). Global climate changes as forecast by Goddard Institute for Space Studies, three-dimensional model. *Journal of Geophysical Research, 93:* 9341–64.

Hansen, J., A. Lacis, and M. Prather (1989). Greenhouse effect of chlorofluorocarbons and other trace gases. *Journal of Geophysical Research, 94:* 16417–21.

Hansen, J., and S. Lebedeff (1987). Global trends of measured surface air temperature. *Journal of Geophysical Research, 92:* 13345–72.

(1988). Global surface air temperatures: update through 1987. *Geophysical Research Letters, 15:* 323–6.

* Hart, R., ed. (1989). *Global climatic change.* Saskatoon: University of Saskatchewan, Extension Division. 217 pp.

Henderson-Sellers, B. (1987). Modelling sea surface temperature rise resulting from increasing atmospheric carbon dioxide concentrations. *Climatic Change, 11:* 349–59.

Holling, C. S. (1986). The resilience of terrestrial ecosystems: local surprise and global change. In *Sustainable development of the biosphere,* ed. W. C. Clark and R. E. Munn. Cambridge: IIASA/Cambridge University Press. 292–316.

* Houghton, J. T., G. J. Jenkins, and J. J. Ephraums, eds. (1990). *Climate change: the IPCC scientific assessment.* Cambridge: Cambridge University Press. 365 pp.

Huntley, B. (1990). Lessons from climates of the past. In *Global warming: the Greenpeace report,* ed. J. Leggett. Oxford: Oxford University Press. 133–48.

* Independent Commission on International Humanitarian Issues (1986a). *The encroaching desert: the consequences of human failure.* London: Zed Books. Harare, Zimbabwe: College Press. 132 pp.

* (1986b). *The vanishing forest: the human consequences of deforestation.* London: Zed Books. Harare, Zimbabwe: College Press. 89 pp.

* Intergovernmental Panel on Climate Change (1990a). *Potential impacts of climate change: Report from Working Group II.* Geneva: World Meteorological Organization/United Nations Environmental Program. 46 pp.

*   (1990b). *Scientific assessment of climate change.* Geneva: World Meteorological Organization/United Nations Environmental Program. 365 pp. (Note: see also Houghton et al. 1990.)

Jones, P. D., M. L. Wigley, and P. B. Wright (1986). Global temperature variations between 1861 and 1984. *Nature, 322:* 430–4.

Karl, T. R., and R. R. Heim (1990). Are droughts becoming more frequent or severe in the United States? *Geophysical Research Letters, 17:* 1921–4.

Karl, T. R. and P. M. Steurer (1990). Increased cloudiness in the United States during the first half of the twentieth century: fact or fiction? *Geophysical Research Letters, 17:* 1925–8.

Karl, T. R., J. D. Tapley, R. G. Quayle, H. F. Diaz, D. A. Robinson, and R. S. Bradley (1989). The recent climate record: what it can tell us. *Reviews of Geophysics, 27:* 405–30.

Karl, T. R., R. R. Heim and R. G. Quayle (1991). The greenhouse effect in central North America – if not now, when? *Science, 251,* 1058–61.

Karoly, D. J. (1989). Northern hemisphere temperature trends: a possible greenhouse gas effect. *Geophysical Research Letters, 16:* 465–6.

* Kates, R. W., J. H. Ausubel, and M. Berberian, eds. (1985). *Climate impact assessment: studies of the interaction of climate and society,* SCOPE 27. Chichester: Wiley. 625 pp.

Kerr, R. A. (1988). Is the greenhouse here? *Science, 239:* 559–61.

  (1990). Global warming continues in 1989. *Science, 247:* 521.

* King, C. (1984). *Immigrant killers: introduced predators and the conservation of birds in New Zealand.* Auckland: Oxford University Press. 224 pp.

Koerner, R. M., and D. A. Fisher (1990). A record of Holocene summer climate from a Canadian high-Arctic ice core. *Nature, 343:* 630–1.

Kohlmaier, G. H., E.-O. Sire, A. Janacek, C. D. Keeling, S. C. Piper, and R. Revelle (1989). Modelling the seasonal contribution of a CO_2 fertilization effect of the terrestrial vegetation on the amplitude increase in atmospheric CO_2 at Mauna Loa observatory. *Tellus, 41B:* 487–510.

Kuo, C., C. Lindberg, and D. J. Thomson (1990). Coherence established between atmospheric carbon dioxide and global temperature. *Nature, 343:* 709–14.

Lachenbruch, A. H., and B. V. Marshall (1986). Changing climate: geothermal evidence from permafrost in the Alaskan Arctic. *Science, 234:* 689–96.

Lashof, D. A., and D. R. Ahuja (1990). Relative contributions of greenhouse gas emissions to global warming. *Nature, 344:* 529–31.

Latham, J., and M. H. Smith (1990). Effect on global warming of wind-dependent aerosol generation at the ocean surface. *Nature, 347:* 372–3.

Laxon, S. (1990). Seasonal and inter-annual variations in Antarctic sea ice extent as mapped by radar altimetry. *Geophysical Research Letters, 17: 1553–6.*

* Liss, P. S., and A. J. Crane (1983). *Man-made carbon dioxide and climatic change – a review of scientific problems.* Norwich: Geobooks. 127 pp.

Liu, Q., and C. J. E. Schuurmans (1990). The correlation of tropospheric and stratospheric temperatures and its effect on the detection of climate changes. *Geophysical Research Letters, 17:* 1085–8.

* MacDonald, G. J., ed. (1982). The long-term impacts of increasing atmospheric carbon dioxide levels. Cambridge, MA: Ballinger. 252 pp.

MacDonald, G. J. (1990). Role of methane clathrates in past and future climates. *Climatic Change, 16:* 247–81.

Manabe, S., and K. Bryan (1985). CO_2-induced change in a coupled ocean-atmosphere model and its palaeoclimatic implications. *Journal of Geophysical Research, 90:* 11689–707.

Manabe, S., K. Bryan, and M. J. Spelman (1990). Transient response of a global ocean-atmosphere model to a doubling of atmospheric carbon dioxide. *Journal of Physical Oceanography, 20:* 722–49.

Manabe, S., and R. J. Stouffer. (1980). Sensitivity of a global climate model to an increase of CO_2 concentration in the atmosphere. *Journal of Geophysical Research, 85:* 5529–54.

McElroy, M. B. (1986). Change in the natural environment of the Earth: the historical record. In *Sustainable development of the biosphere,* ed. W. C. Clark and R. E. Munn. Cambridge: IIASA/Cambridge University Press. 199–211.

Mikolajewicz, U., B. D. Santer, and E. Maier-Reimer (1990). Ocean response to global warming. *Nature, 345:* 589–93.

Mitchell, J. F. B. (1988). Local effects of greenhouse gases. *Nature, 332:* 399–400.
    (1989). The greenhouse effect and climate change. *Reviews of Geophysics, 27:* 115–39.

Mitchell, J. F. B., and G. Lupton (1984). A 4xCO_2 integration with prescribed changes in sea surface temperatures. *Progress in Biometeorology, 3:* 353–74.

Mitchell, J. F. B., C. A. Senior, and W. J. Ingram (1989). CO_2 and climate: a missing feedback? *Nature, 341:* 132–4.

Mitchell, J. F. B., and D. A. Warrilow (1987). Summer dryness in northern mid-latitudes due to increased CO_2. *Nature, 330:* 238–40.

* Myers, N., ed. (1984). *Gaia – an atlas of planet management.* New York: Anchor/Doubleday. 272 pp.
    (1989). The future of forests. In *The fragile environment,* ed. L. Friday and R. Laskey. Cambridge: Cambridge University Press. 22–40.

* Norton, B. G., ed. (1986). *The preservation of species – the value of biological diversity.* Princeton, NJ: Princeton University Press. 305 pp.

Overpeck, J. T., E. Rind, and R. Goldberg (1990). Climate-induced changes in forest disturbance and vegetation. *Nature, 343:* 51–3.

Peltier, W. R. (1988). Global sea level and earth rotation. *Science, 240:* 895–901.

Peltier, W. R., and A. M. Tushingham (1989). Global sea level rise and the greenhouse effect: might they be connected? *Science, 244:* 806–10.

Plantico, M. J., T. R. Karl, G. Kukla, and J. Gavin (1990). Is recent climate change across the United States related to rising levels of anthropogenic greenhouse gases? *Journal of Geophysical Research, 95:* 16617–37.

Plumb, A. (1990). Atmospheric dynamics: ozone depletion in the Arctic. *Nature, 347:* 20–1.

Polarstern Shipboard Party (1988). Breakthrough in Arctic deep-sea research: the R/V *Polarstern* expedition 1987. *EOS, 69:* 665–78.

Proffitt, M. H., J. J. Margitan, K. K. Kelly, M. Loewenstein, J. R. Podolski, and K. R. Chan (1990). Ozone loss in the Arctic polar vortex inferred from high-altitude aircraft measurements. *Nature, 347:* 31–6.

Ramanathan, V. (1988). The radiative and climatic consequences of the changing atmospheric composition of trace gases. In *The changing atmosphere,* ed. F. S. Rowland and I. S. A. Isaksen. Chichester: Wiley. 159–86.

Ramanathan, V., L. Callis, R. Cess, J. Hansen, I. Isaksen, W. Kuhn, A. Lacis, F. Luther, J. Mahlman, R. Reck, and M. Schlesinger (1987). Climate-chemical interactions and effects of changing atmospheric trace gases. *Reviews of Geophysics, 25,* 1441–82.

Rind, D. (1988). The doubled CO_2 climate and the sensitivity of the modeled hydrologic cycle. *Journal of Geophysical Research, 93:* 5385–412.

Robin, G. deQ. (1986). Changing the sea level. In *The greenhouse effect, climatic change and ecosystems,* SCOPE 29, ed. B. Bolin, B. R. Doos, J. Jager, and R. A. Warrick. Chichester: Wiley. 323–62.

Robinson, D. A., and K. F. Dewey (1990). Recent secular variations in the extent of Northern Hemisphere snow cover. *Geophysical Research Letters, 17:* 1557–60.

Rocken, C., T. M. Kelecy, G. H. Born, L. E. Young, G. H. Purcell, and S. K. Wolf (1990). Measuring precise sea level from a buoy using the global positioning system. *Geophysical Research Letters, 17:* 2145–8.

Rosenzweig, C. (1987). Potential CO_2-induced climate effects on North American wheat-producing regions. *Climatic Change, 7:* 367–89.

* Rosswall, T., R. G. Woodmansee, and P. G. Risser, eds. (1988). *Scales and global change,* SCOPE 35. Chichester: Wiley. 355 pp.

Schindler, D. W., K. G. Beaty, E. J. Fee, D. R. Cruikshank, E. R. DeBruyn, D. L. Findlay, G. A. Linsey, J. A. Shearer, M. P. Stainton, and M. A. Turner (1990). Effects of climatic warming on lakes of the central boreal forest. *Science, 250:* 967–70.

Schlesinger, M. E., and X. Jiang. (1988). The transport of CO_2-induced warming into the ocean: an analysis of simulations by the OSU coupled atmosphere-ocean general circulation model. *Climate Dynamics, 3:* 1–17.

Schneider, S. (1987). Climate modelling. *Scientific American, 256:* 72–89.

* (1989). *Global warming.* San Francisco: Sierra Books. 317 pp.

(1990). The science of climate modelling and a perspective on the global warming debate. In *Global warming: the Greenpeace report,* ed. J. Leggett. Oxford: Oxford University Press. 44–67.

Slingo, A. (1990). Sensitivity of the Earth's radiation budget to changes in low clouds. *Nature, 343:* 49–51.

Sonka, S. T., and P. J. Lamb (1987). On climate change and economic analysis. *Climatic Change, 11:* 291–311.

Spencer, R. W., and J. R. Christy (1990). Precise monitoring of global temperature trends from satellites. *Science, 247:* 1558–62.

Stamnes, K., J. Slusser, M. Bowen, C. Booth, and T. Lucas (1990). Biologically effective ultraviolet radiation, total ozone abundance and cloud optical depth at McMurdo station, Antarctica, September 15, 1988 through April 15, 1989. *Geophysical Research Letters, 17:* 2181–4.

Stolarski, R. S., M. R. Schoeber, P. A. Newman, R. D. McPeters, and A. J. Krueger (1990). The 1990 Antarctic ozone hole as observed by TOMS. *Geophysical Research Letters, 17:* 1267–70.

Stone, P. H., and J. S. Risbey (1990). On the limitations of general circulation climate models. *Geophysical Research Letters, 17:* 2173–6.

Stouffer, R. J., S. Manabe, and K. Bryan (1989). Interhemispheric asymmetry in climate response to a gradual increase of atmospheric CO_2. *Nature, 342:* 660–2.

Strauss, S. (1991). 1990 weather picture: hottest year on record, world figures show. *The Globe and Mail* (Toronto), 5 January: A1–5.

Wahr, J. M., and A. S. Trupin (1990). New computation of global sea level rise from 1990 tide gauge data. *EOS, 71:* 1267.

Wang, W., Z. Zeng, and T. R. Karl (1990). Urban heat islands in China. *Geophysical Research Letters, 17:* 2377–80.

Warrick, R. A., H. H. Shugart, M. J. Antonovsky, J. R. Tarrant, and C. J. Tucker (1986). The effects of increased CO_2 and climatic change on terrestrial ecosystems. In *The greenhouse effect, climatic change and ecosystems*, SCOPE 29, ed. B. Bolin, B. R. Doos, J. Jager, and R. A. Warrick. Chichester: Wiley. 363–92.

Washington, W. M., and G. A. Meehl. (1989). Climate sensitivity due to increased CO_2: experiments with a coupled atmosphere and ocean general circulation model. *Climate Dynamics, 4:* 1–38.

Wigley, T. M. L., and S. C. B. Raper (1987). Thermal expansion of sea water associated with global warming. *Nature, 330:* 127–31.

(1990). Natural variability of the climate system and detection of the greenhouse effect. *Nature, 344:* 324–7.

Williams, G. D. V., R. A. Fautley, K. H. Jones, R. B. Stewart, and E. E. Wheaton (1987). Estimating effects of climatic change on agriculture in Saskatchewan, Canada. In *The impact of climatic variations on agriculture*. Vol. 1: *Assessments in cool temperate and cold regions*, ed. M. L. Parry, T. R. Carter, and N. T. Konjin. Dordrecht: D. Reidel.

Wilson, C. A. and J. F. B. Mitchell (1987). A doubled CO_2 climate sensitivity experiment with a GCM including a simple ocean. *Journal of Geophysical Research, 92:* 13,315–43.

(1987). Simulated climate and CO_2-induced climate change over Western Europe. *Climatic Change, 10:* 11–42.

Woodroffe, C., R. McLean, H. Polach, and E. Wallensky (1990). Sea level and coral atolls: Late Holocene emergence in the Indian Ocean. *Geology, 18:* 62–6.

Woodwell, G. M. (1990). The effects of global warming. In *Global warming: the Greenpeace report*, ed. J. Leggett. Oxford: Oxford University Press. 116–32.

5: To manage the planet

Ausubel, J. H. (1991). Does climate still matter? *Nature, 350:* 649–51.

Bingham, T. H. (1989). Social values and environmental quality. In *Changing the global environment*, ed. D. B. Botkin, M. F. Caswell, J. E. Estes, and A. A. Orio. San Diego: Academic Press. 369–82.

Broccoli, A. J., and S. Manabe (1990). Can existing climate models be used to study anthropogenic changes in tropical cyclone climate? *Geophysical Research Letters, 17:* 1917–20.

Brokensha, D., and Riley, B. W. (1989). Managing natural resources: the local level. In *Changing the global environment*, ed. D. B. Botkin, M. F. Caswell, J. E. Estes, and A. A. Orio. San Diego: Academic Press. 341–56.

d'Arge, R. C. (1989). Ethical and economic systems for managing the global commons. In *Changing the global environment*, ed. D. B. Botkin, M. F. Caswell, J. E. Estes, and A. A. Orio. San Diego: Academic Press. 237–337.

Darmstadter, J. (1987). The role of policy in managing natural resource problems. In *Resources and world development*, ed. D. J. McLaren and B. J. Skinner. Chichester: Wiley. 473–82.

Dasgupta, P. (1989). Exhaustible resources. In *The fragile environment*, ed. L. E. Friday and R. A. Laskey. Cambridge: Cambridge University Press. 107–26.

* Department of the Environment (1991). *The potential effects of climate change in the United Kingdom*. London: Her Majesty's Stationery Office. 124 pp.

* Deutscher Bundestag (1988). *Schutz der Erdatmosphar*. Report of Commission of Inquiry of the Eleventh West German Parliament, B. Schmidbauer, chair. Dt. Bundestag, Referat Offentlichkeitsarbeit, Zur Sache; 88, 5, Bonn.

* Fisher, D. E. (1990). *Fire and ice: the greenhouse effect, ozone depletion and nuclear winter*. New York: Harper and Row.

* Friday, L. E., and R. A. Laskey, eds. (1989). *The fragile environment*. Cambridge: Cambridge University Press. 198 pp.

Goldemberg, J. (1990). Policy responses to global warming. In *Global warming: the Greenpeace report*, ed. J. Leggett. Oxford: Oxford University Press. 166–84.

Goudie, A. (1989). The changing human impact. In *The fragile environment*, ed. L. E. Friday and R. A. Laskey. Cambridge: Cambridge University Press. 1–21.

Hardin, G. (1968). The tragedy of the commons. *Science, 162:* 1243–8.

Hartwick, J. M. (1989). Market processes and sustainable economic development. Dept. of Economics, Queen's University, Kingston, Canada.

* Hartwick, J. M., and Nancy D. Olewiler (1986). *The economics of natural resource use*. New York: Harper & Row. 530 pp.

* Institute for Research on Public Policy (1990). *The social challenge of global change*. Environment and Sustainable Development Program, 275 Slater St., Ottawa, Canada: IRPP. 50 pp.

International Development Research Centre (1990). *The global research agenda: a south-north perspective*. Ottawa: IDRC, P.O. Box 8500, Ottawa K1G 3H9.

Intergovernmental Panel on Climate Change (1990). *Formulation of response strategies: summary of Working Group III*. Geneva: World Meteorological Organization/United Nations Environmental Program. 46 msp.

* International Geosphere-Biosphere Program (1988). *A plan of action*. Global Change, Report No. 4. IGBP Secretariat. Stockholm: Royal Swedish Academy of Sciences. 200 pp.

* International Geosphere-Biosphere Program (1990). *The initial core projects*. Global Change, Report No. 12. IGBP Secretariat. Stockholm: Royal Swedish Academy of Sciences.

Krupp, H.-J. (1987). Economic growth in a world of limited resources. In *Resources and world development*, ed. D. J. McLaren and B. J. Skinner. Chichester: Wiley. 49–63.

* Leggett, J., ed. (1990a). *Global warming: the Greenpeace report*. Oxford: Oxford University Press. 554 pp.

(1990b). The nature of the greenhouse threat. In *Global warming: the Greenpeace report*, ed. J. Leggett. Oxford: Oxford University Press. 14–43.

* Leonard, H. J. (1988). *Pollution and the struggle for the world product*. Cambridge: Cambridge University Press. 254 pp.

* McLaren, D. J., and B. J. Skinner, eds. (1987). *Resources and world development*. Chichester: Wiley. 940 pp.

* Mendell, W. W. ed. (1985). *Lunar bases and space activities of the 21st century*. Lunar and Planetary Institute, 3303 NASA Road One, Houston, TX. 865 pp.

Nisbet, E. G. (1988). The business of planet management. *Nature, 333:* 617.

* Nisbet, Robert A. (1969). *Social change and history*. Oxford: Oxford University Press.

* Pearson, C. S. (1985). *Down to business: multinational corporations, the environment and development*. Washington, DC: World Resources Institute. 107 pp.

Rasool, S. I. (1987). Understanding the global change: an opportunity to seize. *EOS, 68:* 1609–11.

* Redclift, M. (1987). *Sustainable development: exploring the contradictions*. New York: Routledge.

* Repetto, R., ed. (1985). *The global possible*. New Haven: Yale University Press.

Roberts, P. C. (1987). Malthus and after: a retrospective look at projection models and resources concerns. In *Resources and world development*, ed. D. J. McLaren and B. J. Skinner. Chichester: Wiley. 101–11.

* Rowe, S. (1990). *Home place: essays on ecology*. Edmonton, Alberta: NeWest.

Schneider, S. H. (1989). The greenhouse effect: science and policy. *Science, 243:* 771–81.

(1990). The costs of cutting – or not cutting – greenhouse gas emissions. In *Global warming: the Greenpeace report*, ed. J. Leggett. Oxford: Oxford University Press. 185–92.

Sun, M. (1990). Emissions trading goes global. *Science, 247:* 520–1.

Thatcher, M. (1989). Speech to the United Nations General Assembly, 8 November. Press release, United Kingdom embassy to the United Nations. Edited version in *The Guardian* (London, Manchester), 9 November: 7.

Umana Quesada, A. U. (1989). Greenhouse economics: global resources and the political economy of climate change. San José, Costa Rica: Ministry of Natural Resources, Energy and Mines.

Wallis, M. K. (1990). Leaky answer to greenhouse gas. *Nature, 344:* 25–6.

* Weiner, J. (1990). *The next one hundred years: shaping the fate of our living Earth*. New York: Bantam.

Wigley, T. M. L. (1991). Could reducing fossil-fuel emissions cause global warming? *Nature, 349:* 503–6.

Wiman, I. M. B. (1990). Expecting the unexpected: some ancient roots to current perceptions of Nature. *Ambio, 19:* 62–9.

* World Commission on Environment and Development (1987). *Energy 2000: a global strategy for sustainable development*. London: Zed Books. 76 pp.

* World Meteorological Office (1989). *The changing atmosphere: implications for global security*. Toronto, Canada, 27–30 June 1988. Geneva: WMO. 483 pp.

* World Resources Institute (1988). *World resources 1988–89*. New York: Basic Books. 372 pp.

Zerbe, J. I. (1990). Biomass energy and global warming. *World resource review, 2:* 66–81.

6: Reducing our impact

* Advisory Group of Hydrogen Opportunities (1987). *Hydrogen, National Mission for Canada*. M27-86/1987E. Ottawa: Ministry of Supply and Services, Canada. 66 pp.

Anderson, C. (1990). Global warming: targets work – up to a point. *Nature, 348:* 668.

Anonymous (1987a). China's coal output. *Mining Journal, 309* (23 Oct.): 331.

* (1987b). *Technical Advisory Committee on the nuclear fuel waste management program*. Atomic Energy of Canada Ltd. 86 pp.

(1988). U.S. coal production forecasts. *Mining Magazine* (April): 249.

* (1989). *Clean coal technology demonstration program*. DOE/EIS-0146D, Office of Clean Coal Technology. Washington, DC: U.S. Department of Energy.

* Armstead, H. C. H. (1978). *Geothermal energy*. London: Spon/Wiley. 357 pp.

Ausubel, J. H., A. Grubler, and N. Nakicenovic (1988). Carbon dioxide emissions in a methane economy. *Climatic Change, 12:* 245–63.

Barnaby, F. (1988). Acid rain: U.K. policies. *Ambio, 17:* 161–2.

Baum, V. (1988). Electricity. *Petroleum Economist,* November: 359–61.

* Bennett, D. J., and J. R. Thomson (1989). *The elements of nuclear power.* 3rd ed. London: Longman. New York: Wiley. 293 pp.

Berry, R. D., and J. J. Colls (1990). Atmospheric carbon dioxide and sulphur dioxide on an urban/rural transect – I. Continuous measurement at the transect ends. *Atmospheric environment, 24a:* 2681–8.

Bilham, R. (1988). Earthquakes and urban growth. *Nature, 336:* 625–6.

Boardman, N. K. (1980). Energy from the biological conversion of solar energy. *Philosophical transactions of the Royal Society, London, A295:* 477–89.

Bodlund, B., E. Mills, T. Karlsson, and T. Johansson (1990). The challenge of choices: technology options for the Swedish economy. In *Global warming: the Greenpeace report.,* ed. J. Leggett. Oxford: Oxford University Press. 317–71.

Boeing Commercial Airplane Group (1990). High-speed civil transport study: special factors. NASA contractor report 181881. Hampton, VA: NASA Langley Research Center. (Note: report details ozone impact of airplane fleets.)

Bondi, Sir H. (1980). Indirect utilization of solar energy. *Philosophical transactions of the Royal Society, London, A295:* 501–6.

Brady, M. (1985). The relation of acid rain to energy policy. *Energy, 10:* 1113–18.

Brasseur, G. P., C. Granier, and S. Walters (1990). Future changes in stratospheric ozone and the role of heterogeneous chemistry. *Nature, 348,* 626–8.

Brown, A. C., C. E. Canosa-Mas, A. D. Parr, K. Rothwell, and R. P. Wayne (1990). Tropospheric lifetimes of three compounds for possible replacement of CFC and halons. *Nature, 347:* 541–3.

* Brown, G. C., and E. Skipsey (1986). *Energy resources.* Milton Keynes, U.K.: Open University Press. 213 pp.

Brundtland, G. H. (1989). Global change and our common future – the Benjamin Franklin lecture. Washington, DC, 2 May 1989. Unpublished release.

Bunyard, P. (1988). The myth of France's cheap nuclear electricity. *The Ecologist, 18:* 1, 4–13.

Burnett, W. M., and S. D. Ban (1989). Changing prospects for natural gas in the United States. *Science, 244:* 305–11.

Caplan, H. S. (1979). A consumer's guide to nuclear reactors. *The Musk-ox, 24:* 68–74.

Capobianco, G. (1988). A western Canadian coal producer's viewpoint. *Canadian Institute of Mining and Metallurgy Bulletin, 81:* 36–8.

* Carter, L. J. (1987). *Nuclear imperatives and public trust: dealing with radioactive waste.* Washington, DC: Resources for the Future. 473 pp.

* Chapman, N. A., and I. G. McKinley (1987). *The geological disposal of nuclear waste.* Chichester: Wiley. 280 pp.

Clark, M. J. and Smith, F. B. (1988). Wet and dry deposition of Chernobyl releases. *Nature, 332:* 245–9.

Clarke, R. H., and Southwood, T. R. E. (1989). Risks from ionizing radiation. *Nature, 338:* 197–8.

Comly, J. B. (1980). Solar heating and air-conditioning. *Philosophical transactions of the Royal Society, London, A295:* 415–22.

Corelli, R. (1990). Fateful consequences: development could take a heavy toll. *Maclean's* magazine (Toronto), 21 May: 56–8. (Note: article discusses environmental impact of the James Bay hydroelectric project.)

Darmstadter, G. J. (1986). Energy patterns – in retrospect and prospect. In *Sustainable development of the biosphere*, ed. W. C. Clark and R. E. Munn. Cambridge: IIASA/ Cambridge University Press. 140–67.

Darnley, A. G. (1987). Resources for nuclear energy. In *Resources and world development*, ed. D. J. McLaren and B. J. Skinner. Chichester: Wiley. 187–210.

Derwent, R. (1988). A better way to control pollution. *Nature, 331:* 575–8.

Deuser, W. G. (1988). Whither organic carbon? *Nature, 332:* 396–7.

* Deutscher Bundestag (1988). *Schutz der Erdatmosphar.* Report of Commission of Inquiry of the Eleventh West German Parliament, B. Schmidbauer, chair. Dt. Bundestag, Referat Offentlichkeitsarbeit, Zur Sache; 88, 5, Bonn.

Dickman, S. (1990). Greenhouse gases: Swiss cut it to the bone. *Nature, 348:* 189.

* Dienes, L., and T. Shabad (1979). *The Soviet energy system: resource use and politics.* New York: Wiley.

* Doern, G. B. (1980). *Government intervention in the Canadian nuclear industry.* Montreal: Institute for Research on Public Policy.

* Doll, Sir R., and A. E. M. McLean, eds. (1979). *Long-term hazards from environmental chemicals. Philosophical transactions of the Royal Society, London,* B205: 1–197.

* Dostrovsky, I (1988). *Energy and the missing resource.* Cambridge: Cambridge University Press. 182 pp.

* Duffie, J. A., and W. A. Beckman (1980). *Solar engineering of thermal processes.* New York: Wiley.

* Economides, M., and P. Ungemach (1987). *Applied geothermics.* New York: Wiley.

* Edwards, L. M., G. V. Chilingar, H. H. Riecke, and W. H. Fertl (1982). *Handbook of geothermal energy.* Houston: Gulf Publishing.

El Osery, I. A. (1984). A nuclear-electric-hydrogen energy system. *Energy, 9:* 709–11.

* Elsom, D. M. (1987). *Atmospheric pollution.* Oxford: Blackwell. 319 pp.

Enz, R. (1988). Prices and earnings around the globe. Publication 108. Union Bank of Switzerland.

Evans, H. J. (1990). Leukaemia and nuclear facilities: data in search of explanation. *Nature, 347:* 712–3.

Fairhall, D. (1989). So ironic – the letter that led to historic U-turn on nuclear power. *The Guardian* (London, Manchester), 1 December: 17.

Financial Times, London (1990). Vehicles and the environment: Financial Times Survey. 27 July: 1–6.

Fisher, D. A., C. M. Hales, W.-C. Wang, M. K. W. Ko, and N. D. Sze (1990). Model calculations of the relative effects of CFC's and their replacements on global warming. *Nature, 344*: 513–16.

Gittus, J. H. (1986). Degraded core analysis for the pressurized-water reactor. Review Lecture. London: Royal Society.

* Glasstone, S., and A. Sesonke (1981). *Nuclear reactor engineering.* 3d ed. New York: Van Nostrand.

Golay, M. W., and N. E. Todreas (1990). Advanced light water reactors. *Scientific American, 262:* 58–65.

Golombek, A., and R. G. Prinn (1989). Global three-dimensional model calculations of the budgets and present-day atmospheric lifetimes of $CF_2ClCFCl_2$ (CFC-113) and $CHClF_2$ (CFC-22). *Geophysical Research Letters, 16:* 1153–6.

* Gordon, R. L. (1987). *World coal.* Cambridge: Cambridge University Press. 145 pp.

Goreau, T. J. (1987). The other half of the global carbon dioxide problem. *Nature, 328:* 581–2.

Gorst, I. (1988). Soviet Union: continuing gas boom. *Petroleum Economist,* May, 147–9.

Gould, P. (1988a). Fire in the rain: the democratic consequences of Chernobyl. Baltimore, MD: Johns Hopkins University Press.

(1988b). Tracing Chernobyl's fallout. *Earth and Mineral Sciences, 57:* 57–65.

Grace, J. D., and G. F. Hart (1986). Giant gas fields of northern West Siberia. *American Association of Petroleum Geologists Bulletin, 70:* 830–52.

Hafele, W., H. Barnert, S. Messner, M. Strubegger, and Anderer (1986). Novel integrated energy systems: the case of zero emissions. In *Sustainable development of the biosphere,* ed. W. C. Clark and R. E. Munn. Cambridge: IIASA/Cambridge University Press. 171–92.

Hall, D. O., and P. J. de Groot (1987). Biomass energy: the parallel requirements for fuel and food from plants. In *Resources and world development,* ed. D. J. McLaren and B. J. Skinner. Chichester: Wiley. 157–85.

Hammitt, J. K., F. Camm, P. S. Connell, W. E. Mooz, K. A. Wolf, D. J. Wuebbles, and A. Bamezai (1987). Future emission scenarios for chemicals that may deplete stratospheric ozone. *Nature, 330:* 711–16.

Hammond, A. L., E. Rodenburg, and W. Moomaw (1990). Accountability in the greenhouse. *Nature, 347:* 705–6.

* Harrison, R. M., ed. (1990). *Pollution: causes, effects, control.* Cambridge: Royal Society of Chemistry. 393 pp.

Heaton, T. H. E., J. C. Vogel, G. von la Chevallerie, and G. Collett (1986). Climatic influence on the isotopic composition of bone nitrogen. *Nature, 322:* 822–3.

Hill, C., and A. Laplanche (1990). Overall mortality and cancer mortality around French nuclear sites. *Nature, 347:* 755–7.

Holt, A. R. (1988). The use of uranium and coal at Ontario Hydro. *Canadian Institute of Mining and Metallurgy Bulletin, 81:* 39–42.

Hubbard, H. M. (1989). Photovoltaics today and tomorrow. *Science, 244:* 297–304.

James, C. G. (1990). Natural gas and the greenhouse. *Nature, 347:* 720–1.

* Johansson, T. B., and P. Steen (1981). *Radioactive waste from nuclear power plants.* Berkeley: University of California Press.

Jovanovich, J. V., and R. M. Rotty (1986). A plausible future of rapid energy growth in China. *Energy, 11:* 829–41.

Keepin, Bill (1990). Nuclear power and global warming. In *Global warming: the Greenpeace report,* ed. J. Leggett. Oxford: Oxford University Press. 295–316.

Keepin, W., and G. Kats (1988a). Global warming – discussion. *Science, 241:* 1027.

(1988b). Greenhouse warming: comparative analysis of two abatement strategies. *Energy Policy, 16:* 538.

Kellogg, W. W. (1987). Mankind's impact on climate: the evolution of an awareness. *Climatic Change, 10:* 113–36.

Kelly, P. M. (1990). *Halting global warming.* Amsterdam: Greenpeace International.

LaPorta, C. (1990). Renewable energy: recent commercial performance in the USA as an index of future prospects. In *Global warming: the Greenpeace report,* ed. J. Leggett. Oxford: Oxford University Press. 224–59.

Leahey, D. M., and Hansen, M. C. (1990). Observation of ozone depletion by nitric oxide at 40 km downwind of a medium-sized city. *Atmospheric environment, 24a:* 2533–40. (Note: city is Calgary, Canada.)

Ledic, M. (1987). China: energy, key to economic growth. *Petroleum Economist* (November): 405–8.

Legg, J. F. (1990). Carbon dioxide extraction and disposal and its potential in Canada. OERD 90-06. Ottawa: Office of Energy, Research and Development, Ministry of Energy, Mines and Resources.

Licht, S. (1987). A description of energy conversion in photoelectrochemical solar cells. *Nature, 330:* 1148–51.

Lovins, A. (1990). The role of energy efficiency. In *Global warming: the Greenpeace report,* ed. J. Leggett. Oxford: Oxford University Press. 193–223.

* Lovins, A., L. H. Lovins, F. Krause, and W. Bach (1989). Least-cost energy: solving the CO_2 problem. 2nd ed. Snowmass, CO: Rocky Mountain Institute. 184 pp.

* MacDonald, G. J., ed. (1982). *The long-term impacts of increasing atmospheric carbon dioxide levels.* Cambridge, MA: Ballinger, 252 pp.

Maclean's Weekly Magazine (1990). Power to burn: the James Bay battle. Toronto: 21 May: 50–58.

Maddox, J. (1990). Two gales do not make a greenhouse. *Nature, 343:* 407.

* McGown, L. B., and J. O'M. Bockris (1980). *How to obtain abundant clean energy.* New York: Plenum Press.

McKay, C. P. (1987). Making an Earth of Mars. *Planetary Report, 7* (Nov.–Dec.): 26–7.

* Milnes, A. G. (1985). *Geology and radwaste.* London: Academic Press. 328 pp.

* Ohta, T., ed., (1979). *Solar-hydrogen energy systems.* Oxford: Pergamon.

* Okken, P. A., R. J. Swart, and S. Zwerver, eds. (1989). *Climate and energy: the feasibility of controlling CO_2 emissions.* Dordrecht: Kluwer. 267 pp.

* Page, A. L., T. G. Logan, and J. A. Ryan (1987). *Land application of sludge: food chain implications.* Chelsea, MI: Lewis.

Pool, R. (1988). Solar cells turn 30. *Science, 241:* 900–1.

Porter, Sir G. (1989). Anniversary address by the President. Royal Society News. Supplement, *5:* 6i–vi.

* Porter, Sir G., and Sir W. Hawthorne, eds. (1980). *Solar energy. Philosophical transactions of the Royal Society, London,* A295: 343–511.

Prinn, R. G., and A. Golombek (1990). Global atmospheric chemistry of CFC-123. *Nature, 344:* 47–9.

Pye, K., and N. Schiavon (1989). Cause of sulphate attack on concrete, render, and stone indicated by sulphur isotope ratios. *Nature, 342:* 663–4.

Rao, R. (1988). What price Tehri dam? *Ambio, 17:* 246–7.

Righelato, R. C. (1980). Microbial production of energy sources from biomass. *Philosophical transactions of the Royal Society, London,* A295: 491–500.

* Rippon, S. (1984). *Nuclear energy.* London: Heinemann.

* Roberts, L. E. J., P. S. Liss, and P. A. H. Saunders (1990). *Power generation and the environment.* Oxford: Oxford University Press. 240 pp. (added in proof: not read).

Ross, M. (1989). Improving the efficiency of electricity use in manufacturing. *Science, 244:* 311–17.

* Roxburgh, I. S. (1987). *Geology of high-level nuclear waste disposal.* London: Chapman and Hall.

Rusk, J. (1991). Ontario may be groping in the dark with its energy plan. *The Globe and Mail* (Toronto), 2 January: B1–4.

Russell, A. G., D. St. Pierre, and J. B. Milford (1990). Ozone control and methanol fuel use. *Science, 247:* 201–5.

* Sawyer, S. W. (1986). *Renewable energy: progress, prospects.* Washington, DC: Association of American Geographers. 102 pp.

Schilling, H.-D., and D. Wiegand (1987). Coal resources. In *Resources and world development,* ed. D. J. McLaren and B. J. Skinner. Chichester: Wiley. 129–56.

Smith, K., and D. R. Ahuja (1990). Toward a greenhouse equivalence index: the total exposure analogy. *Climatic change, 17:* 1–7.

Stewart-Patterson, D. (1988). Coal giant shrinks slowly but surely. *The Globe and Mail* (Toronto), 20 June: B18.

Tabor, H. (1980). Non-convecting solar ponds. *Philosophical transactions of the Royal Society, London, A295:* 423–33.

Taylor, J. J. (1989). Improved and safer nuclear power. *Science, 244:* 318–25.

* U.S. Department of Energy (1989). Clean coal technology demonstration program. DOE/EIS – 0146D. Washington, DC: U.S. Department of Energy.

van der Veen (1987). Ice sheets and the CO_2 problem. *Surveys in Geophysics, 9:* 1–42.

Victor, D. G. (1990). Calculating greenhouse budgets. *Nature, 347:* 431.

* Walker, C. A., L. C. Gould, and E. J. Woodhouse, eds. (1983). *Too hot to handle? – social and policy issues in the management of radioactive wastes.* New Haven, CT: Yale University Press. 209 pp.

Walsh, M. P. (1990). Motor vehicles and global warming. In *Global warming: the Greenpeace report,* ed. J. Leggett. Oxford: Oxford University Press. 260–94.

Wang, W.-C., M. P. Dudek, X.-Z. Liang, and J. T. Kiehl (1991). Inadequacy of effective CO_2 as a proxy in simulating the greenhouse effect of other radiatively active gases. *Nature, 350:* 573–7.

White, D. A. (1987). Conventional oil and gas resources. In *Resources and world development,* ed. D. J. McLaren and B. J. Skinner. Chichester: Wiley. 129–56.

White, W. M., E. S. Macias, D. F. Miller, D. E. Schorran, T. E. Hoffer, and D. P. Rogers (1990). Regional transport of the urban workweek: methylchloroform cycles in the Nevada-Arizona desert. *Geophysical Research Letters, 17:* 1081–4.

Whitelegg, J. (1989). Transport in turmoil. *The Guardian* (London, Manchester), 24 November: 25.

* Wieder, S. (1982). *An introduction to solar energy for scientists and engineers.* New York: Wiley.

Williams, R. H. (1987). Exploring the global potential for more efficient use of energy. In *Resources and world development,* ed. D. J. McLaren and B. J. Skinner. Chichester: Wiley. 211–44.

* Wilson, C. L. (1980). *Coal – bridge to the future. Report of the World Coal Study.* Cambridge, MA: Ballinger.

* World Resources Institute (1988). *World Resources 1988–9.* New York: Basic Books. 372 pp.

    (1990). *World Resources 1990–1.* New York: Oxford University Press. 383 pp.

Yunzhen, J. (1986). China's coal industry: present status and prospects for the future. *Energy, 11:* 1059–65.

7: Managing the Earth's vegetation

Adams, R. M., C. Rosenzweig, R. M. Peart, J. T. Ritchie, B. A. McCarl, J. D. Glyer, R. B. Curry, J. W. Jones, K. J. Boote, and L. H. Allen, Jr. (1990). Global climatic change and U.S. agriculture. *Nature, 345:* 219–24.

Anonymous (1988). Water 2020. Sustainable use for water in the 21st century. Science

Council of Canada Report 40. Ottawa: Ministry of Supply and Services Canada. 40 pp.

Arnold, J. E. M. (1987). Deforestation. In *Resources and world development*, ed. D. J. McLaren and B. J. Skinner. Chichester: Wiley. 711–25.

Ball, T. F. (1986). Historical evidence and climatic implications of a shift in the boreal forest – tundra transition in central Canada. *Climatic Change, 8:* 121–34.

Blank, L. W., T. M. Roberts, and R. A. Skeffington (1988). New perspectives on forest decline. *Nature, 336:* 27–30.

* Body, R. (1982). *Agriculture: the triumph and the shame.* London: Temple Smith. 139 pp.

Breen, C., and G. Begg (1987). Wetlands – a valuable natural resource. *African Wildlife, 41:* 219–21.

* Brett, M. R. (1989). *Pilanesberg.* Sandton, South Africa: Frandsen Publishers. 140 pp.

Burgess, P. F. (1989). Asia. In *No timber without trees,* ed. D. Poore. London: Earthscan Publications. 117–53.

* Campbell, B. M., R. F. Du Toit, and C. A. M. Attwell (1989). *The Save study.* Harare: Univ. of Zimbabwe Press, P.O. Box M.P. 45, Mount Pleasant.

* Carson, R. (1962) *Silent spring.* 1982 reprint. London: Pelican Books. 317 pp.

Casey, J., and L. K. Muir (1988). Institutional responsibility for social forestry in Zimbabwe. Working Paper AE 1/88, Dept. of Agricultural Economics, University of Zimbabwe.

Christiansen, C. (1988). Degradation and rehabilitation of agropastoral land: perspective on environmental change in semi-arid Tanzania. *Ambio, 17:* 144–52.

Clark, W. C. (1986). Sustainable development of the biosphere: themes for a research program. In *Sustainable development of the biosphere,* ed. W. C. Clark and R. E. Munn. Cambridge: IIASA/Cambridge University Press. 5–48.

* Cottrell, R. (1987). *The sacred cow.* London: Grafton (Collins). 192 pp.

Crosson, P. (1986). Agricultural development – looking to the future. In *Sustainable development of the biosphere,* ed. W. C. Clark and R. E. Munn. Cambridge: IIASA/Cambridge University Press. 104–35.

Currey, B., A. S. Fraser, and K. L. Bardsley (1987). How useful is Landsat monitoring? *Nature, 328:* 587–9.

Dregne, H. E. (1987). Desertification. In *Resources and world development,* ed. D. J. McLaren and B. J. Skinner. Chichester: Wiley. 697–710.

Drohan, M. (1990). Farm supports laid out: [Canada's] annual programs worth $8.8 billion. *The Globe and Mail* (Toronto), 3 October: B1.

Ehrenfeld, D. (1988). Life in the next millenium: who will be left in Earth's community? In *The last extinction,* ed. L. Kaufman and K. Mallory. Cambridge, MA: MIT Press. 167–86.

Ehrlich, A. (1990). Agricultural contributions to global warming. In *Global warming: the Greenpeace report,* ed. J. Leggett. Oxford: Oxford University Press. 400–21.

Fearnside, P. M. (1989). The charcoal of Carajas: a threat to the forests of Brazil's Eastern Amazon region. *Ambio, 18:* 141–3.

Food and Agriculture Organization, U.N. (1990). The state of world food and agriculture (compiled by T. J. Aldington and F. Zegarra). *World resource review, 2:* 7–41.

Foose, T. J. (1988). Riders of the lost ark: the role of captive breeding in conservation strategies. In *The last extinction,* ed. L. Kaufman and K. Mallory. Cambridge, MA: MIT Press. 141–66.

* Franklin, M. (1988). *Rich man's farming: the crisis in agriculture.* London: Royal Institute of International Affairs/Routledge. 103 pp.

Fung, I. Y., C. J. Tucker, and K. C. Prentice (1987). Application of advanced very high resolution radiometer vegetation index to study atmosphere-biosphere exchange of CO_2. *Journal of Geophysical Research, 92:* 2999–3015.

Galt, V. (1987). Bank's bosses brazen out flood of red ink. *The Globe and Mail,* (Toronto), 26 August: B1.

Gleick, P. H. (1989). Climate change, hydrology and water resources. *Reviews of Geophysics, 27:* 329–44.

Goreau, T. J. (1990). Balancing atmospheric carbon dioxide. *Ambio, 19:* 230–7.

* Grigg, D. (1984). *An introduction to agricultural geography.* London: Unwin Hyman. 204 pp.

    (1987). The industrial revolution and land transformation. In *Land transformation in agriculture,* SCOPE 32, ed. M. G. Wolman and F. G. A. Fournier. Chichester: Wiley.

Gunn, J. M., and Keller, W. (1990). Biological recovery of an acid lake after reductions in industrial emissions of sulphur. *Nature, 345:* 431–3.

* Harris, L. D., ed. (1984). *The fragmented forest.* Chicago: University of Chicago Press. 211 pp.

* Havel, V. (1988). *Letters to Olga, June 1979–September 1982,* trans. P. Wilson. New York: Knopf. London: Faber and Faber.

Henderson-Sellers, A., and V. Gornitz (1984). Possible climatic impacts of land cover transformations, with particular emphasis on tropical deforestation. *Climatic Change, 6:* 231–57.

Hooley, J. (1988). The level of provisions. *Banking World* (March): 34.

Houghton, R. A. (1990). The future role of tropical forests in affecting the carbon dioxide concentration of the atmosphere. *Ambio, 19:* 204–9.

Ittekkot, V. (1988). Global trends in the nature of organic matter in river suspensions. *Nature, 332:* 436–8.

Janzen, D. H. (1988). Tropical ecological and biocultural restoration. *Science, 239:* 243–4.

Johns, A. D. (1988). Economic development and wildlife conservation in Brazilian Amazonia. *Ambio, 17:* 302–6.

Justice, C. O., ed., (1986). Monitoring the grasslands of semi-arid Africa using NOAA-AVHRR data. *International Journal of Remote Sensing, 7:* 1383–1622.

Kawasaki, T. (1985). Fisheries. In *Climatic impact assessment,* SCOPE 27, ed. R. W. Kates, J. H. Ausubel, and M. Berberon. Chichester: Wiley. 131–54.

* Kingdon, J. (1990). *Island Africa: the evolution of Africa's rare animals and plants.* London: Collins. 287 pp.

Knox, P. (1991). Experts try new ideas to save forests. *The Globe and Mail* (Toronto), 2 January: A6.

* Koopowitz, H., and H. Kaye (1983). *Plant extinction: a global crisis.* Washington, D.C.: Stonewall Press. 239 pp.

Kuusela, K. (1987). Forest products – world situation. *Ambio, 16:* 81–5.

Lal, R. (1989). Soil degradation and conversion of tropical rainforests. In *Changing the global environment,* ed. B. Botkin, M. F. Caswell, J. E. Estes, and A. A. Orio. San Diego: Academic Press. 137–54.

Lamont-Doherty Geological Observatory (1989). New greenhouse paradox: northern land plants, not southern oceans, proposed as the major sink for fossil fuel CO_2. *Lamont, 22* (fall): 1–2.

Lascelles, D. (1987). Third world loss provision hits British banks. *The Globe and Mail* (Toronto), 8 October.

Leader-Williams, N., and S. D. Albon (1988). Allocation of resources for conservation. *Nature, 336:* 533–5.

Lean, J., and D. A. Warrilow (1989). Simulation of the regional climatic impact of Amazon deforestation. *Nature, 342:* 411–13.

Le Houeron, H. N. (1985). Pastoralism. In *Climate impact assessment*, SCOPE 27, ed. R. W. Kates, J. H. Ausubel, and M. Berberon. Chichester: Wiley. 155–86.

Lovejoy, T. E. (1989). Deforestation and the extinction of species. In *Changing the global environment*, ed. D. B. Botkin, M. F. Caswell, J. E. Estes, and A. A. Orio. San Diego: Academic Press. 89–98.

Lovelock, J. E., and E. G. Nisbet (1987). A modest proposal. *Geology, 15:* 983.

Makuku, S. J. (1991). Indigenous timber woodlots resuscitate natural ecology, provide more socio-economic benefits and are easier to establish and cheaper to maintain than exotic timber woodlots. *Zimbabwe Science News, 24:* 91–3.

Matthews, E. (1983). Global vegetation and land use: new high-resolution data bases for climate studies. *Journal of Climate and Applied Meteorology, 22:* 474–87.

May, R. (1989). How many species? In *The fragile environment*, ed. L. E. Friday and R. A. Laskey. Cambridge: Cambridge University Press. 61–81.

Mayo, E. G., and K. D. Casteel (1986). The changing face of Latin America's tin industry and its effect on the world tin market. In *Mining Latin America*. London: Institution of Mining and Metallurgy. 257–70.

McKie, A. B. (1988). Taxing the banks – fable and fact. *Canadian Banker, 95:* 22–4.

McQuaig, L. (1987). The big boondoggle. *The Globe and Mail* (Toronto), *Report on Business,* December: 57–71.

Micklin, P. P. (1988). Dessication in the Aral Sea: a water management disaster for the Soviet Union. *Science, 241:* 1170–6.

Mittlestaedt, M. (1990). Photos reveal huge clearcut. *The Globe and Mail* (Toronto), 11 December: A1–2.

Moreno, A. M., and K. D. Casteel (1986). Latin American gold mining: a growing force in the world market. In *Mining Latin America*. London: Institution of Mining and Metallurgy. 285–97.

Muir-Leresche, K. (1988a). Marketing wildlife products and services. International Conference on Wildlife Management in Sub-Saharan Africa. Harare: Dept. of Agricultural Economics, University of Zimbabwe.

   (1988b). Wildlife utilisation as a sustainable land use system for arid lands. Programme synopsis. Harare: Dept. of Agricultural Economics, University of Zimbabwe.

Myers, N. (1989). The future of forests. In *The fragile environment*, ed. L. E. Friday and R. A. Laskey. Cambridge: Cambridge University Press. 22–40.

   (1990). Tropical forests. In *Global warming: the Greenpeace report*, ed. J. Leggett. Oxford: Oxford University Press. 372–99.

Nix, H. A. (1985). Agriculture. In *Climate impact assessment*, SCOPE 27, ed. R. W. Kates, J. H. Ausubel, and M. Berberon. Chichester: Wiley. 105–30.

* Organization for Economic Cooperation and Development (1989). *Agricultural and environmental policies*. Paris: OECD. 200 pp.

Palmer, J. (1989). Management of natural forest for sustainable timber production: a commentary. In *No timber without trees*, ed. D. Poore. London: Earthscan. 154–89.

Parry, M. L. (1986). Some implications of climatic change for human development. In *Sustainable development of the biosphere*, ed. W. C. Clark and R. E. Munn. Cambridge: IIASA/Cambridge University Press. 378–406.

Pastor, J., and W. M. Post (1988). Response of northern forests to CO_2-induced climatic change. *Nature, 334:* 55–8.

Peters, C. M., A. H. Gentry, and R. O. Mendelsohn (1989). Valuation of an Amazonian rainforest. *Nature, 339:* 655–7.

Poore, D. (1989a). The sustainable management of tropical forest: the issues. In *No timber without trees*, ed. D. Poore. London: Earthscan. 1–27.

* Poore, D. ed. (1989b) *No timber without trees*. London: Earthscan. 252 pp.

Prance, G. T. (1986). The Amazon: paradise lost? In *The Last Extinction*, ed. L. Kaufman and K. Mallory. Cambridge, MA: MIT Press. 63–106.

(1990). Consensus for conservation. *Nature, 345:* 384.

Reganold, J. P., L. F. Elliott, and Y. L. Unger (1987). Long term effects of organic and conventional farming on soil erosion. *Nature, 330:* 370–3.

Regier, U. A., and G. L. Baskerville (1986). Sustainable development of the biosphere. In *Sustainable development of the biosphere*, ed. W. C. Clark and R. E. Munn. Cambridge: IIASA/Cambridge University Press. 75–100.

Repetto, R. (1990). Deforestation in the tropics. *Scientific American, 282:* 18–25.

Richards, J. F. (1986). World environmental history and economic development. In *Sustainable development of the biosphere*, ed. W. C. Clark and R. E. Munn. Cambridge: IIASA/Cambridge University Press. 53–70.

Rietbergen, S. (1989). Africa. In *No timber without trees*, ed. D. Poore. London: Earthscan. 40–73.

Rocha, J. (1988). Ozone fears as Amazon forest burns. *Manchester Guardian Weekly, 138* (24 April): 9.

Ruddle, K. (1987). The impact of wetland reclamation. In *Land transformation in agriculture*, SCOPE 32, ed. M. G. Wolman and F. G. A. Fournier. Chichester: Wiley. 171–201.

Schoener, T. W., and D. A. Spiller (1987). High population persistence in a system with high turnover. *Nature, 330:* 474–7.

Shugart, H. H., M. Y. Antonovsky, P. G. Jarvis, and A. P. Sandford (1986). CO_2, climatic change and forest ecosystems. In *The greenhouse effect, climatic change and ecosystems*, SCOPE 29, ed. B. Bolin, B. R. Doos, J. Jager, and R. A. Warrick. Chichester: Wiley. 475–521.

Shukla, J., C. Nobre, and P. Sellers (1990). Amazon deforestation and climate change. *Science, 247:* 1322–5.

* Simmonds, N. W. (1976, 1986). *Evolution of crop plants*. Harlow, U.K.: Longmann. 339 pp.

Simmons, I. G. (1987). Transformation of the land in pre-industrial time. In *Land transformation in agriculture*, SCOPE 32, ed. M. G. Wolman and F. G. A. Fournier. Chichester: Wiley. 45–78.

Singh, T., and J. M. Powell (1986). Climatic variation and trends in the boreal forest region of western Canada. *Climatic Change, 8:* 267–78.

* Stewart, J. W. B., ed. (1990). *Soil quality in semi-arid agriculture*. Saskatoon: Saskatchewan Institute of Pedology, University of Saskatchewan. Vol. 1: 292 pp; Vol. 2: 391 pp.

Synott, T. (1989). South America and the Caribbean. In *No timber without trees*, ed. D. Poore. London: Earthscan. 74–116.

Timmerman, P. (1986). Mythology and surprise in the sustainable development of the biosphere. In *Sustainable development of the biosphere*, ed. W. C. Clark and R. E. Munn. Cambridge: IIASA/Cambridge University Press. 435–52.

* Tyson, P. D. (1986). *Climatic change and variability in southern Africa.* Oxford: Oxford University Press. 220 pp.
* Visser, M. (1986). *Much depends on dinner.* Toronto: McClelland and Stewart.
  Voytek, M. A. (1990). Addressing the biological effects of decreased ozone on the Antarctic environment. *Ambio, 19:* 52–61.
  Warrick, R. A., R. M. Gifford, and M. L. Parry (1986). CO_2, climatic change and agriculture. In *The greenhouse effect, climatic change and ecosystems,* SCOPE 29, ed. B. Bolin, B. R. Doos, J. Jager, and R. A. Warrick. Chichester: Wiley. 393–474.
* Westoby, J. (1988). *Introduction to world forestry.* Oxford: Blackwell. 228 pp.
  Wheaton, E. E., T. Singh, R. Dempster, K. D. Higgenbotham, J. P. Thorpe, G. C. Kooten, and J. S. Taylor (1987). An exploration and assessment of the implications of climatic change for the boreal forest and forestry economics of the Prairie Provinces and Northwest Territories. Saskatchewan Research Council Technical Report No. 211. Saskatoon: Saskatchewan Research Council, Innovation Place. 282 pp.
  Whelan, T. (1988). Central American environmentalists call for action on the environment. *Ambio, 17:* 72–5.
  Whitlow, J. R. (1988). Soil erosion – a history. *Zimbabwe Science News, 22:* 83–5.
* Wolman, M. G., and F. G. A. Fournier, eds. (1987). *Land transformation in agriculture,* SCOPE 32. Chichester: Wiley. 531 pp.
* World Commission on Environment and Development (chair: G. H. Brundtland) (1987). *Our common future.* Oxford: Oxford University Press. 400 pp.
* World Resources Institute (1988). *World Resources 1988–9.* New York: Basic Books. 372 pp.
*   (1990). *World Resources 1990–1.* New York: Oxford University Press. 383 pp.
  Wright, D. M. (1990). Human impacts on energy flow through natural ecosystems, and implications for species endangerment. *Ambio, 19:* 189–94.

8: The management of man

  Anderson, R. M., R. M. May, and A. R. McLean (1988). Possible demographic consequences of AIDS in developing countries. *Nature, 32:* 228–34.
* Billings, E. L., and A. Westmore (1981). *The Billings method.* New York: Random House.
* Billings, J. J. (1983). *The ovulation method.* 7th ed. Melbourne, Australia: Advocate Press.
* Blaikie, P., and H. Brookfield, eds. (1987). Land degradation and society. London: Methuen. 296 pp.
  Caccia, C. (1989). Political leadership and the Brundtland report: what are the implications for public policy? Forum on global change and our common future. Washington, DC: National Academy of Sciences.
  Caldwell, J. C., and P. Caldwell (1990). High fertility in Sub-Saharan Africa. *Scientific American, 262:* 82–9.
* Campbell, B. M., R. F. Du Toit, and C. A. M. Attwell, eds. (1989). *The Save study.* Harare: University of Zimbabwe Press, P.O. Box M.P. 45, Mount Pleasant.
  Cherfas, J. (1990). FAO proposes "new" plan for feeding Africa. *Science, 250:* 748–9.
  Coleman, D., and R. Schofield, eds. (1986). The state of population theory. Oxford: Blackwell. 311 pp.

Demeny, P. (1987a). Population change: global trends and implications. In *Resources and world development*, ed. D. J. McLaren and B. J. Skinner. Chichester: Wiley. 29–48.

(1987b). Population change in rural and urban areas. In *Resources and world development*, ed. D. J. McLaren and B. J. Skinner. Chichester: Wiley. 753–66.

Dorfman, R., and W. P. Falcon (1987). Food for a developing world. In *Resources and world development*, ed. D. J. McLaren and B. J. Skinner. Chichester: Wiley. 767–85.

Dudal, R. (1987). Land resources for plant production. In *Resources and world development*, ed. D. J. McLaren and B. J. Skinner. Chichester: Wiley. 29–48.

Haines, A. (1990). The implications for health. In *Global warming: the Greenpeace report*, ed. J. Leggett. Oxford: Oxford University Press. 149–62.

Hayter, T. (1989). *Exploited Earth: Britain's aid and the environment*. London: Earthscan. 276 pp.

Jones, H. (1990). *Population geography*. 2nd ed. London: Paul Chapman Publishing, 144 Liverpool Road, London N1 1LA. 321 pp.

Pearce, D. (1989). Sustainable futures: some economic issues. In *Changing the global environment*, ed. D. B. Botkin, M. F. Caswell, J. E. Estes, and A. A. Orio. San Diego: Academic Press. 311–23.

Pimental, D. (1987). Technology and natural resources. In *Resources and world development*, ed. D. J. McLaren and B. J. Skinner. Chichester: Wiley. 29–48.

Ramakrishna, K. (1990). Third world countries in the response to global climate change. In *Global warming: the Greenpeace report*, ed. J. Leggett. Oxford: Oxford University Press. 42.

Schwarz, W. (1990). The dirty men of Europe: USSR. *The Guardian* (London, Manchester), 19 January: 23.

Sinha, S. K., N. H. Rao, and M. S. Swaminathan (1989). Food security in the changing global climate. In *The changing atmosphere*. Geneva: World Meteorological Organization. 167–90.

Thapa, S., R. V. Short, and M. Potts (1988). Breast feeding, birth spacing and their effects on child survival. *Nature, 335:* 679–82.

Timmer, C. P. (1987). Role of markets and international trade in national food security. In *Resources and world development*, ed. D. J. McLaren and B. J. Skinner. Chichester: Wiley. 819–29.

Torrey, B. B., and W. W. Kingkadi (1990). Population dynamics of the United States and the Soviet Union. *Science, 247:* 1548–52.

Traynor, I. (1990). On the front-line of filth: Eastern Europe. *The Guardian* (London, Manchester), 19 January: 25.

Whitehead, R. (1989). Famine. In *The fragile environment*, ed. L. E. Friday and R. A. Laskey. Cambridge: Cambridge University Press. 82–106.

Wittwer, S. H. (1989). Food problems in the next decades. In *Changing the global environment*, ed. D. B. Botkin, M. F. Caswell, J. E. Estes, and A. A. Orio. San Diego: Academic Press. 119–34.

* World Commission on Environment and Development (chair: G. H. Brundtland) (1987). *Our common future*. Oxford: Oxford University Press. 400 pp.

* World Resources Institute (1987). *World resources 1987*. New York: Basic Books. 369 pp.

* (1988). *World resources 1988–89*. New York: Basic Books. 372 pp.

* (1990). *World resources 1990–91*. New York: Oxford University Press. 383 pp.

Wrong, M., and C. Mica (1990). The black snows of Transylvania: Romania. *The Guardian* (London, Manchester), 19 January: 23.

9: Summary: a new global economy

* Ellison, H. L. (1967). *The Psalms.* London: Scripture Union.
  George, S. (1990). Managing the global house: redefining economics in a greenhouse world. In *Global warming: the Greenpeace report,* ed. J. Leggett. Oxford: Oxford University Press. 438–56.
* Havel, V. (1987). *Letters to Olga,* trans. Paul Wilson. Cambridge: Granta No. 21. London: Faber and Faber. New York: Knopf.
  John Paul II (1990). Peace with God the creator, peace with all of creation. Pontiff's New Year message, 1990. Vatican.
* Leggett, J. (1990). Global warming: a Greenpeace view. In *Global warming: the Greenpeace report,* ed. J. Leggett. Oxford: Oxford University Press. 457–80.

Index

Cropland, 228–9
Cuba, 230, 273–4, 277–80
Cucumbers, 228, 231
Cumbria, 202
Curie, 204
Cycle tracks, 162–6
Czechoslovakia, 134, 212, 221
    coal, 176

Daisy planet, 15–16, 16f
Darwin, C., 91
Darwinian selection, 16
Debt, international, 254–6
Debt, poor nations, 292
Deforestation, 94–108, 296
    corporations, 107
    by country, 249
    worldwide scope, 106
Denitrification, flue-gas, 176
Denmark, 245f, 269
    CFC, 170
Denver, 200
Deodars, 260
Desertification, 129, 137–8, 239
Desulfurization, flue-gas, 175
Deuterium, 193–5
Dew-point curve, 24f
Diesel emissions, 305
Diesel smoke, 166
Dimethyl sulfide, *see* DMS
Dinitrogen pentoxide, 306
Dinosaurs, 5
DMS, 36, 83, 85, 234, 306, 307
Dobson, G. M. B., 75
Domesday Book, 92
Doppler broadening, 199–200
Dry tropical forest, 45
Dublin, 266
Ducrot, General, xiv
Dumbutshena, E., 104
Dylan, Bob, 265

Eagle pub, 164
Earth
    albedo, 15
    incoming radiation, 8f, 10f, 12f, 14f
    infrared emission, 8f, 10f, 12f, 14f
    outgoing radiation, 8f, 10f, 12f, 14f
    surface temperature, 29
    temperature, 5–15, 114f
Earthquakes, 30f, 31, 32f, 159
East Asia, cropland, 229–30
East Indies
    climate, 23
    forest, 39
Eastern Europe, coal, 83–4, 174–7
Economic growth, 60, 62f, 148
Ecosystem productivity, 42
Ecosystems, area, 42
Ecuador, 106, 230, 245, 249, 255

Education, 272–7
Educational expenditure, 276
Eemian interglacial, 124
Egypt, 137, 150, 188, 276, 278–80
Einkorn wheat, 231
El Chichon, 33, 87f, 114
El Niño, 18, 21f, 33, 98, 136
El Salvador, 191, 278, 280
Electric
    cars, 180–1
    transport, 162–7
    vehicles, 159
Electricity, 153f
    sources, 182
    transport, 179
Electrification, rural, 241–4
Ellison, H. L., 294
Elm, 94
Elsom, D., 170
Emmer wheat, 231
Energy, 151–226, 296–8
    in atmosphere, 7–9, 8f, 10f, 12f, 14f, 16f
    composition, 167–8
    conservation, 156–67, 226
    consumption, 55, 57
    consumption, by region, 146
    cost, 148–50
    currency, 179–80, 297
    efficiency, 159–60
    goethermal, 151
    regional production, 152f, 153f
    renewable, 152
    solar, 7–16, 8f, 10f, 12f, 14f, 16f, 22
England, 92–4, 140, 161
Enrichment, 201, 213
    nuclear, 195
ENSO event, 18
Environmental boycott, 252
Eocene, 124
Ethics, definition, 143–4
Ethiopia, 137–8, 231, 241, 266, 276, 278–80
Etna, Mt., 87
Eucalyptus, 38, 241, 254, 261, 263, 264
Eurasia
    climate, 22, 124
    fauna, 90–1
    warming, 123
Europe, 150, 188, 242, 272, 284
    agriculture, 227, 234–7
    CFC, 170–1
    climate change, 133–4
    CO_2 emissions, 58–60, 221
    coal, 174, 176, 289
    conservation, 160–1
    cropland, 228–30
    energy, 55, 57
    fauna, 90–2
    forest, 107, 230
    gas, 181
    GDP poor nations, 277